ASTRONOMY

A VISUAL GUIDE

ASTRONOMY

A VISUAL GUIDE

Mark A. Garlick

Firefly Books

Contents

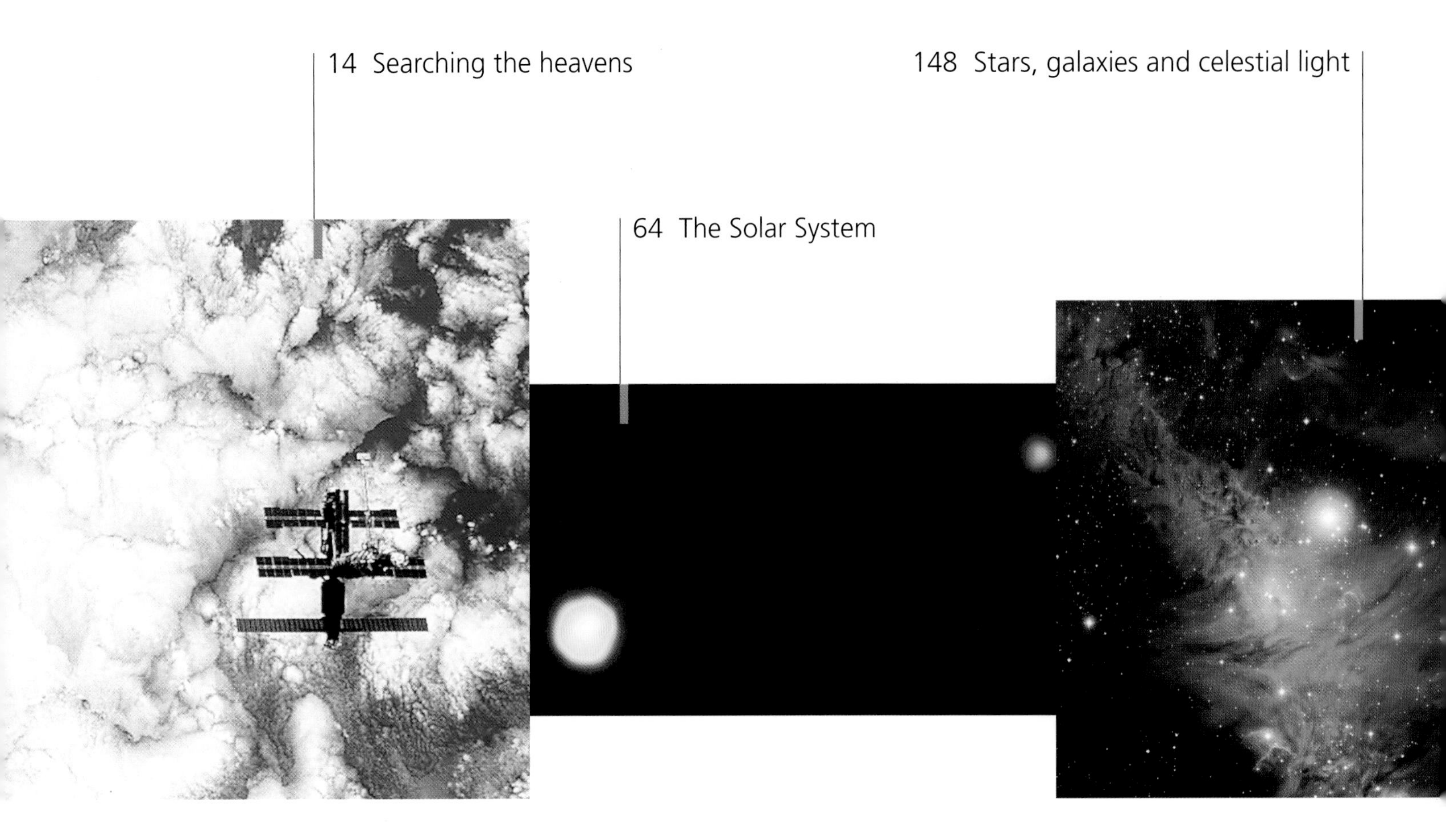

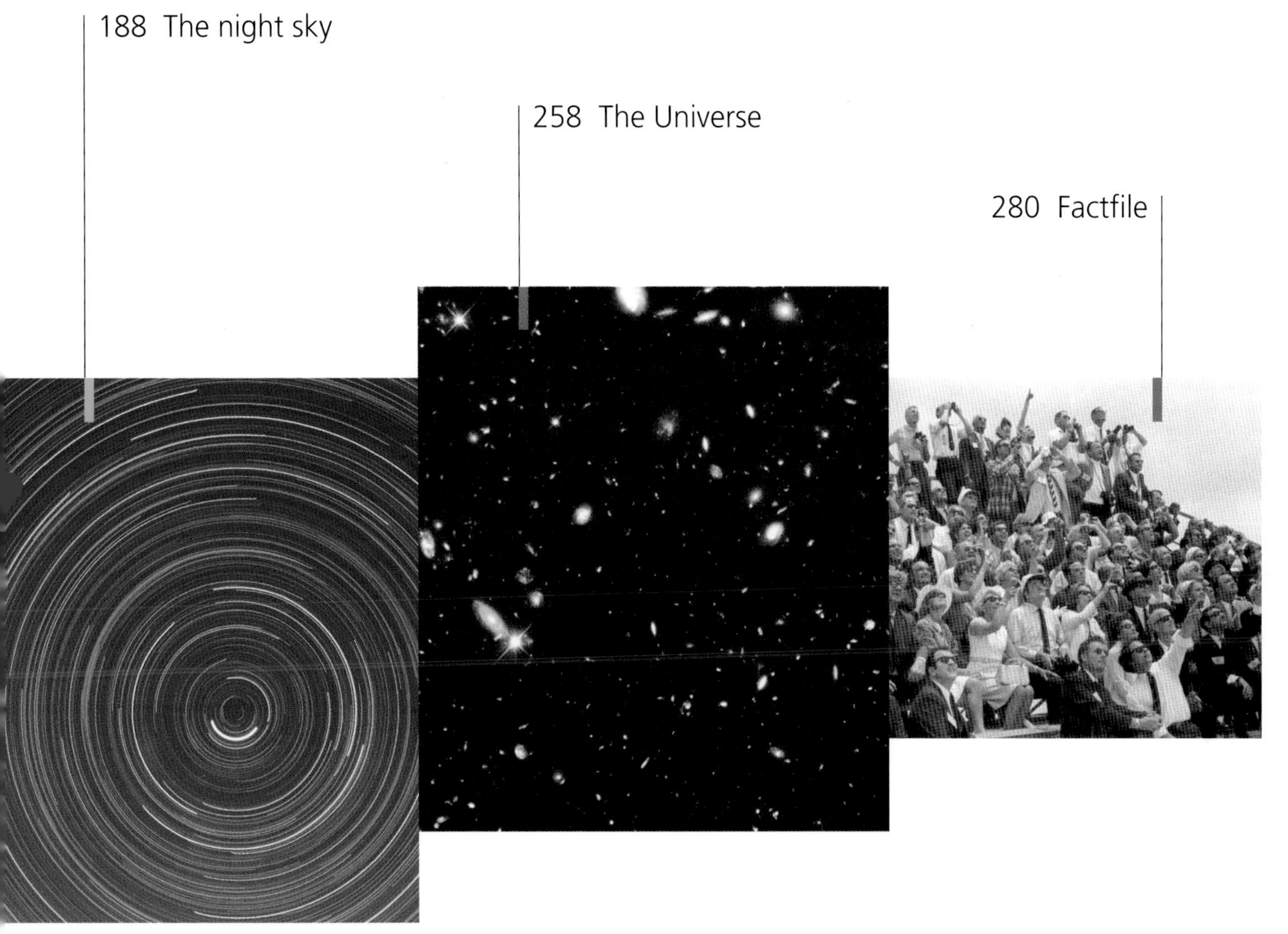

Introduction

Astronomy has fascinated us as far back in our history as we care to explore. At first, the interest stemmed from religious beliefs. The stars and planets were considered deities—indeed, this is true even today. Later, as we learned to domesticate animals and grow crops, the sky served as a celestial agricultural calendar, its constant motions reminding us when to sow and reap.

Today, our interest in astronomy is no less profound, and some might argue it is even more deep-rooted, perhaps simply because we know so much more about it. Or perhaps it's the human desire to explore, and to understand that which we currently do not. Whatever the reason, astronomy is a subject of research at universities across the globe. The more time our scientists spend peering into the depths of space, the greater the wonders—and mysteries—they uncover. Exploding stars called supernovae shine, just briefly, with the power of hundreds of billions of our Sun. They produce superdense neutron stars, shiny balls of matter the size of a city, and yet containing the mass of an entire planetary system, spinning hundreds of times every second. Enigmatic black holes twist time and space and suck in light. At the edge of the known Universe, quasars spew jets of charged particles millions of light-years into space. And then there is the material we cannot see but which we know must exist—dark matter and the newly discovered, bizarre and poorly understood dark energy.

Join us on a cosmic journey in which we will meet all of these marvels and many more, in stunning pictures and words.

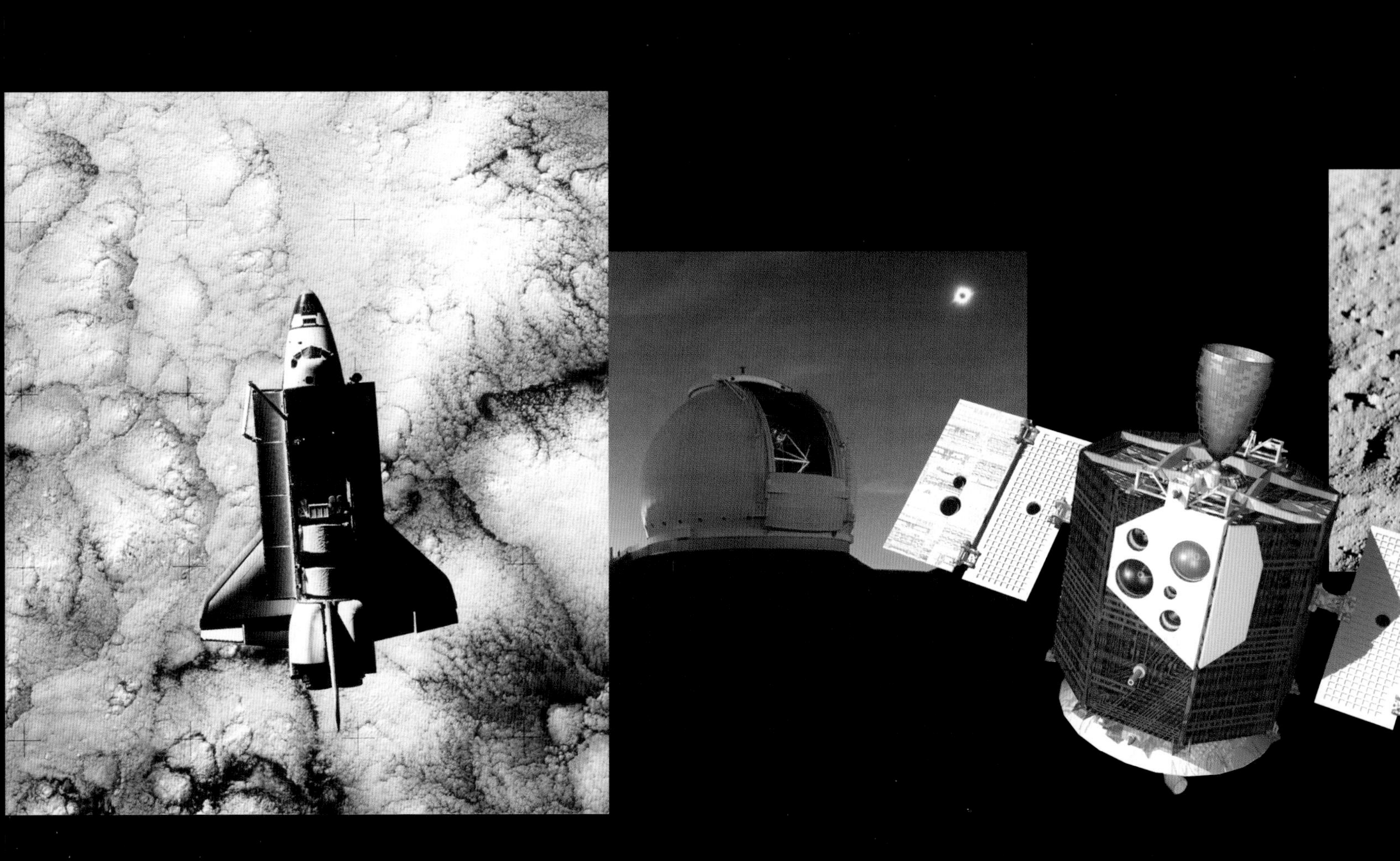

Previous page The Russian space station Mir hovers gently far above the cloud-tops of Earth.

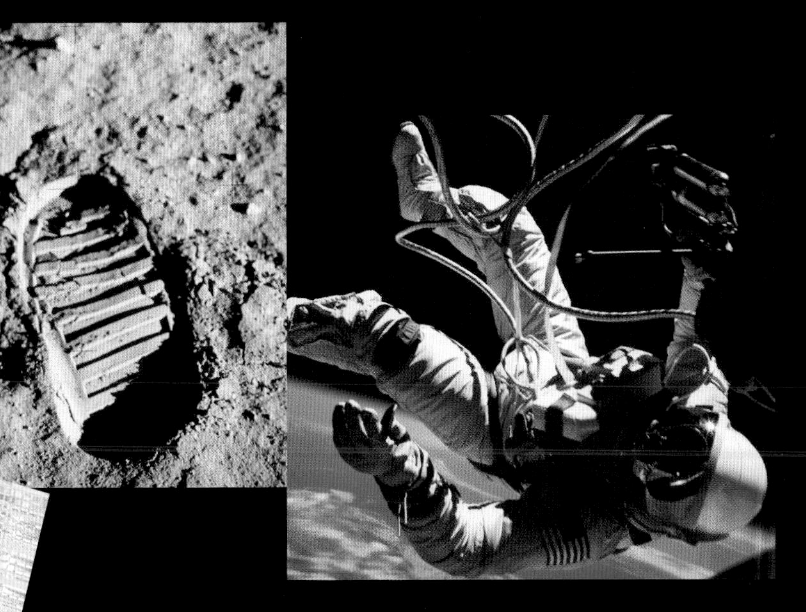

Searching the heavens

Ever since humans first gathered in groups, people have searched the heavens. Our ancestors looked to the skies for messages from the gods to guide them, celestial signs telling when to reap and sow. And even today, thousands of years later, we retain a fascination with the sky, though for different reasons.

Astronomy through the ages

c. 32,000 BC Stone-age people score bones with what appear to be phases of the Moon.

c. 4000 BC Sumerians found the city of Ur and name the earliest constellations.

c. 3000 BC Egyptians build Giza pyramids. First stage of Stonehenge erected in England.

c. 2000 BC Stone circles added at Stonehenge. Birth of the Babylonian civilization.

c. 1300 BC The Chinese invent possibly the world's first calendar.

c. 600 BC True science begins, in Greece, with Thales of Miletus the first scientist. He proposes that Earth is a disk floating on water.

520 BC Anaximander of Miletus suggests that the Earth's surface is curved, and that Earth is a cylinder.

500 BC First suggestion that Earth is a sphere.

350 BC Aristotle proposes that Earth is at the center of the Universe, and this holds sway for more than 1800 years.

300 BC Aristarchus of Samos becomes the first to suggest the quickly forgotten idea that the Sun is at the center of the Universe.

240 BC China: first recorded sighting of the comet now attributed to Halley.

235 BC Eratosthenes of Cyrene accurately measures the diameter of Earth for the first time.

165 BC China: first observations of sunspots.

c. 150 BC Hipparchus of Nicea measures the length of the year to an accuracy of minutes, makes the first fairly accurate measurements of the distances to the Moon and Sun, discovers Earth's precession, compiles the first star catalog and invents the magnitude scale of star brightness.

c. AD 150 Ptolemy further develops Aristotle's idea of an Earth-centered Universe and includes the concept of epicycles to explain the motion of the planets.

1054 Chinese astronomers witness a supernova in Taurus—the explosion that created the Crab Nebula.

1066 The comet now attributed to Halley makes a return to the skies.

1543 Nicolaus Copernicus publishes, on his deathbed, his concept of the Sun-centered Universe. The Church considers the idea fiercely heretical but the idea soon takes hold.

1576–1597 Tycho Brahe makes accurate measurements of the movement of the planets.

1609 Johannes Kepler finds that the orbit of Mars is elliptical. During 1618–1621, he proposes the now axiomatic three laws of planetary motion.

1610 Galileo Galilei becomes the first person to use a telescope for astronomy. He discovers, among other things, the large moons of Jupiter and the craters on the Moon.

1655 Christiaan Huygens discovers Titan, Saturn's largest moon. He goes on to correctly deduce the particulate nature of Saturn's rings.

1675 Giovanni Cassini discovers the gap in Saturn's rings that still bears his name.

1676 The speed of light is found to be finite.

1687 Sir Isaac Newton publishes his law of gravity.

1705 Edmund Halley suggests that the comet that now bears his name will return (which it does) in 1758.

1755 Immanuel Kant correctly proposes that stars and planets condense from interstellar clouds.

1771 Charles Messier publishes the first version of his catalog of celestial objects. He completes the list in 1781.

1781 Sir William Herschel discovers Uranus. He goes on to invent the term "planetary nebula," discovers several Saturnian moons, and publishes the catalog that becomes the basis for the *New General Catalog*.

1801 Ceres becomes the first known asteroid.

1814 Joseph von Fraunhofer discovers dark lines in the spectrum of the Sun for the first time.

1838 Astronomers measure the first stellar distance, for the star 61 Cygni.

1842 Christian Doppler first explains the Doppler effect, now widely used to determine distances and speeds in astronomy.

c. 3000 BC The Egyptian pyramids are the oldest structures built using astronomical knowledge.

1066 Comet Halley, as it is later named in 1758, is recorded when it makes an appearance in the skies.

1655 Christiaan Huygens correctly suggests that the rings of Saturn are particulate in nature.

1845 Lord Rosse discovers the first spiral galaxy, the Whirlpool, then called a spiral nebula.

1846 Neptune becomes the second new planet found since antiquity. Its largest moon, Triton, is also found.

1849 The speed of light is measured to lie within 5 percent of the modern value.

1859 James Maxwell proves mathematically that the rings of Saturn must be particulate, as Huygens proposed 204 years earlier.

1860s Spectroscopy, used for the first time, revolutionizes astronomy, and reveals the hitherto only guessed at make-up of astronomical objects.

1873 Moon craters attributed to meteorite impacts.

1877 The two moons of Mars are discovered.

1884 The Greenwich Prime Meridian is established.

1887 The Michelson–Morley experiment proves that the "ether"—the medium through which it was assumed light propagated—is fictitious.

1888 Johan Dreyer publishes the *New General Catalog Nebulae and Clusters of Stars* (NGC). It includes 7840 objects.

1895 The first extension to the NGC, called the *Index Catalog* (IC I), adds another 1529 objects.

1901 Annie Jump Cannon invents stellar spectral classification and classifies 225,000 stars.

1905 Albert Einstein publishes *The Special Theory of Relativity.*

1906 The spiral structure of the Milky Way Galaxy is first suggested.

1908 The second extension to the NGC is published, IC II, adding 3587 objects.

1908–1912 Henrietta Leavitt discovers the period–luminosity relationship for Cepheid variables that makes them invaluable distance yardsticks.

1912 Cosmic rays are discovered on balloon flights.

1915 Einstein publishes *The General Theory of Relativity*. Sirius B becomes first known white dwarf star.

1920s Edwin Hubble finds that "spiral nebulae" are galaxies in their own right, that they are all rushing apart, and that the Universe is therefore expanding.

1930 Clyde Tombaugh discovers Pluto.

1931 The first detection of radio waves from space, specifically from the Milky Way.

1938 Nuclear fusion inside stars is explained.

1940s Fred Hoyle sarcastically coins the term "Big Bang." Cygnus A becomes first known radio galaxy. Fred Whipple correctly deduces the nature of comets.

1950s Many radio galaxies are discovered. Stellar nucleosynthesis explains where elements heavier than helium come from.

1963 Taurus X-1 is the first known X-ray source found aside from the Sun. First quasar found.

1965 The discovery of the cosmic microwave background radiation provides real support for the Big Bang theory.

1967 Jocelyn Bell and collaborators discover pulsars.

Early 1970s US satellites discover the first gamma-ray bursters.

1971 Cygnus X-1 becomes the first strong candidate black hole.

1977 Astronomers detect the rings of Uranus from Earth.

1978 Pluto's satellite, Charon, is discovered.

Late 1970s The idea of inflation, which explains the first stages immediately following the Big Bang, is outlined in detail.

1981 Astronomers discover the first void in space, empty regions now known to exist between superclusters.

1982 The first millisecond pulsar is discovered.

Late 1980s Evidence for "dark matter" first announced. Supernova 1987A becomes the first naked-eye supernova in nearly 400 years.

1995 The first extrasolar planet is discovered.

1996 Announcement of possible fossilized Martian microbes inside a meteorite from Mars.

1998 The Universe's expansion found to be accelerating.

2002 The number of known extrasolar planets exceeds 100.

1873 Astronomers recognize the origin of the Moon's craters are due to meteorite impacts.

1915 Einstein publishes *The General Theory of Relativity*—which greatly influences astronomy.

1987 Supernova 1987A in the Large Magellanic Cloud creates a stellar explosion on our doorstep.

Sky gods

Ever since we dwelt in caves, the skies have kept us in awe. The earliest evidence of this dates to the end of the last Ice Age, when our stone-age ancestors scored animal bones with apparent phases of the Moon. Perhaps they kept watch on the Sun and the Moon and the patterns in the stars to record the seasons. Certainly this was the practice when agriculture was developed and animals were domesticated around 10,000 BC in Mesopotamia (part of modern-day Iraq). The sky became a giant farming calendar, telling these people when to sow and when to reap. It was these early Mesopotamian societies, specifically the Sumerians around 4000 BC, who named the oldest surviving constellations. They noted the groups of visible stars in the sky at important times of the year on the agricultural calendar, and gave them names. And because the skies controlled these people's way of living, the heavens were deified. Soon, more constellations were added to represent various ancient gods. The 12 constellations where the gods of the Sun, Moon and planets resided (the zodiac) were especially important. Thus the night sky became not just a farming tool—it was the home of the gods and a picture book relating stories of important figures to a people who at first, lacking writing, had no other means of keeping records.

Meanwhile, in other parts of the world, other civilizations were developing their own independent astronomy and skylore, always with religious associations.

↑ **A Babylonian tablet** containing astronomical records describing the motions of the stars and planets. It is around 2500 years old.

ASTRONOMY IN THE NEW WORLD

The New World was equally fascinated with the sky. The Maya (3rd century BC to 9th century AD) were keen observers of the heavens. Venus, to them, represented their rain god. And the Aztecs (14th–16th centuries) also worshiped Venus, this time as the feathered snake god Quetzalcoatl, who demanded regular human sacrifice.

This Aztec sunstone is a primitive form of calendar, weighing some 25 tons and measuring 13 feet (4 m) across. At the center is the Aztec god Tonatiuh.

LATER MESOPOTAMIAN ASTRONOMY

Building on the knowledge amassed by the earliest Mesopotamian civilizations, successive societies in the same region developed a greater understanding of the motions of the heavenly bodies. The Babylonians and the Assyrians invented sophisticated calendars to aid them with their agriculture, similar to those that also arose in Egypt. The Babylonians learned how to predict eclipses of the Moon and the Sun, and they could forecast the motions of the five then-known planets with some precision. The Babylonians were even responsible for inventing the unit still used to measure distances on the sky—the degree. This, the last of the great Mesopotamian civilizations, lasted until around the first century BC. By that time the Babylonians had begun to transform astronomy into a true science, but they lacked the necessary tools—geometry and trigonometry—to do the job well. Thus, it fell to the Greeks, who inherited this wealth of astronomical data and ideas, to free astronomy of its history of mystical hand-waving once and for all.

EGYPTIAN ASTRONOMY

In Egypt, as in other early civilizations, astronomy was seen more as a religion than a science. Astronomer-priests propagated tales of Egyptian gods, and used primitive calendars based on the skies to predict important agricultural events. For example, the first appearance of Sirius—the brightest star in the sky—after months of absence was an event known as the heliacal rising. Upon its occurrence, astronomer-priests could predict when the Nile's next annual flood was due to occur, important because it was needed to irrigate their crops. These people also saw patterns in the stars—constellations. Orion was, to them, Osiris, the god of the dead, for example. And in the band of light that we now know to be the Milky Way Galaxy, they saw the goddess of the sky, Nut, giving birth to Ra, the god of the Sun.

← **This ceremonial mask** pays homage to the Sun God of La Tolita—a pre-Columbian culture from Ecuador (around 500 BC).

↓ **In this hand-colored** engraving taken from an Egyptian mummy case, the god of the atmosphere, Shu, is seen lifting his daughter Nut, goddess of the sky, to separate her from Earth's perils.

Archaeoastronomy

We tend to think of early civilizations as ignorant of science. Indeed, by many standards they were very primitive. But they were also remarkable in that their acute knowledge of astronomy influenced their architecture. Around 3000 BC, construction began on one of the most famous monuments in the world, Stonehenge, now thought to have been an early astronomical observatory. Around the same time, the Egyptians carved out their massive pyramids on the Giza plateau, which were constructed to align perfectly with certain stars in the sky. Two millennia later, the Maya were doing the same thing in the New World. These are just some of the many civilizations separated by oceans of water and vast rivers of time, each ignorant of the other, and yet each independently developing their architecture to conform to the motions of the lights (and gods) in the sky.

GIZA PYRAMIDS
The ancient Egyptians had a great interest in the sky. The pyramids on the Giza plateau (*below*) are aligned to the compass cardinal points to very high accuracy, each side facing north, south, east or west to within a few tenths of a degree. Moreover, at the time the Giza pyramids were constructed (about 3000 BC), the sky had a different north polar star (Thuban in Draco) owing to the precession of Earth's spin axis. Passages within the pyramids emerging from their north faces pointed almost directly to this star at the time. One theory, as yet unsubstantiated, suggests the Giza site was laid out to mirror the night sky: the three pyramids represent the belt stars in the constellation of Orion (*right*)—known to the Egyptians as Osiris, the God of the Dead—while the Sphinx is Leo and the Nile is the Milky Way.

STONEHENGE

Stonehenge (*left*) is probably the best-known megalithic monument in Europe, situated in the English countryside and now a major tourist attraction. No one is sure how it was built, by whom or even when. But the best thinking is that it was constructed in several different stages by the Druids. The earliest work may have begun 5000 years ago, with the creation of the original henge—an arrangement of earth banks, holes and ditches. The first circle of stones was erected possibly 1000 years later, around 2000 BC, with the monument completed around 1500 BC. Some think Stonehenge was used for human sacrifice or as a burial place for high-ranking citizens. But a popular theory is that it was some kind of observatory, used to predict the motion and eclipses of the Moon and Sun. The evidence is in the monument's alignment, which is such that on the summer solstice every year, the Sun rises above one of the main stones, the so-called Heel Stone.

MAYAN PYRAMIDS

The Maya civilization occupied what is now modern-day Mexico, Guatemala, Belize, Honduras and El Salvador, and flourished between 1500 BC and AD 900. Like the ancient Greeks, the Maya were scientifically minded and drawn to the sky through their religion. Evidence of their fascination with the heavens is clear, particularly in the layout of many of their cities. Some were aligned with the main compass points, or with the spring and autumn equinoxes, or especially with the rising and setting cycles of Venus. A good example is the famous pyramid at the Mayan site of Palenque (*below*). Windows in its top and sides are arranged so that the rooms they lead to are fully illuminated by the Sun on the morning before Venus becomes visible. Another example is the pyramid at Chichén Itzá. On the equinoxes, the illumination of the Sun on the stairs and top of the pyramid creates the illusion of a serpent—the Mayan snake god Quetzalcoatl, personifying the planet Venus.

Scientific beginnings

Around 600 BC in Greece, science was born. Here were a people who sought to understand the cosmos by observation, theory and experiment—the cornerstones of the scientific method. Aristotle (384–322 BC) was perhaps the first person to attempt to explain the Universe scientifically. He saw the cosmos as a series of spheres with Earth in the center. Several hundred years later, in AD 150, the last of the great Greek philosophers, Ptolemy (about AD 100–170), adapted Aristotle's idea. His geocentric model predicted the motions of the heavens to high precision, and held sway for an astonishing 13 centuries—despite the fact that it could not be more wrong.

Astronomy received a rude awakening in 1543, courtesy of a Polish churchman named Nicolaus Copernicus (1473–1543, *left*). In his posthumous publication, he revealed that the Sun, not Earth, was at the center of the Solar System. Despite Church opposition, the new idea slowly took hold. Seven decades later, when Galileo Galilei (1564–1642) turned a new invention—the telescope—to the skies for the first time, he found that a small family of four moons orbited Jupiter. It was a Solar System in miniature. Here at last was hard observational evidence that not everything went around the sacred Earth, despite the strong theological opposition.

PTOLEMY'S UNIVERSE

According to Ptolemy, the unmoving Earth lay at the center of the Universe. The Sun, Moon and the five planets (Mercury, Venus, Mars, Jupiter and Saturn) all went around it. But while the Sun and Moon went around in circles—the perfect shape—the planets moved in epicycles, small circles superimposed on the rims of their main orbits (*not shown*).

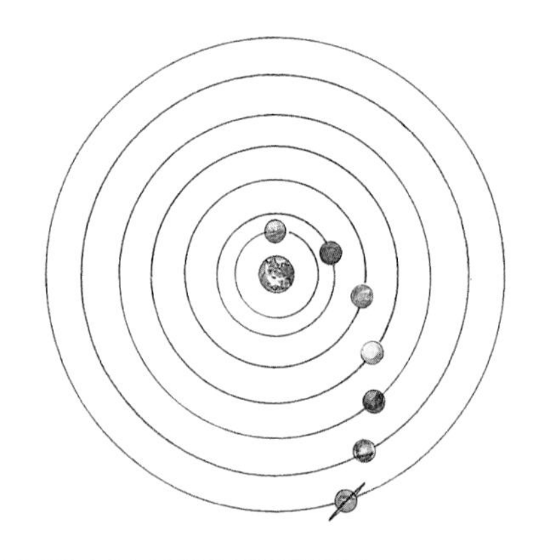

TYCHO BRAHE'S UNIVERSE

Dutch astronomer Tycho Brahe (1546–1601) accurately recorded the positions and motions of the planets. Brahe realized the data did not exactly fit the model of Ptolemy, nor that of Copernicus. So he formulated his own model. He had the planets orbiting the Sun, but he had the Sun orbiting Earth. Apparently he was unwilling to accept the notion that Earth was not important enough to occupy the middle.

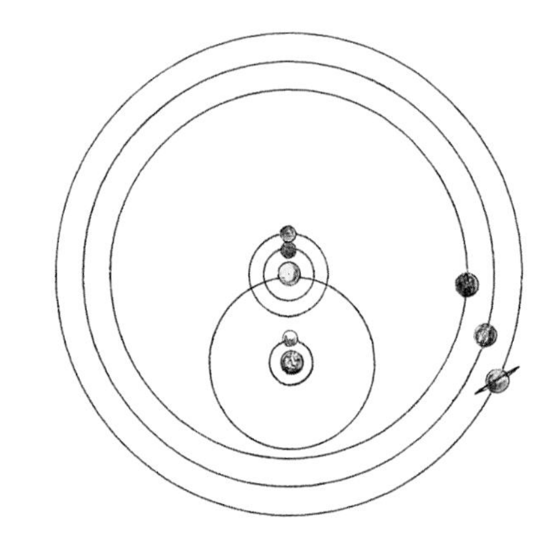

← **Italian-born Galileo** was the first person to use the telescope for astronomical observation in 1609.

COPERNICUS' UNIVERSE Copernicus revolutionized astronomy—bogged down with an Earth-centered cosmos for centuries largely because of religious opposition to any alternatives—when he dared to propose a heliocentric model (*left*). Around the Sun went Earth and the other planets, while the stars were placed far beyond the planets, their motions across the sky correctly attributed to Earth's rotation. His model was still flawed, in that it retained circular orbits and epicycles. Nevertheless, it was certainly the kick-start that astronomy needed. Copernicus knew well that the fanatical Church would regard his heliocentric model as heretical, so he wisely withheld its open publication until the year of his death.

↓ **Galileo built** a number of telescopes, such as the one below, and discovered many previously invisible objects—helping to expand greatly the knowledge of astronomy.

The development of astronomy tools

c. 3000 BC First stage of Stonehenge erected.

c. 2000 BC Babylonians invent the degree.

c. 1300 BC The Chinese invent possibly the world's first calendar.

c. 210 BC The magnetic compass is invented in China.

c. 1555 The first telescope, a reflector, is invented by Leonard Digges. (Uncertain.)

1582 Pope Gregory XIII invents the Gregorian calendar, used throughout the world today.

1608 Dutchman Hans Lippershey possibly invents, or at least first constructs and markets, the refracting telescope. (Who really invented the telescope remains a mystery.)

1610 Galileo uses a telescope to view the heavens for the first time.

1668 Isaac Newton constructs a reflecting telescope and applies it to the sky. It has a magnification of 35 times and an aperture of just 1 inch (25 mm).

1731 John Hadley invents the octant, a predecessor of the sextant—a device to measure the angular distances of celestial objects above the horizon.

1757 John Campbell invents the sextant.

1770s Sir William Herschel designs, builds and uses a reflecting telescope with an unprecedented aperture of 48 inches (1.2 m)—then the largest in the world.

1826 Joseph Niepce invents the camera.

1860s Spectroscopy used for the first time.

1880s Photography used as an astronomical aid.

1887 Albert Michelson and Edward Morley use interferometry for the first time—the technique of combining signals from more than one detector to make an instrument with a larger effective aperture.

1930 Radio astronomy begins.

1945 Founding of the Nuffield Radio Astronomy Laboratory at Jodrell Bank in the United Kingdom.

1948 The Palomar Observatory in San Diego, USA, officially opens, with the famous 200-inch (5 m) Hale Telescope taking pride of place. It remains the world's largest telescope until 1975.

1949 A rocket-borne experiment detects the first known astronomical X-ray source—the Sun.

1961 Opening of the 208-foot (64 m) radio telescope at the Parkes Observatory in New South Wales, Australia. The observatory participated in the Apollo 11 Moon landing, relaying messages to NASA.

1962 The European Southern Observatory (ESO) organization is founded.

1963 The Arecibo Observatory in Puerto Rico opens. The telescope is built into a natural bowl in the ground, and measures 1000 feet (305 m) across. It is still the world's largest single-dish radio telescope.

1964 La Silla Observatory is founded in Chile.

Late 1960s Ultraviolet astronomy begins.

1970 X-ray astronomy begins with the first dedicated X-ray satellite, Uhuru—Swahili for "freedom"—launched off the coast of Kenya.

1972 The South African Astronomical Observatory (SAAO) is founded near Sutherland, South Africa. It remains the most important observatory in Africa. In November, NASA launches its second Small Astronomy Satellite, SAS-2, after Uhuru, opening up the field of gamma-ray astronomy.

1974 The 13-foot (3.9 m) Anglo–Australian Telescope (AAT) is opened at the Anglo–Australian Observatory in New South Wales, Australia. The United Kingdom launches the long-lived X-ray satellite Ariel 5.

1975 The Russian 20-foot (6 m) Bolshoi Telescope overtakes the Hale Telescope as the world's largest. In May, the USA launches the last of its SAS satellites, SAS-3. In August, the European Space Agency (ESA) launches the gamma-ray space observatory COS-B.

1977 NASA launches the first of its High-Energy Astrophysical Observatories—HEAO-1. It had unprecedented sensitivity.

1978 In January, the International Ultraviolet Explorer (IUE) is launched. This joint NASA–ESA–UK satellite proved very successful for UV astronomy, operating for 18 years. Later, in November, NASA launches its second high-sensitivity X-ray satellite, HEAO-2, later renamed Einstein. It was the first X-ray telescope with grazing-incident optics, and was the first mission to take X-ray photographs of actual sources.

1770s Sir William Herschel's telescope becomes a marvel of eighteenth-century engineering.

1948 The 200-inch (5 m) Hale telescope is completed. Its mirror is resurfaced every two years.

1975 Russia's Bolshoi Telescope is the largest in the world when built—overtaking the Hale Telescope.

1979 NASA launches the last HEAO satellite, HEAO-3. The 12½-foot (3.8 m) United Kingdom Infrared Telescope (UKIRT) opens at the Mauna Kea Observatory, Hawaii. The Roque de los Muchachos Observatory opens on La Palma, in the Canary Islands. It currently houses several large telescopes and is the premier European observatory.

1980 Completion of the Very Large Array (VLA) of radio dishes in New Mexico. The Array consists of 27 individual radio dishes—each 82 feet (25 m) in diameter—that can be moved along rails into four separate configurations.

1983 In January, the Infrared Astronomical Satellite (IRAS) makes the first infrared survey of the sky, covering 95 percent. In May, ESA launches the highly successful X-ray satellite Exosat, which operated until 1986.

1987 The Japanese launch the last of three X-ray satellites, Ginga. It operated until 1990.

1988 The Australia National Telescope Facility—a group of eight radio antennas in New South Wales—begins operation.

1989 Opening of the New Technology Telescope (NTT) at La Silla Observatory, Chile. Launch of NASA's Cosmic Background Explorer (COBE) satellite, which measured the background temperature of space to high precision and provided solid evidence for the Big Bang.

1990 In April, the 8-foot (2.4 m) Hubble Space Telescope (HST) is launched after many years of delay—its mirror is later found to be defective. In June, the joint American–German–British X-ray satellite, Rosat, is put into space to perform the first all-sky imaging survey in X-rays.

1991 Launch of NASA's Compton Gamma-Ray Observatory (GRO).

1992 NASA launches the Extreme Ultraviolet Explorer (EUVE) satellite to probe high-energy ultraviolet sources. Completion of the Keck I telescope at the Keck Observatory, Hawaii.

1993 Astronauts repair the HST's mirror during the first servicing mission. In the USA, the Very Long Baseline Array (VLBA) begins operation—a series of 10 radio dishes spread across the mainland, the Virgin Islands and Hawaii. In unison, it amounts to a single instrument 5000 miles (8000 km) across.

1994 Construction begins on the Laser–Interferometer Gravitational Wave Observatory (LIGO) in California. It will search for gravitational waves from violent astronomical phenomena.

1995 ESA launches the Infrared Space Observatory (ISO). This is the second important IR detector in space. It carried a 2-foot (0.6 m) infrared telescope, two spectrometers and a photopolarimeter, for measuring the degree of polarization of astronomical light sources.

1996 The Italians launch the X-ray satellite BeppoSAX, with financial aid from ESA and the Netherlands. Completion of the Keck II telescope at the Keck Observatory, Hawaii. Keck I and Keck II each have 33-foot (10 m) mirrors composed of 36 individual segments. They are currently the largest individual optical telescopes in the world.

1999 July sees the launch of the Chandra X-ray Observatory (formerly known as AXAF), the successor to the Einstein Observatory. In December, ESA follow suit with an X-ray observatory called XMM-Newton. Both telescopes are still in operation and their extreme sensitivity is revolutionizing X-ray astronomy.

2000 The Very Large Telescope begins operations at Paranal Observatory, Chile. It consists of four separate telescopes, each with a diameter of 27 feet (8.2 m), which can be combined to make an instrument with an effective diameter of 52 feet (16 m)—the largest effective optical telescope in the world. Also this year, the Green Bank Telescope—with its slightly elliptical 330-foot (100 m) radio dish—is opened in West Virginia, USA.

2002 The International Gamma-Ray Astrophysics Laboratory (INTEGRAL) spacecraft is launched by the European Space Agency.

2003 The Mount Stromlo Observatory in Australia is tragically destroyed by a bushfire.

1980 The 27 radio dishes that form the Very Large Array (VLA) in New Mexico are completed.

1990 The Hubble Space Telescope is launched. Three years later, astronauts repair Hubble's faulty mirror.

1996 The Keck telescopes are completed and are sited at the highest observatory in the world.

Observatories

Our observatories have come a long way since their humble beginnings back in the seventeenth and eighteenth centuries. Nowadays, the largest ground-based observatories have telescopes with mirrors up to 33 feet (10 m) across. Virtually all such professional observatories use CCDs (charged-coupled devices) to capture their data—chips exactly like those found inside digital cameras that turn images into digital data. Looking through the telescope itself is a thing of the past. In most cases the astronomer relaxes in the comfort of the "warm room" adjacent to the dome, from where the observing is fully automated using a computer. Radio observatories have grown too, especially since individual radio dishes can be combined (using "interferometry") to make larger effective collecting areas.

↑ **Some 13,700 feet (4200 m)** above sea level atop a mountain in Hawaii lies the highest observatory in the world—Mauna Kea. It is home to several telescopes, including the Canada–France–Hawaii Telescope with an aperture of 11¾ feet (3.6 m).

← **The Very Large Array (VLA)** in New Mexico, USA, consists of a total of 27 radio dishes, each 82 feet (25 m) in diameter. They can be moved independently, and the entire array can conform to four different configurations. Using the technique of interferometry, the signals from these individual instruments can be combined to create, in effect, a single large radio telescope with an effective baseline up to several miles wide.

← **The Keck telescopes,** housed at the W. M. Keck Observatory in Mauna Kea, Hawaii, are currently the largest single optical telescopes in the world. Each of the two matching telescopes has a 33-foot (10 m) mirror composed of several individual segments. These "adaptive optics" segments can be modulated independently of one another to counteract the blurring effect of Earth's atmosphere.

↓ **This photo shows one** of several telescopes at the Crimean Astrophysics Observatory (CrAO), situated at an altitude of 1960 feet (600 m) in the Ukraine. It is a solar telescope with an aperture of 3¼ feet (1 m), equipped with a coronagraph for photographing the outer atmosphere of the Sun, the corona. As well as the solar telescope, the CrAO houses one 8½-foot (2.6 m) and two 4-foot (1.25 m) telescopes, and also manages several others on a different mountain.

Major space centers

Our globe is dotted with hundreds of observatories, with every major continent having at least a handful of large telescopes and radio dishes for astronomical research. The greatest concentration is in North America, where many of the observatories are privately owned by universities. But space research is not only carried out at observatories and universities. NASA institutions, such as the Goddard Space Flight Center or the Johnson Space Center, are also active hives of space research, as are some launch sites such as the Kennedy Space Center at Cape Canaveral.

NORTH AMERICA
Dominion Astrophysical Observatory (Canada)
Goddard Space Center (Maryland)
Jet Propulsion Lab (California)
Johnson Space Center (Texas)
W. M. Keck Observatory (Hawaii)
Kennedy Space Center (Florida)
Kitt Peak Observatory (Arizona)
Lowell Observatory (Arizona)
Mount Hopkins Observatory (Arizona)
Mount Wilson Observatory (California)
Palomar Observatory (California)
Very Large Array (New Mexico)
Yerkes Observatory (Wisconsin)

CENTRAL AND SOUTH AMERICA
Arecibo Telescope (Puerto Rico)
Cerro Tololo Inter-American Observatory (Chile)
Paranal Observatory (Chile)
La Silla Observatory (Chile)
European Space Agency launch site (French Guiana)

EUROPE
Bolshoi Telescope (Russia)
Effelsberg Telescope (Germany)
European Space Agency (France)
European Space Operations Center (ESOC) (Germany)
Jodrell Bank (England)
Roque de los Muchachos Observatory (La Palma, Canary Islands)
Pic du Midi (France)
Star City (Cosmonaut training, Russia)

AUSTRALASIA
Anglo–Australian Observatory (Australia)
Australia Telescope (Australia)
Baikonur Cosmodrome (Kazakhstan)
Kagoshima Space Center (Japan)
Parkes Observatory (Australia)

AFRICA
South African Astronomical Observatory (South Africa)

The summit of Mauna Kea, Hawaii, is home to the Keck telescopes as well as many other observatories.

The Jodrell Bank radio telescope, constructed in 1957, remains one of the largest radio dishes in the world.

The Paranal Observatory, Chile, is most famous for the Very Large Telescope—four linked optical telescopes.

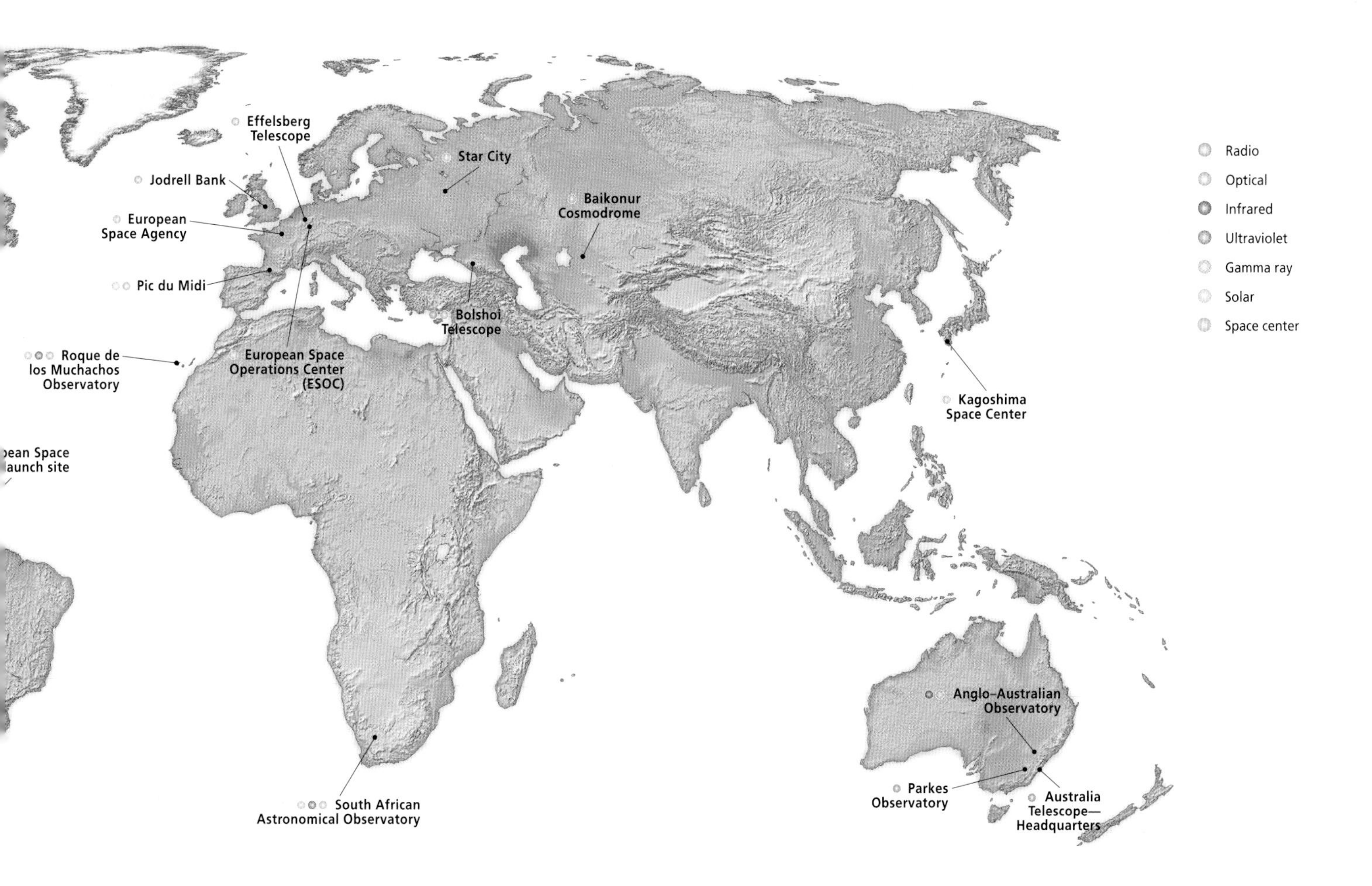

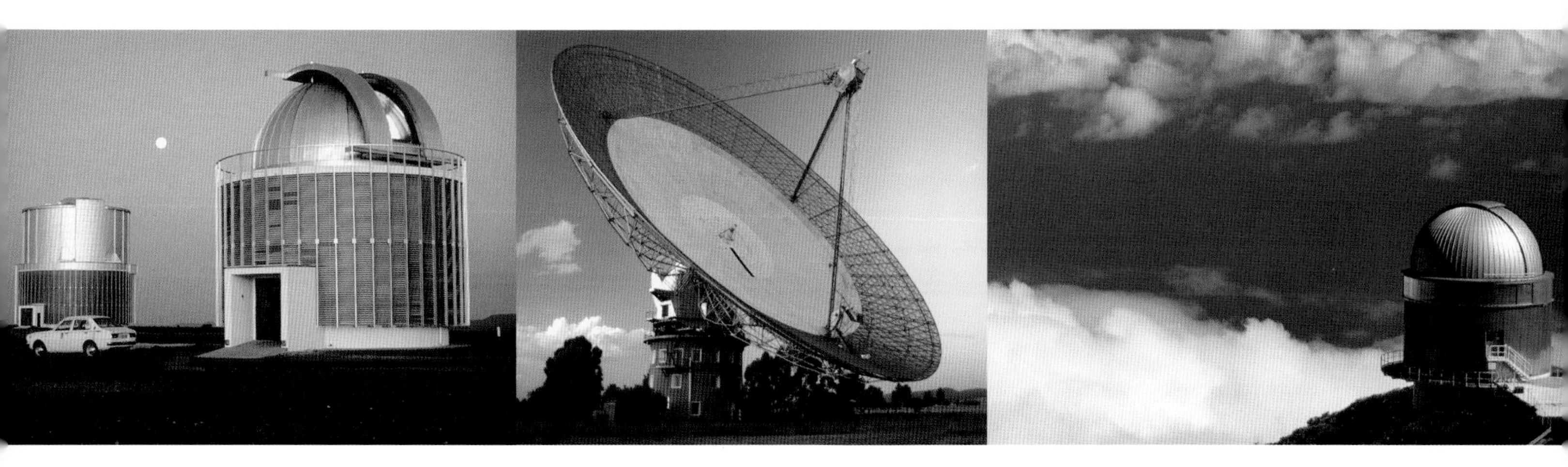

The South African Astronomical Observatory remains the most significant observatory in Africa.

The Parkes Observatory in Australia houses a radio telescope measuring 210 feet (64 m) in diameter.

Roque de los Muchachos Observatory, lies above the clouds on the island of La Palma in the Canary Islands.

Space observatories

Think of space observatories and the Hubble Space Telescope (HST) will almost certainly come to mind. However, Hubble is by no means the only space telescope. Far from it. Since the space age began, there have been dozens of space telescopes, predominately American and Russian in origin, observing in all regions of the electromagnetic spectrum. On Earth, our atmosphere, which protects us so well from harmful ultraviolet radiation, X-rays and gamma rays, also makes it impossible for ground-based telescopes to observe in these regions of the spectrum. Putting the telescope in space—where, as a bonus, there is no atmospheric blurring—is the natural answer. The Hubble Space Telescope is perhaps most famous because it observes partly in the optical band, which is sensitive to human eyes. But Hubble is a relatively old telescope—launching into space in early 1990. Its replacement, the James Webb Space Telescope, is already on the drawing board and set for a 2011 launch.

↑ **This photo is an unusual** combination of an optical and a ROSAT X-ray photograph. The magenta background is the X-ray image, showing that the three optical galaxies are immersed in a giant cloud of very hot, X-ray-emitting gas.

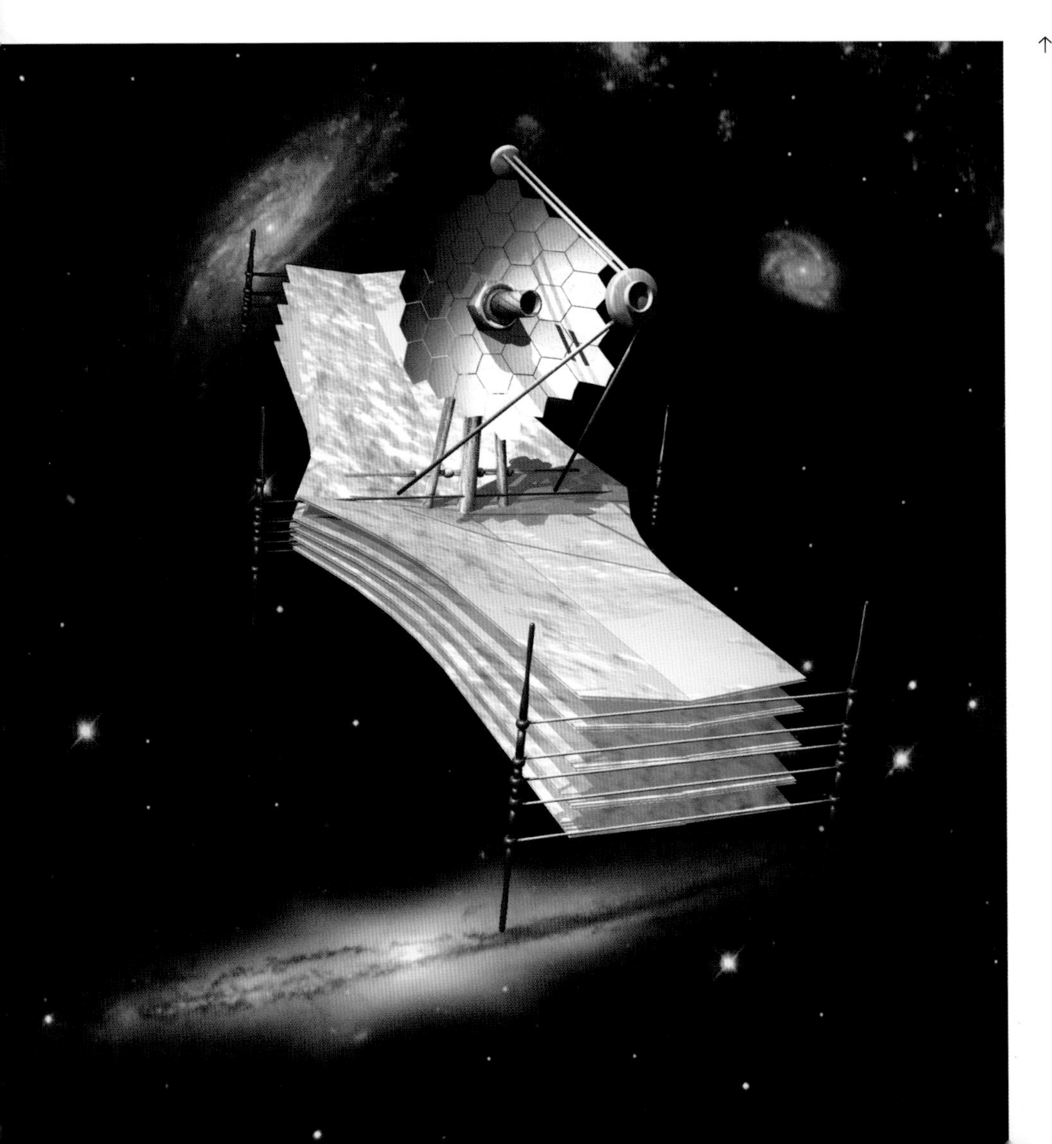

JAMES WEBB SPACE TELESCOPE

When it is time for the Hubble Space Telescope (HST) to come out of service, a new and more powerful space observatory will claim its crown. The eagerly anticipated successor is known as the James Webb Space Telescope (JWST, *left*), named in honor of the NASA administrator who led the space agency from 1961 to 1968. The JWST was formerly known as the Next Generation Space Telescope, as it will be constructed using technology that did not exist when the HST was engineered. The most radical application of this technology concerns its main mirror. While the HST mirror is a solid block weighing about one ton and spanning 8 feet (2.4 m), the JWST mirror will consist of much smaller, thinner and lighter hexagonal segments, and will measure a total of 20 feet (6 m) across. A solid mirror of this size will not fit into a shuttle cargo bay, so the new mirror will be folded for its trip and will then unfurl itself when deployed. This makes the observatory very lightweight. It is anticipated that the JWST, although much larger than the HST, will have only half of its mass.

The JWST is designed to observe in the region of the spectrum running from visible green light to the mid-infrared, where it will study the formation of planets, stars and galaxies. If all goes well, the JWST will see first light in 2011, and it should be 10 to 100 times more sensitive than the HST. It will be the most powerful and largest space telescope ever built.

↑ **This photo shows** the mirror system of the X-ray satellite ROSAT. The mirrors are a series of cylinders with a slight taper to them, designed to catch X-rays that are incident at very shallow angles.

→ ↓ **A team of technicians** rigorously examine various components of the Infrared Space Observatory (ISO) satellite, during testing at the European Space Agency's ground station in Kourou, French Guiana. ISO was launched in 1995 and carried a 2-foot (0.6 m) infrared telescope.

Hubble Space Telescope

In April 1990, the lauded and much-delayed Space Telescope was finally launched—and renamed Hubble, in honor of the American astronomer Edwin Hubble (1889–1953), who discovered the expansion of the Universe. Today, still going strong, the Hubble Space Telescope (HST) is probably the most famous telescope ever, courtesy of the plethora of breathtaking images that the satellite continues to beam to Earth. Hubble is by no means large. Its main mirror, which measures 8 feet (2.4 m) across, is only a quarter of the size of the biggest ground-based instruments. But it is the Hubble Space Telescope's position some 380 miles (600 km) above the blurry atmosphere that gives it its remarkable power and insight.

↑ **Just one of many** photos taken by the HST, this shows a close-up of the Swan Nebula, M17, in Sagittarius, where new stars are being born.

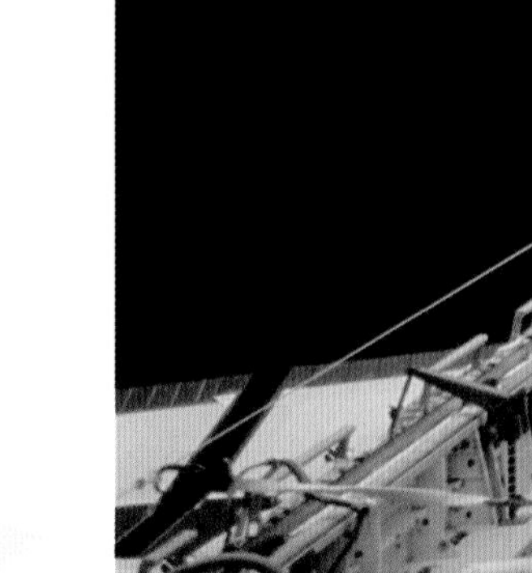

Communications antenna

Telescope directional system

Spacecraft compartment

Aperture door

Secondary mirror

Primary mirror

Scientific instruments and cameras

Solar array

HUBBLE IN CLOSE-UP
This cross-section of the HST shows the locations of the main parts, including the primary and secondary mirrors. Power comes primarily from the solar panels, but there are also batteries as back-up. When the telescope is moving between targets, the aperture door is closed over the end of the tube to protect the instruments from accidental exposure to the effects of the Sun and also to Earth's bright day side.

↑ **After the HST was launched,** astronomers realized that its main mirror had been made perfectly—but to the wrong specification. Its shape was wrong, and this resulted in blurred images beamed back to Earth. So in December 1993, the space shuttle Endeavour intercepted the telescope, and a team of astronauts carried out the much-needed repair in a series of spacewalks. Since then, there have been two additional "service missions," in 1997 and 1999. NASA took these opportunities to effect small upgrades, and to replace some of the instruments with newer ones.

→ **The Hubble Space Telescope** floats above a sea of blue, approximately 380 miles (600 km) skyward, in this NASA photograph. From here, the telescope has an uninterrupted view of the sky, 24 hours a day, free from the constraints of weather and atmospheric turbulence that disrupt ground-based instruments.

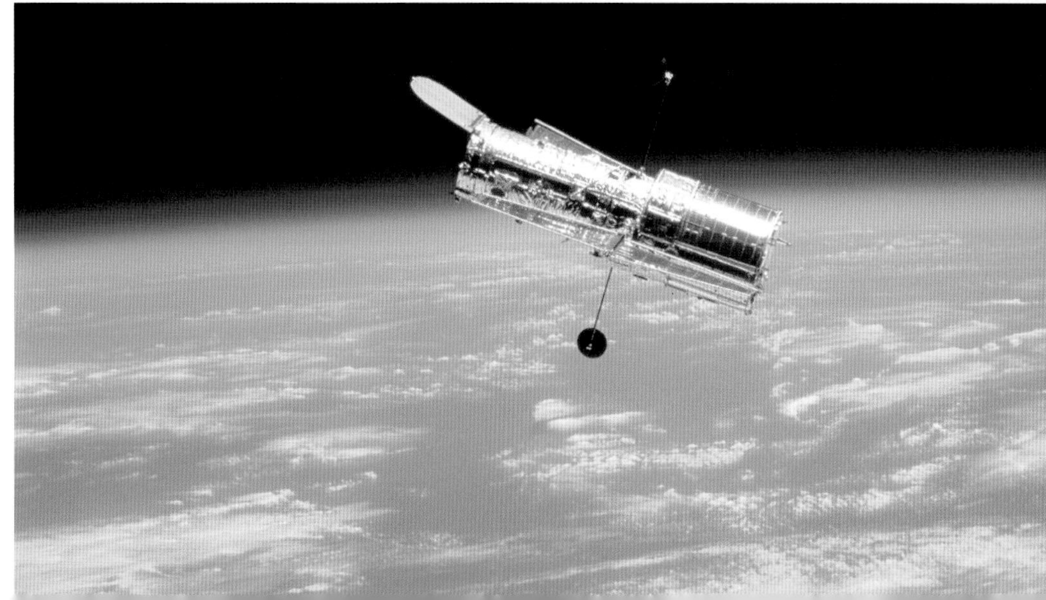

Radio astronomy

The radio part of the electromagnetic spectrum is the widest, having very long wavelengths in the range 1/25 inch to 100 feet (1 mm to 30 m) and therefore situated beyond the far infrared. Radio astronomy had its origins in 1931/32, when the American radio engineer Karl Jansky found to his surprise that the sky itself seemed to be emitting radio waves—in fact the emission was coming from the Milky Way Galaxy. Nowadays, radio astronomy is a huge area of research. Not only are there many phenomena that emit radio waves—radio galaxies, pulsars, the cosmic microwave background—but the entire radio spectrum is accessible from the ground, since our atmosphere, day and night, is transparent to radio waves.

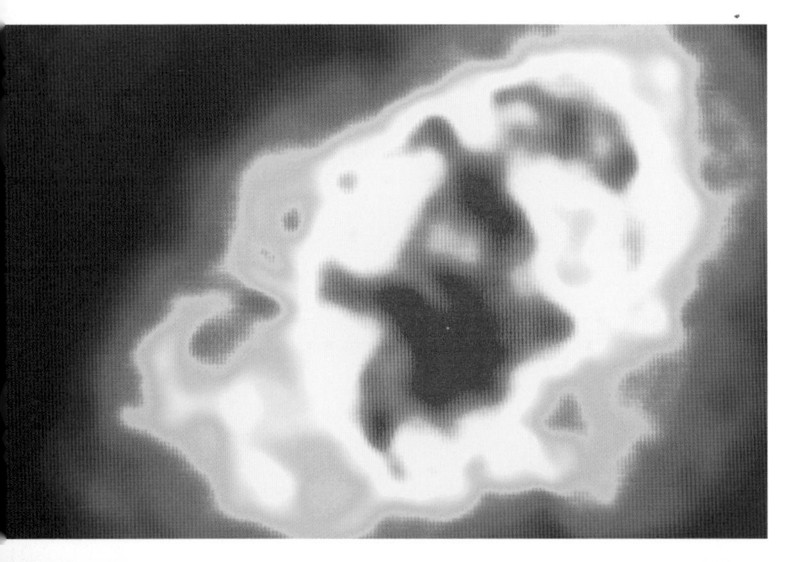

← **The Crab Nebula (M1)** is seen here as it appears in the radio region of the spectrum at a wavelength of 1½ inches (3.6 cm). The photo was obtained using the Very Large Array (VLA) in New Mexico, USA.

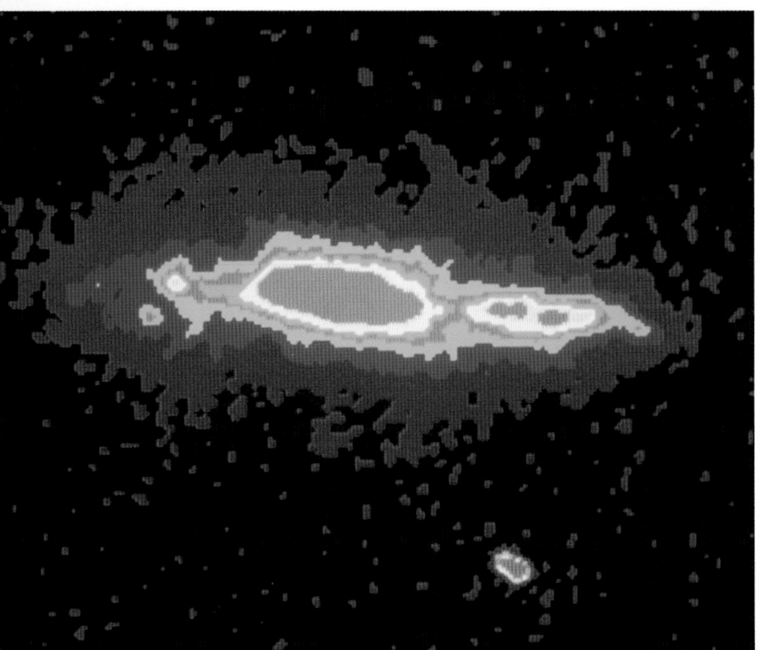

← **Radio telescopes** are usually dishes with parabolic curvature. The radio waves bounce off the dish and are focused at the receiver, situated on a framework above the dish surface.

⇇ **This photo**, again taken using the VLA, is a false-color radio map of the spiral galaxy NGC 4631, which is seen edge-on from Earth.

MEASURING RADIO WAVES: ARECIBO

Opened in 1963, the radio dish in Puerto Rico near Arecibo is still the largest single astronomical dish in the world. Measuring 1000 feet (305 m) across, 167 feet (51 m) deep and covering about 20 acres (8 hectares), the dish is suspended from wires above a natural bowl in Earth's surface. Its surface is a mesh of 40,000 aluminum panels curved into a spherical shape—not the usual parabolic shape often used for radio dishes. Because it is so large and fixed to the ground, the dish is not steerable. Naturally, this limits the objects it can see. Instead, the powerful telescope scans a strip of the sky as the heavens rotate around it. The Arecibo dish is used not only for astronomy, but also for studies of the atmosphere. And in 1973, the radio dish was famously employed to beam a radio message to the globular cluster M13, in an attempt to contact any extraterrestrials who might be listening in. Ironically, since this publicity stunt, good theoretical reasons have emerged for believing that globular clusters are extremely unlikely places to find planets. In any case, since M13 is around 23,000 light-years away, it will take at least 46,000 years before we receive any potential reply.

↑ **Suspended 450 feet (138 m)** above the surface of the Arecibo radio dish is a 900-ton (915 tonne) platform somewhat like a bridge. Here the platform is seen tracing a circle of lights in a time-lapse image.

↓ **The electromagnetic spectrum (EMS)** is a type of radiation that includes visible, or optical, light. Radio waves have the longest wavelengths (lowest frequency), while gamma rays have the shortest wavelengths.

ELECTROMAGNETIC SPECTRUM

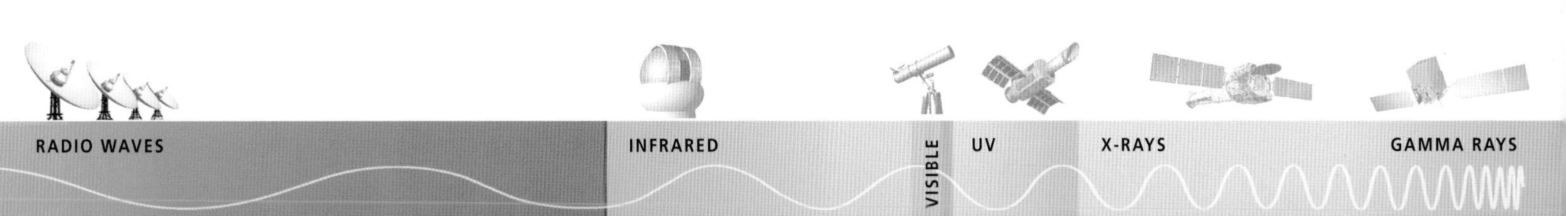

Infrared astronomy

Just beyond the range of human visibility is the domain of the infrared part of the electromagnetic spectrum. Infrared means "below red," because this region has wavelengths slightly longer than that of visible light—in the range 1/25,000 to 1/80 inch (1 to 300 micrometers). Although you cannot see infrared, you can still detect it—as heat. Astronomical objects that emit this heat radiation include very cool objects, such as star-forming regions, protoplanetary disks and the cores of some active galaxies. Infrared astronomy is important, but ground-based research—though possible at certain wavelengths—is hampered by our atmosphere. Therefore, to carry out their research more effectively, astronomers must use infrared satellites.

← **American and Dutch technicians** prepare the Infrared Astronomical Satellite (IRAS) for launch at a United States Air Force base in California. IRAS—a Dutch, US and UK mission—was the original pioneering infrared satellite, launched in 1983 to study the sky at infrared wavelengths, without atmospheric disturbance, for the very first time. It revolutionized infrared astronomy, despite its relatively short lifetime of just 10 months.

GROUND-BASED INFRARED: UKIRT
Like other ground-based infrared observatories, the United Kingdom Infrared Telescope (UKIRT) observes in the near-infrared, adjacent to the red end of the visible spectrum. The observatory, opened in 1979, is situated at an altitude of 13,700 feet (4200 m) at the Mauna Kea Observatory in Hawaii, but it is owned by the Particle Physics and Astronomy Research Council (PPARC) in the UK. A reflecting telescope, UKIRT is the world's largest observatory dedicated solely to the infrared. The mirror is 12½ feet (3.8 m) across, and was the first in such a size class that was constructed to new, thin and compact specifications. Because of the nature of infrared light, the images obtained by UKIRT are sharper than those that can be obtained with a ground-based optical telescope of similar size. Mauna Kea is higher than any other observatory, and as such is generally regarded as the clearest place on Earth for ground-based infrared and optical astronomy.

← **The open dome that houses** the 12-foot (3.8 m) United Kingdom Infrared Telescope is on the left in this photo, next to a smaller telescope belonging to the University of Hawaii.

↓ **The Crab Nebula (M1)** as it appears at a near-infrared wavelength of about 1/25,000 inch (1 micron). At the center lies the famous Crab pulsar, 6000 light-years away.

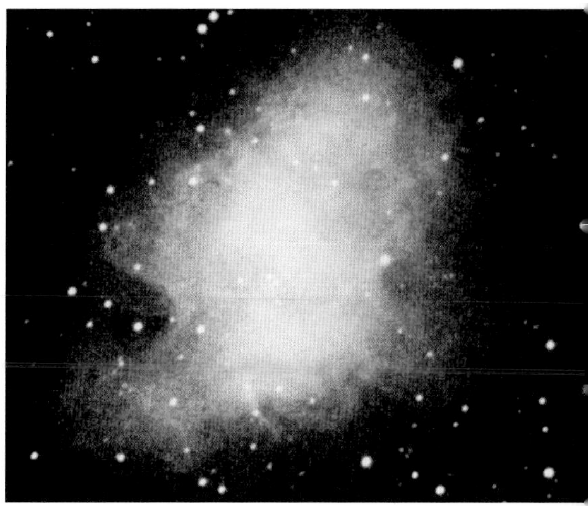

SPACE-BASED INFRARED: IRAS

Prior to 1983, infrared astronomy was extremely limited on Earth. Although it is possible to observe in the infrared through certain wavelength "windows," water vapor and carbon dioxide in our atmosphere absorb the rest of the infrared spectrum. Indeed, this is the reason behind the greenhouse effect: the Sun heats up the Earth, but only part of this heat radiation can escape back into space through the atmosphere, while the rest is trapped and absorbed. In January 1983, infrared astronomy got a major upgrade when a joint British–Dutch–US mission called IRAS—the Infrared Astronomical Satellite—came onto the scene. It was equipped with a 2-foot (0.6 m) telescope and an array of different detectors. To avoid interference from other heat sources, such as the telescope itself, the satellite was supercooled to just a few degrees above absolute zero. IRAS was the first satellite to systematically observe almost the entire sky at infrared energies. It performed four separate all-sky surveys, each at a different infrared wavelength, covered 95 percent of the sky, and observed in total some 250,000 sources—before its planned 10-month lifetime (dictated by the amount of on-board coolant) expired. In its brief time, the IRAS mission has had a major impact on many aspects of astronomy. Among IRAS's discoveries were several new comets, celestial objects such as Beta Pictoris that have since been confirmed to be planetary systems in the earliest stages of formation, infrared emission from active galaxies and the first view of the core of the Milky Way. Occasionally, astronomers refer to a galaxy that is very bright in the infrared as an IRAS galaxy, since it was IRAS that discovered galaxies of this type. They are suspected to be galaxies that are interacting with others and so they contain a great deal of active star formation.

ELECTROMAGNETIC SPECTRUM

RADIO WAVES INFRARED VISIBLE UV X-RAYS GAMMA RAYS

UV astronomy

Ultraviolet, as the name suggests, is a form of electromagnetic radiation having wavelengths just beyond the violet end of the optical spectrum, where the human eye cannot probe. Many objects—including active galaxies, novae, supernovae, and of course the Sun—emit this dangerous radiation. Fortunately for those of us conscious of the dangers of UV-induced skin cancer, the ozone in our atmosphere largely shields us from this constant onslaught. Unfortunately for astronomers, they can only carry out ultraviolet astronomy using satellites. Ultraviolet astronomy began with a series of satellites in the late 1960s, but it wasn't until 1978 that the International Ultraviolet Explorer satellite really showed astronomers how many ultraviolet sources are "out there."

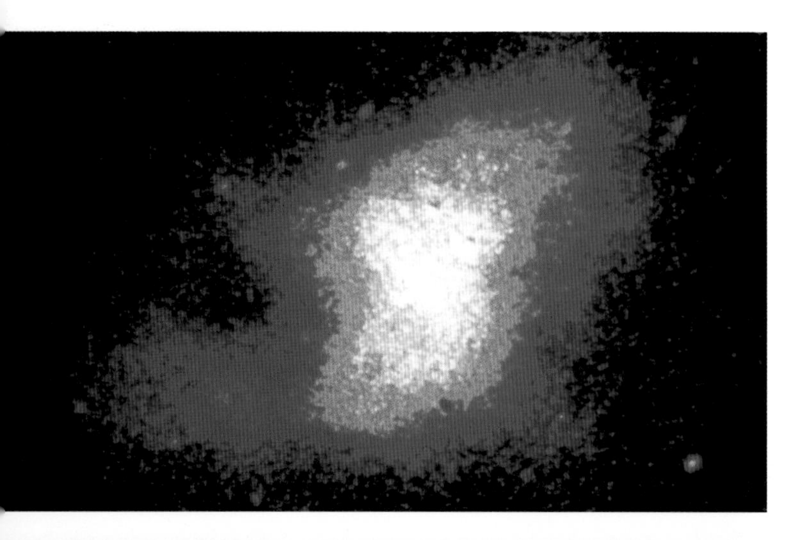

← **The Crab Nebula (M1)** is seen here in ultraviolet light. The colors are false and represent the intensity of the emission, which decreases from the center outward.

← **In ultraviolet light** , the surface of the Sun appears dark because its temperature is much lower than that of the atmosphere. The bright orange loops are ionized gas filaments.

→ **Here, Earth is captured** in ultraviolet light, using an ultraviolet camera on board the Apollo 16 spacecraft. The red emission comes from an otherwise invisible "halo" of hydrogen gas, which surrounds the planet.

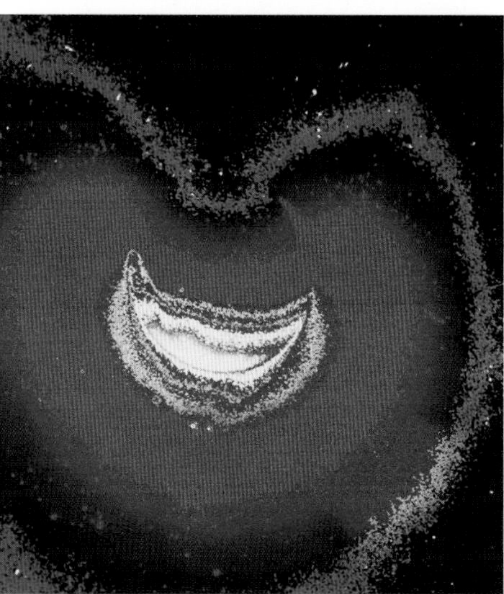

MEASURING UV: IUE

The International Ultraviolet Explorer (IUE) was a joint UK–NASA–ESA mission. Launched on July 26, 1978, IUE was the first satellite dedicated solely to ultraviolet astronomy, and it significantly improved astronomers' understanding of this science, in the Solar System and far beyond. The main telescope aperture was 1½ feet (45 cm), and there were also a pair of spectrometers. These are instruments that split light up into a "rainbow" of wavelengths—in this case, to form a spectrum of ultraviolet light—and which give important information about the temperature, composition and speed of astronomical objects. IUE was placed in a so-called geosynchronous or geostationary orbit, 22,300 miles (35,900 km) above the surface of Earth. At this altitude, all satellites take exactly 24 hours to complete an orbit and so stay above the same point on the surface of Earth at all times, which is why communications satellites are often parked there. From this orbit, IUE observed tens of thousands of ultraviolet-emitting objects in space. Despite having a planned lifetime of three to five years, IUE in fact became the longest-serving astronomical satellite—switched off in September 1996 after a durable 18 years of activity.

↑ **The International Ultraviolet Explorer (IUE)**, portrayed here in an artist's rendering, was the first astronomical satellite to be placed into a high Earth orbit, much farther out in space than previous observatories had orbited.

ELECTROMAGNETIC SPECTRUM

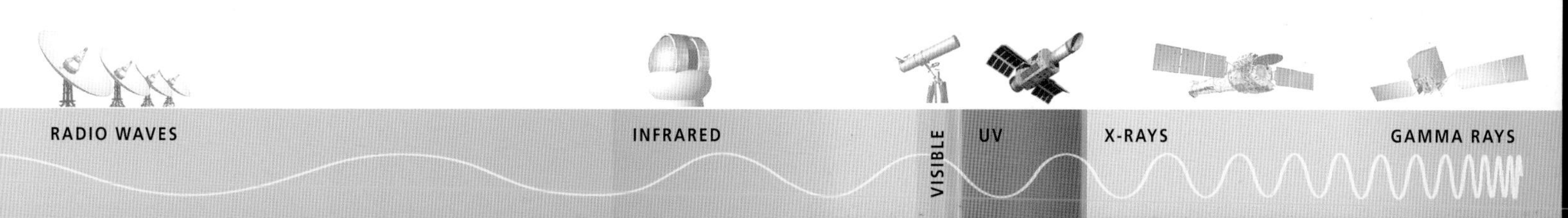

X-ray and gamma-ray astronomy

Blueward of the ultraviolet is the X-ray part of the electromagnetic spectrum, with very small wavelengths down to about 1/2,500,000,000 inch (0.01 nanometer). Merging with the short-wavelength X-ray realm and extending to even smaller wavelengths are the gamma rays. X-rays and gamma rays are the most energetic forms of radiation, and they are emitted by very violent and extreme processes—for example, by the accretion disks that surround black holes in some binary stars. Indeed, the first celestial X-ray source found (in 1962), not including the Sun, was one such binary star called Scorpius X-1. As with ultraviolet and infrared astronomy, researchers can only probe this region of the spectrum with satellites floating above Earth's atmosphere.

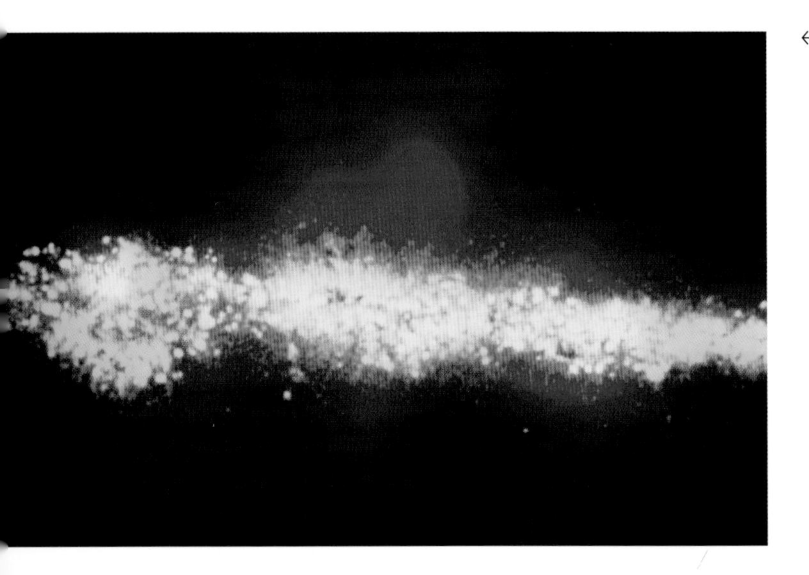

← **This unusual photo** is a combination of images taken in the optical and X-ray parts of the spectrum, showing the spiral galaxy NGC 4631 seen edge-on. The red areas are regions of intense X-ray emission, while the green parts show vibrant optical emission from regions where stars are forming.

← **Eastman Kodak technicians** are here hoisting the partially complete Chandra X-ray Observatory into a vertical position. Chandra, still flying, is the most powerful and successful X-ray observatory ever built.

COMPTON GAMMA-RAY OBSERVATORY

Very high-energy gamma rays can penetrate our atmosphere, enabling a limited form of gamma-ray astronomy to be performed on the ground. But to get the best look at the gamma-ray sky, astronomers use satellites. In 1991, NASA launched the Compton Gamma-Ray Observatory (CGRO). It is named in honor of the American physicist Arthur Compton who was a pioneer of high-energy X-ray physics. CGRO is designed to observe celestial objects in the gamma-ray region of the spectrum, just beyond the X-ray realm. In all, the 16-ton (16.3 tonne) satellite has four instruments. One, called BATSE (Burst and Transient Source Experiment), is specifically used to observe gamma-ray bursts—mysterious blasts of highly intense gamma rays from the edge of the Universe. It has significantly improved astronomers' grasp of these objects. The other three instruments are used to image the sky in three different gamma-ray bands of the spectrum.

← **In the payload bay** of the space shuttle Atlantis, an American astronaut readies the Compton Gamma-Ray Observatory (CGRO) for insertion into Earth orbit in 1991. One of the solar panel arrays that will provide the observatory with power is in the background.

↓ **One of many images** taken by the Compton Gamma-Ray Observatory (CGRO) satellite, this photo shows the Crab Nebula (M1) as it would appear if you could detect gamma rays. The colors are false and represent intensity, with red and yellow the strongest emission regions.

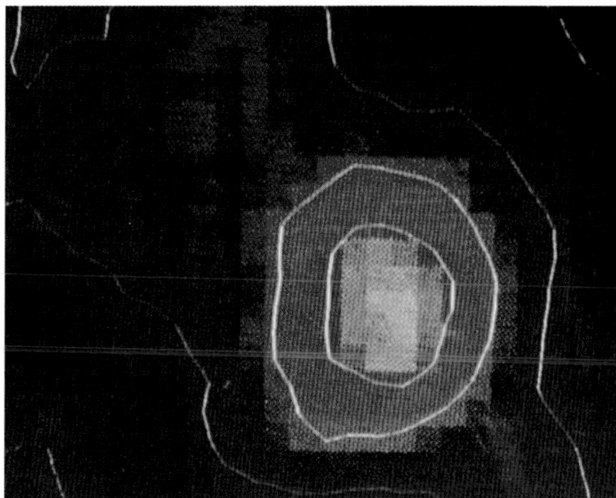

CHANDRA X-RAY OBSERVATORY

X-ray astronomy took a giant leap forward with the launch of NASA's Chandra X-ray Observatory in 1999, the largest satellite ever launched by shuttle. It currently orbits Earth 200 times higher than the Hubble Space Telescope. Chandra's resolving power—which is equivalent to being able to read a road sign from 12 miles (18 km)—is at least 50 times higher than the best X-ray telescope that came before it, the ROSAT Observatory, revealing images of the X-ray Universe in unprecedented clarity. Part of this power stems from Chandra's mirrors, which are the largest, smoothest and most precisely shaped mirrors ever ground. Indeed, if the surface of Earth were as smooth as Chandra's mirrors, the highest mountains would measure just a few inches high. Like other X-ray telescopes, Chandra has so-called grazing-incidence mirrors. Because X-rays have a tendency to pass right through objects rather than bounce off them—which is what makes them so good for medical imaging purposes—normal mirrors cannot focus them. But if the X-rays are incident at a very shallow angle, almost parallel to the surface, they can be focused. So Chandra's mirrors are essentially a series of tapered coaxial tubes. The X-rays enter the end of the tube, bounce off the sides and are focused at the end. Chandra is rapidly improving astronomers' understandings of many high-energy objects such as pulsars, black holes, supernovae and active galaxies. The Chandra X-ray Observatory is named in honor of the Indian astronomer and Nobel laureate Subrahmanyan Chandrasekhar—one of the twentieth century's leading astrophysicists. It is Chandrasekhar who proved theoretically that white dwarf stars can be no more massive than 1.44 times the mass of the Sun—the so-called Chandrasekhar limit.

ELECTROMAGNETIC SPECTRUM

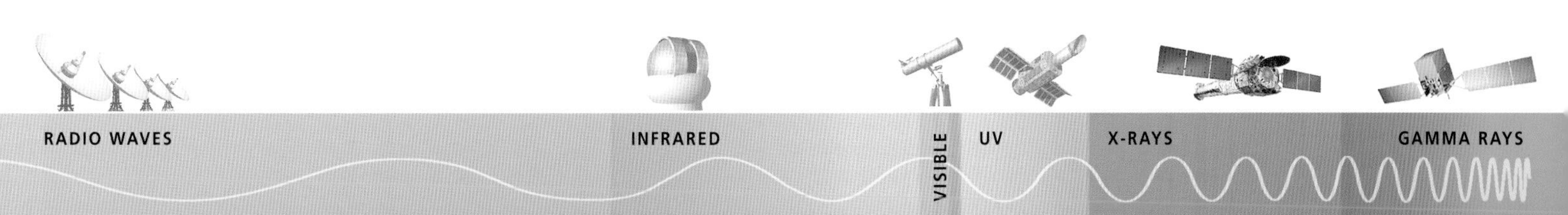

The space race

1946 The German rocket engineer Wernher von Braun, who was in charge of the German long-range missile program during the war, surrenders to the United States. With his help, the Americans build their first copies of the German V-2.

Late 1940s The Cold War between the United States and the Soviet Union begins, a period characterized by mutual mistrust and military competition as each side seeks to better the other with missile technology and nuclear firepower.

1956 Soviet designers Sergei Korolev and Konstantin Feoktistov begin working on the problem of how to build a spacecraft capable of sending a person into space and returning him safely home. This is the beginning of the Vostok project, a spherical craft designed to carry a human into space.

1957 In August, the Soviets launch the first intercontinental ballistic missile (ICBM). Two months later, they launch the first artificial satellite, the bleeping Sputnik 1, aboard a modified ballistic missile called an R-7. Just one month later, in November, they launch the first animal into space, the dog Laika, aboard the substantially larger Sputnik 2. Sputnik—Russian for "traveler"—sparks what popularly becomes known as the "space race."

1958 Construction begins in the USSR on the first Vostok spacecraft. The National Aeronautics and Space Administration (NASA) is created in America.

1960 The Americans, fearing a nuclear attack from the Soviets, launch the first spy satellite, called Corona, disguised as a scientific probe. It successfully returns accurate photographs of Soviet territory taken from orbit.

1961 US President John F. Kennedy makes a speech in which he pledges to put a man on the Moon by "the end of this decade." The Soviets fly the first man into space, Yuri Alekseyevich Gagarin, who orbits Earth once and then returns in his one-man Vostok 1 spacecraft. Three weeks later, Alan Shepherd becomes the first American in space, in a one-man Mercury capsule, but does not orbit Earth.

1962 John Glenn becomes the first American to orbit Earth, which he does three times in his Mercury craft.

1963 Valery Bykovsky becomes the first person to sustain a long-duration spaceflight (five days), aboard Vostok 5. While he is still in space, Soviet cosmonaut Valentina Vladimirovna Tereshkova, launched on the last Vostok mission (number 6), becomes the first woman in space. She orbits Earth 48 times in 71 hours before returning safely.

1964 First two unmanned test flights of the new American spacecraft, Gemini—designed to carry a crew of two. It is essentially a larger version of the more primitive, one-man Mercury capsule. The Soviet Voskhod 1 spacecraft—an expanded Vostok—performs the first two-man flight.

1965 First manned Gemini flight, Gemini 3, with Virgil Grissom and John Young. In March, Soviet cosmonaut Alexei Leonov performs the first "space walk"—drifting in space above Siberia while tethered to his spacecraft Voskhod 2. In June, Edward White becomes the first American to do the same from his Gemini 4 craft.

1966 The Soviets test-fly their Soyuz spacecraft, which is designed to seat three, but is tested without crew.

1967 Disaster for the Americans in January, when a fire on the ground kills three astronauts who were testing a spacecraft that was posthumously named Apollo 1. In April, disaster also strikes the Soviets when the Soyuz 1 spacecraft crashes after a parachute failure, killing its only occupant Vladimir Komarov—the first person ever to die during a spaceflight. Also this year was the first testing of the Americans' Saturn V rocket, designed to carry a payload to the Moon.

1968 Apollo 8, launched by Saturn V rocket, becomes the first craft to carry humans beyond the gravity of the Earth, when it orbits the Moon several times before returning safely home.

1969 In March, Apollo 9 tests the Lunar Lander in orbit. In May, Apollo 10 takes humans to the Moon and goes through the motions of a landing as a rehearsal, but does not actually land.

1957 The launch of Sputniks 1 and 2 stokes the first fires of the space race between the US and the USSR.

1963 Cosmonaut Valentina Tereshkova has reason to smile. She was the first woman in space.

1965 Edward White becomes the first American to tread the vacuum of space above Earth.

1969 In July, Apollo 11 lands the first men on the Moon, Neil Armstrong and Edwin Aldrin. Apollo 12 lands the second pair of humans on the Moon, Alan Bean and Charles Conrad, in November.

1970 Disaster strikes Apollo 13 when an oxygen tank explodes. Because of the ingenuity of the crew and ground staff, the crew is safely returned to Earth after rounding the Moon.

1971 In February, Apollo 14 puts Alan Shepherd and Edgar Mitchell on the Moon. In April, the Soviets launch the first successful space station, Salyut 1, an orbiting platform on which to perform experiments. In July, Apollo 15 touches down on the Moon with David Scott and James Irwin.

1972 In April, Apollo 16 lands John Young and Charles Duke on the Moon. And on December 11, the final Apollo mission, number 17, carries Eugene Cernan and Harrison Schmitt to the lunar surface. No humans have been back since.

1973 NASA launches its first manned space station, Skylab, made from the upper stage of a Saturn V rocket. Three crews occupy it during its lifespan. Also, following the Americans, the Soviets launch their own spy satellite, called Kosmos.

1975 The Apollo–Soyuz Test Project—in which a US Apollo spacecraft docked above Earth with a Soviet Soyuz one—marks a groundbreaking collaboration between the two superpowers.

1979 The now-abandoned Skylab falls from orbit and breaks up above Earth, parts of it famously crashing into Western Australia.

1981 Maiden flight of the first American space shuttle, Columbia.

1983 Sally Kristen Ride becomes the first American woman in space, aboard space shuttle Challenger.

1984 In July, cosmonaut Svetlana Savitskaya becomes the first woman to perform a spacewalk, 18 years after the first man. She is also only the second woman in space (in 1982) despite more than two decades of spaceflight. In October, Kathryn Sullivan becomes the first American woman to walk in space, tethered to space shuttle Challenger.

1986 The Soviet Union launches its space station Mir. The space shuttle Challenger explodes shortly after lift-off, killing all seven astronauts on board. The remaining fleet is temporarily grounded.

1988 The Soviets experiment with their own reusable space shuttle, called Buran. However, the program is scrapped after its first launch, which was unmanned. Cosmonaut Valeri Polyakov begins a record stay on Mir lasting 438 days.

1990 With the fall of the Soviet Union and the end of the Cold War, collaborations between Russia and the United States become more commonplace.

1991 Mir passes its predicted five-year lifespan. Helen Sharman becomes the first British astronaut on Mir.

1995 Arrival at Mir of the first crew to be ferried by a space shuttle, Atlantis. Previous crews had been sent up by Russian Soyuz craft.

1996 After a 188-day stay on Mir, American astronaut Shannon Lucid returns to Earth. She holds the record for the longest stay in space for a woman.

1997 In February, a fire on Mir causes the crew to almost abandon the craft. In June, a cargo ship collides with Mir, causing further damage. These are just the beginnings of many problems with the ailing space station.

1998 The Russians initiate construction on the International Space Station (ISS), with completion expected in 2005. It will hold seven people when complete, and countries participating include the United States, Russia, France, Canada and Japan.

1999 The Mir cosmonaut Sergei Avdeyev becomes the person with the longest stay in space, although not consecutively, with a total of 681 days.

2000 The aging and failing Mir space station is finally deorbited—deliberately crashed to Earth after nearly 15 years in space. Meanwhile, the first occupants of the partially complete International Space Station—two Russians and an American—take up residence.

2003 The Americans lose another shuttle, Columbia (the first to fly). As a result, all flights are grounded, and completion of the ISS delayed.

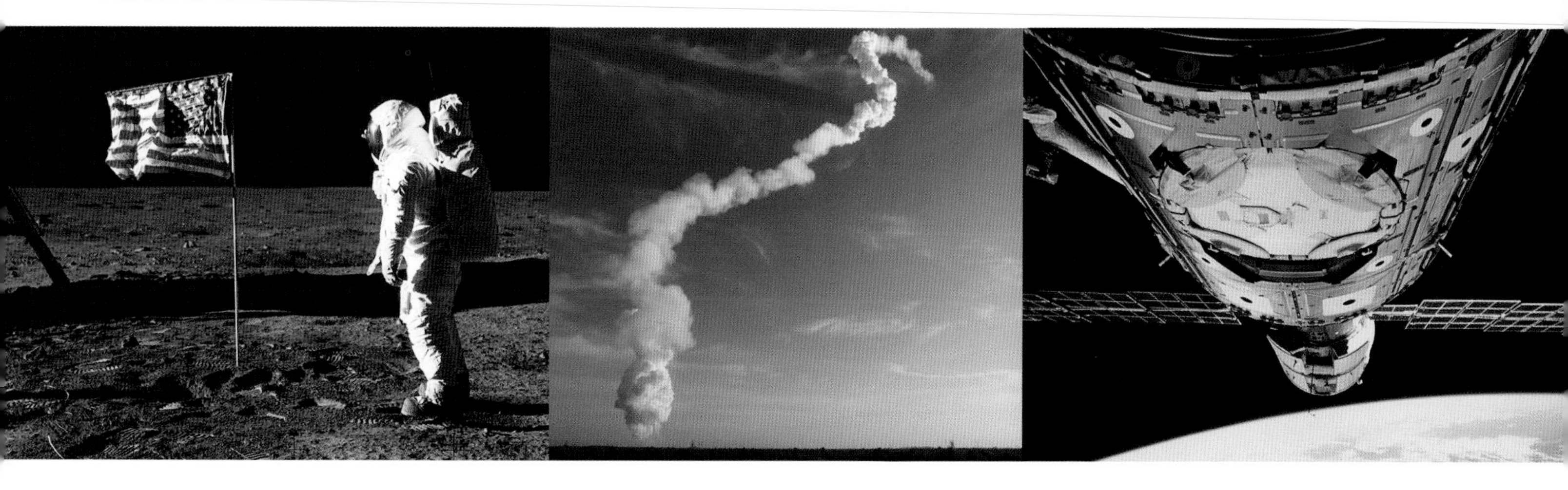

1969 Millions watch as an American astronaut stands beside his country's flag on the surface of the Moon.

1981 The space shuttle Columbia lifts off from Cape Canaveral and trails a pillar of smoke in its wake.

1998 The International Space Station begins to take shape, here photographed from the shuttle Endeavour.

Mission milestones

The world watched in awe as American astronaut Neil Armstrong uttered the now-famous words: "One small step for a man; one giant leap for mankind." The time was July 1969; the place: the Moon's Sea of Tranquility; the mission: Apollo 11. For the Americans it was a triumph, a defeat of their great foe, the Soviet Union, in the race to land men on the Moon and return them safely to Earth. Six other Apollo missions followed, with Apollo 13 the only failure. But although Apollo 11 remains the greatest achievement of the manned space program even now, other milestones must not be forgotten. Indeed, the Soviets beat the Americans in many "firsts." The dog Laika became the first animal in space (1957). The first man was Yuri Alekseyevich Gagarin (1961, *right*). Valentina Vladimirovna Tereshkova was the first woman (1963). And Alexei Leonov was the first man ever to "walk" in space (1965). The space race is now over. The USSR has dissolved and the Cold War has ended. Now, instead of competition, the operative word is "cooperation," with the International Space Station as the perfect example of this change.

TO THE MOON...
After the launch by Saturn V rocket (1), the command-and-service module (CSM) and the lunar module (LM), docked to each other, begin to make their way to the Moon (2 and 3). Once in Moon orbit (4), the LM, carrying two astronauts, breaks away from the CSM (5) and descends to the surface (6), while the CSM waits in orbit with a third astronaut on board.

...AND BACK
To return to Earth, the upper half (ascent stage) of the LM blasts off from the Moon (1), leaving the landing feet behind. The LM docks with the CSM (2) to let the astronauts back in, then the LM is jettisoned (3) and the CSM leaves Moon orbit for Earth. (4) Once there, the command module (CM) separates from the CSM (5) and safely returns the astronaut to Earth (6 and 7).

← **Footprints left** by the 12 Apollo moonwalkers will last millions of years on the Moon, where the only erosion comes from the steady rain of micrometeorites, and where wind is non-existent.

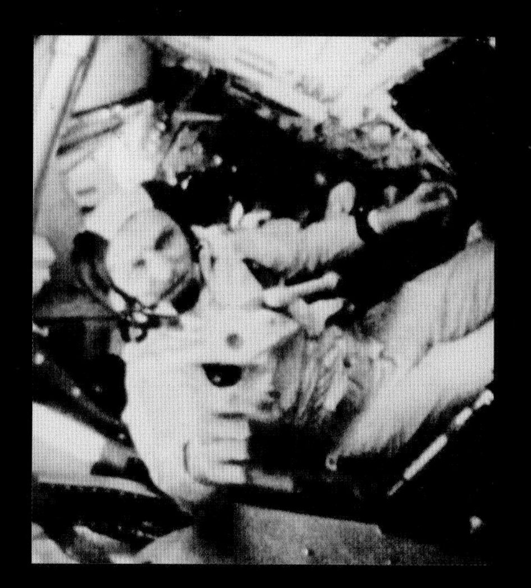

← **In December 1968, Apollo 8** left Earth's gravity and entered into an orbit around the Moon with its crew of three. It was the first time a human had ever escaped the gravity of Earth, orbited another planetary body, and seen Earth as a planet in its own right, gently drifting in space.

LAST MISSION

It is incredible to think that no person has set foot on the Moon since 1972. This last historic mission was Apollo 17, in December of that year, when astronauts Eugene Cernan and Harrison Schmidt set down near the crater Littrow. In total, only 12 men have ever walked on the Moon—and no women. There might have been others, but public interest in the Moon landings rapidly deteriorated after Apollo 11, and NASA finally cut the budget, putting an end to an era.

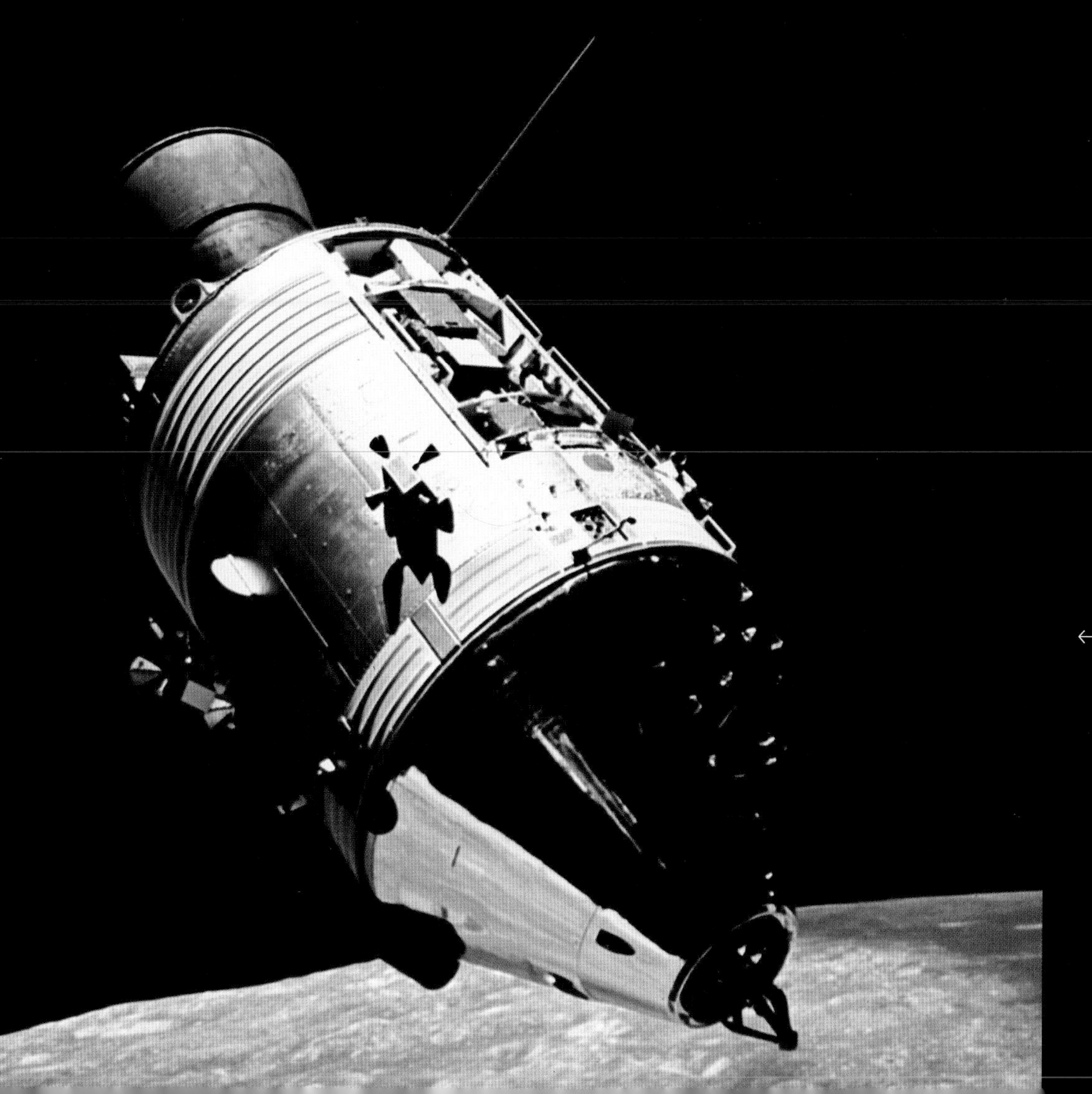

← **The lunar command module** America orbits the Moon during the last manned mission there in 1972—Apollo 17. This photo was taken by an astronaut from the lunar excursion module as it was making its way back to the command module before the final return to Earth.

Space disasters

Space travel may be exotic and heroic, but it is not without danger. Since the manned space program began in the early 1960s, both sides in the space race have suffered a loss of lives. The first person to die as a result of a disaster in space was the Soviet cosmonaut Vladimir Komarov, in April 1967, when the parachute on his Soyuz 1 craft failed to operate after it re-entered the atmosphere. The Americans lost three astronauts that same year during an early Apollo pre-flight test. The three men were sitting in their command module when a rapid fire started—fueled by the cabin's atmosphere of pure oxygen. The hot gases inside the craft had damaged the hatches, so the men could not be evacuated. They died within seconds from smoke inhalation. More recently, the most vividly remembered space disasters are those that befell the Challenger and the Columbia space shuttles, in 1986 and 2003 respectively.

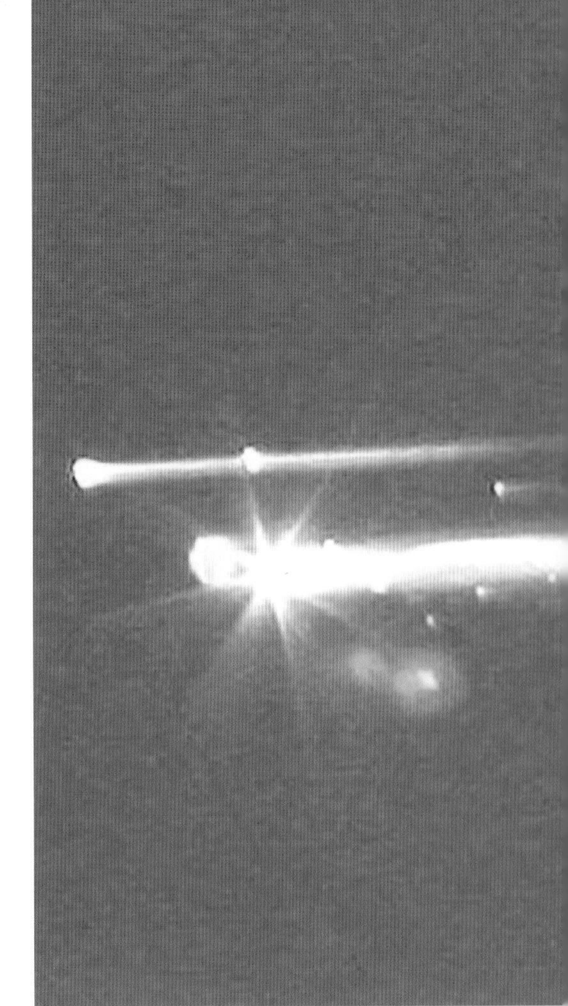

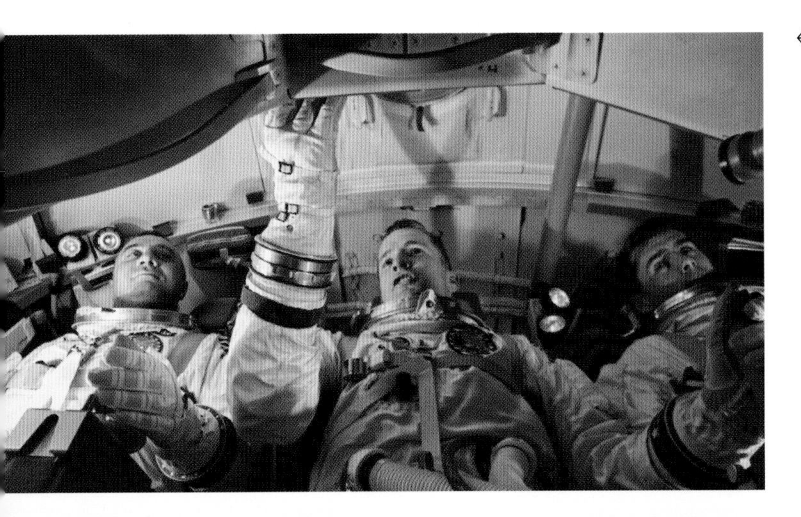

← **In April 1967,** a flash fire swept through the cockpit of the Apollo 1 command module while it was being tested on the launch pad with three astronauts inside—Roger Chaffee, Edward White and Virgil Grissom (*from left to right*). None of them survived. This photograph of the astronauts was taken in a simulator a few months before the deadly accident.

SPACE SHUTTLE CHALLENGER

The world reeled on January 28, 1986, when the space shuttle Challenger exploded in a huge ball of red and yellow flame (*left*) just 73 seconds after launch, 11 miles (17 km) above Earth. The disaster resulted in the death of all seven astronauts on board, including a teacher, 37-year-old Sharon Christa McAuliffe. The cause was traced to a faulty rubber O-ring seal on one of the solid rocket boosters, which had become brittle during a period of extreme cold at Cape Canaveral. Before the disaster, Challenger had flown nine successful missions. Despite stringent safety precautions being implemented afterward, a similar disaster might also have happened in 1995, when the same O-rings that had doomed Challenger were found to have been damaged on the Atlantis shuttle during take-off. NASA feared for the safety of the Discovery shuttle, which was in orbit at the time of the Atlantis find. However, there were no eventualities.

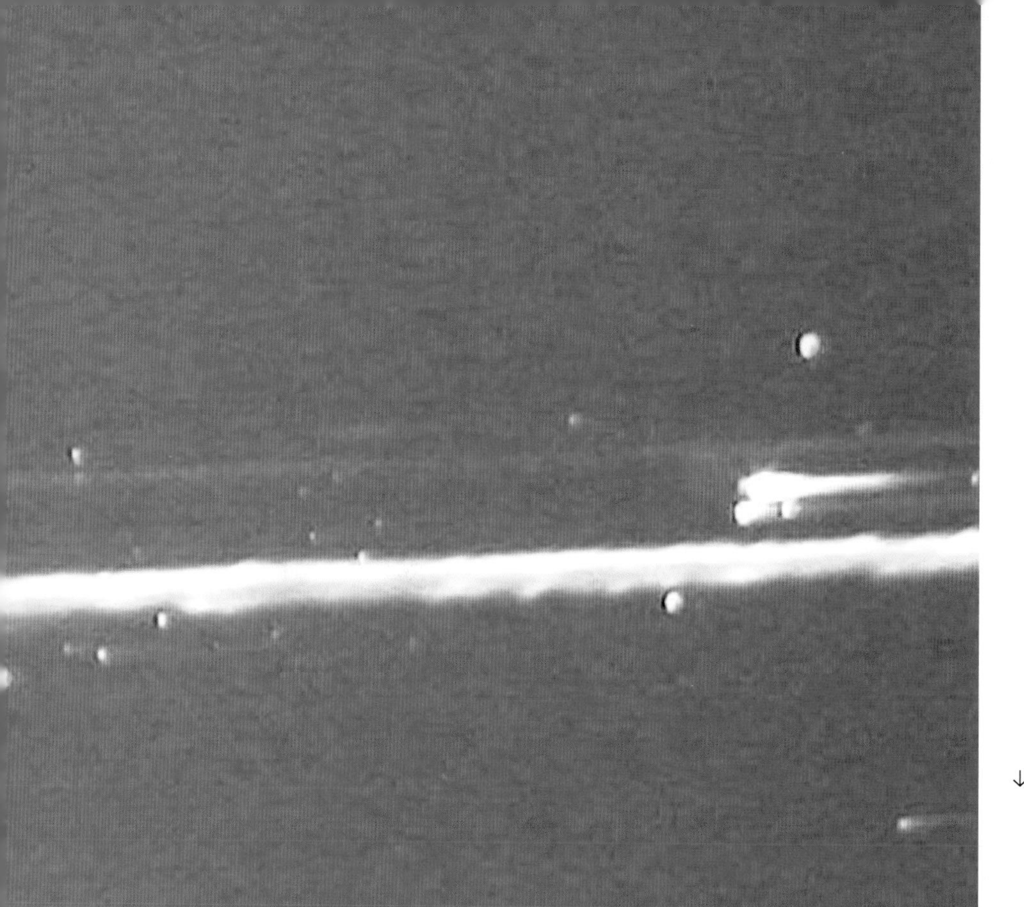

FUTURE OF MANNED MISSIONS

Due to the Columbia shuttle disaster, the American manned space program is currently at a standstill. Meanwhile, construction of the International Space Station has stalled, as the shuttles were ferrying the parts into orbit. The final damning report by Columbia's investigative team found that NASA had not much improved its attitude toward safety since the original Challenger disaster. But NASA has said that it will abide by the recommendations of the investigative team to effect "sweeping changes." In any case, some argue that manned missions waste resources. Why send people when you can send a probe at a fraction of the cost—and with no danger to human life? Indeed, plans for a manned mission to Mars are still embryonic at best. But it is human nature to explore. It is just a question of time before we turn our attention to Mars—and beyond.

↓ **"Houston, we've had a problem,"** went the now-famous words of Jim Lovell, commander of the ill-fated Apollo 13 mission to the Moon. After an oxygen tank exploded, the crew had to construct a CO_2 filter (*below*, seen in the hands of astronaut John Swigert) to ensure that their air remained breathable. Then the ship had to round the Moon before attempting a re-entry into Earth's orbit. Miraculously, all three occupants survived by the narrowest of margins, owing to their ingenuity and that of the staff at Mission Control.

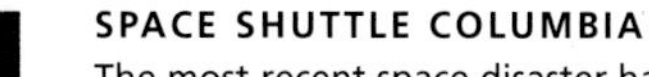

SPACE SHUTTLE COLUMBIA

The most recent space disaster happened on February 1, 2003, when the Americans lost another space shuttle—and with it, seven more astronauts (*left*). This time it was Columbia, the first shuttle to fly, in 1981. When the shuttle was launched on January 16, engineers reported to NASA that a piece of insulation foam had detached from the craft as it left Earth. However, according to the official enquiry into the event, NASA "failed to heed the alarm bells." After an otherwise successful 16-day mission, disaster struck just 15 minutes before Columbia was due to land. While it was re-entering the atmosphere at 18 times the speed of sound, spectators observed as the craft simply disintegrated before their eyes, 39 miles (62 km) above Earth, to form glowing trails of debris (*above*). It is generally believed the insulation foam had damaged the wing as it came off, allowing the super-heated air to enter the shuttle as it was re-entering the atmosphere—with terrible consequences.

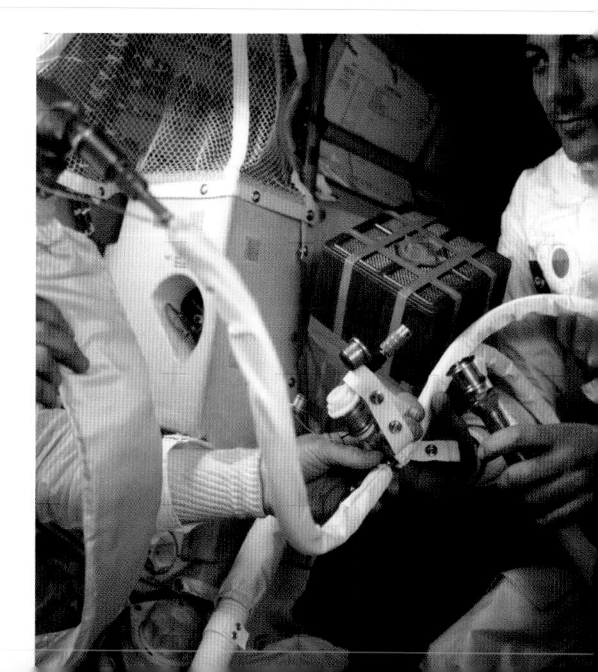

Spacecraft

Space vehicles come in many guises. There are unmanned satellites that orbit Earth, used for weather monitoring, communications and other purposes. There are manned space vehicles. There are space stations. And, of course, to get manned and unmanned payloads into space in the first place, the most important kind of spacecraft is the launch vehicle—in all cases, a form of rocket. The first satellites were lifted into orbit using rockets that were little more than intercontinental ballistic missiles. Later, as the technology improved, the rockets grew more complex and more reliable. The largest launch vehicle ever was the Saturn V—a three-stage monster used to launch the Apollo missions to the Moon. Today, the two major launch vehicles are the reusable American space shuttle (*below*) and the three-stage Russian Soyuz rocket.

LAUNCHING THE SPACE SHUTTLE
During take-off, the shuttle is piggybacked to a red external tank (ET) and two solid rocket boosters (SRBs). Once the fuel is spent, the SRBs drop away and are recovered and reused, but the ET is discarded. The three engines at the rear of the shuttle provide the final thrust to get the craft into orbit.

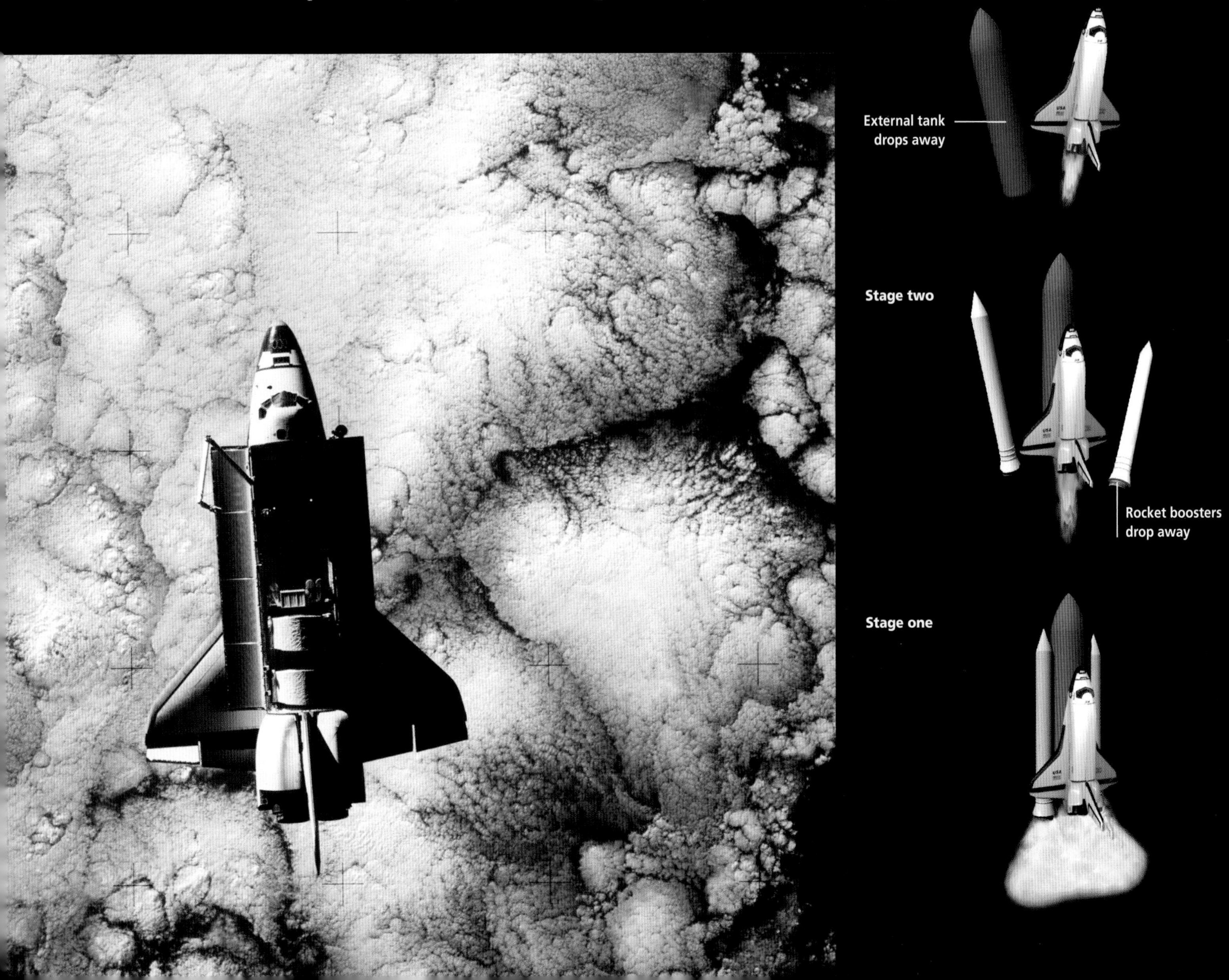

SOYUZ SPACECRAFT
The longest-serving spacecraft in the world is the Soyuz, developed by the Soviet Union in the early 1960s. Soyuz is a three-man spacecraft, and the latest version—Soyuz TM—was most recently used to ferry cosmonauts to and from the Russian space station, Mir. Unlike the American space shuttle, the Soyuz spacecraft cannot be reused. A new one has to be built for every mission, which dramatically increases the cost of launching into space. This photo shows the vehicle used to send the Soyuz into space.

SPACE SHUTTLE
The American space shuttle is still the world's only fully reusable launch vehicle. Originally there were six members of the fleet—Enterprise, Columbia, Challenger, Discovery, Atlantis and Endeavour—although not all were intended as actual launch vehicles. The first one to be built, Enterprise, was flown only during tests, never into space, while the Columbia space shuttle flew the first five real missions, beginning in 1981. Following the losses of Challenger in 1986 and Columbia in 2003, only three active shuttles now remain. They are designed to carry observatories and other satellites into low-Earth orbit, typically at altitudes between 190 and 350 miles (350 to 560 km), where the orbital speed is around 17,500 mph (28,000 km/h).

→ **This time-lapse photo** shows the space shuttle Discovery blasting off from launch pad 39-B at Cape Canaveral on December 19, 1999. This was the third of three missions so far to upgrade and repair the Hubble Space Telescope.

↓ **China has been late** to develop a manned space program. Its Shenzhou 5 manned was finally launched, after several unmanned tests, on October 15, 2003. This image shows the recovered re-entry vehicle from one of the test flights before the launch.

Space stations

Since the early 1970s, more than 100 men and women have worked above Earth on board various space stations. A space station is a little like an ordinary satellite, not necessarily larger, but differs in that it has a life-support system to keep its human crew in tip-top condition. The first space station was the Soviets' Salyut 1, launched on April 19, 1971. Since then there have been several others, most notably Skylab, the Soviet Mir (*below*) and the International Space Station (*right*).

INTERNATIONAL SPACE STATION (ISS)
The International Space Station is the most ambitious space collaboration, pooling resources of no fewer than 16 countries—USA, Canada, Japan, Russia, Brazil and 11 European nations. ISS is still under construction but is already being used. When completed, it will be four times larger than Mir—measuring 356 feet (109 m) across, 290 feet (89 m) long, and orbiting 250 miles (402 km) above Earth.

← **Mir floats gently** above Earth, while a hurricane brews below.

MIR SPACE STATION

Possibly the best-known space station, and the longest lived by far, is Mir. Mir (Russian for "peace," "world" and "village") was launched by the Soviet Union in 1986. Fifteen years later, having outlived the Soviet Union itself, Mir was finally de-orbited, having lasted three times longer than anticipated. Despite the bad press that Mir received during its angst-ridden flight—fires, power cuts, a near collision—its status as a milestone of peace and cooperation between two former enemies, America and the Soviet Union, cannot be overstated. The image above shows cosmonaut Valeriy V. Ryumin (*left*), fresh from the space shuttle Discovery, greeting Mir commander Talgat A. Musabayev after a successful docking.

SKYLAB SPACE STATION

Skylab was launched by Saturn V rocket—and in fact manufactured from the converted upper stage of a Saturn V—in May 1973. In all, three separate crews manned Skylab during its flight, which lasted until July 1979, when the station fell to Earth, parts of it famously landing in Australia. Skylab played host to a series of UV experiments and some X-ray studies of the Sun. This photo (*left*) shows Skylab above Earth, as photographed by the last crew to occupy her.

A history of unmanned probes

1957 The Soviets' Sputnik 1 and 2 become the first artificial satellites of Earth.

1958 The US launches its first satellite, Explorer 1. It also launches Pioneer 0—the first spacecraft to attempt to leave Earth orbit. It is destroyed when its first stage explodes. NASA, then new, takes over with Pioneers 1 and 2, but they fail too. Pioneer 3 also fails to leave Earth orbit, but discovers a second radiation belt around Earth.

1959 The Soviets launch Luna 1, the first successful space probe, which sails past the Moon. Luna 2 crashes onto the Moon (deliberately), and becomes the first man-made object to hit another planetary body. Luna 3 becomes the first probe to see the lunar farside. Meanwhile, the US also finally succeeds in flying a probe past the Moon—Pioneer 4.

1961 The US launches the first two Ranger probes, designed to study interplanetary space. Both fall to Earth. Meanwhile, the Soviet Union's first Venus probe, Venera 1, also fails when contact is lost.

1962 The Americans also attempt to reach Venus with Mariner 1, but it is deliberately destroyed when it veers off course. The successful follow-up, Mariner 2, becomes the first probe to reach another planet.

1964 The first successful probe in the Ranger series, Ranger 7, photographs the Moon in close-up.

1965 Mariner 4 becomes the first probe to fly past Mars. It reveals an unexpectedly cratered surface. Meanwhile, Rangers 8 and 9 photograph the Moon in close-up.

1966 The Soviets' Luna 9 makes the first soft-landing on the Moon and sends back pictures. Shortly afterward, the Americans' Surveyor 1 becomes the first US probe to land safely on the Moon.

1967 The Soviets finally achieve success at Venus with Venera 4, which ejects a capsule into Venus' atmosphere. Mariner 5 also flies past Venus. Surveyors 3, 5 and 6 land on the Moon and perform tests as part of the preparation for the Apollo missions.

1968 Surveyor 7, the last, lands on the Moon.

1969 Veneras 5 and 6 repeat the experiment carried out at Venus by Venera 4. Mariners 6 and 7 fly past Mars and provide the first photos of its polar cap.

1970 Venera 7 becomes the first craft to land on another planet—Venus.

1971 Mariner 9 becomes the first probe to enter into an orbit around Mars. It also photographs the Martian moons in close-up.

1972 Venera 8 becomes the second probe to land on Venus, following Venera 7.

1973 The successful NASA probe Pioneer 10 becomes the first to reach a gas giant planet, Jupiter, flying by at a distance of 80,000 miles (130,000 km).

1974 The final Mariner probe, 10, becomes the first to reach two planets (Venus and Mercury) in a single mission. Meanwhile, the last Pioneer probe, number 11, flies even closer to Jupiter than its predecessor, and takes the first close-up photographs of three of its largest moons.

1975 Veneras 9 and 10, landing on Venus, become the first craft to return pictures from the surface of another planet, via radio transmissions.

1976 NASA lands the first two craft on Mars, Vikings 1 and 2. They provide the first photos of the Red Planet's landscape, and test its soil for evidence of chemistry that might indicate life. None is found.

1978 Veneras 11 and 12 touch down on Venus.

1979 Pioneer 11 becomes the first probe to reach Saturn and observe its rings and satellites up close. Meanwhile, NASA's highly successful Voyager probes (1 and 2) reach Jupiter and send back thousands of photos of its cloud tops and satellites.

1980 Voyager 1 reaches Saturn and discovers six new moons before leaving the plane of the Solar System.

1981 Voyager 2 reaches Saturn and traverses the ring plane, and also passes closer to its satellites than had Voyager 1. Veneras 13 and 14 touch down on Venus.

1983 The last two Venera missions, 15 and 16, enter orbit around Venus and map its surface. Pioneer 10, leaving the Solar System, passes Pluto's orbit.

1957 The Soviet Union's bleeping Sputnik 1, our first artificial satellite, was no larger than a basketball.

1965 A scientist puts together a primitive mosaic of Mars from strip photos taken by the Mariner 4 probe.

1976 The Viking spacecraft returned a whole series of photos of the Martian surface, such as this one.

1986 Voyager 2 reaches Uranus—the first probe to do so. Meanwhile, the European Space Agency's (ESA) Giotto and the Soviets' Vega 1 and Vega 2 become the first probes to intercept a comet—the Halley's Comet encounter.

1989 Voyager 2 reaches the last giant planet, Neptune, and photographs it and Triton, its largest moon, in close-up. It also discovers Neptune's rings and several new moons. NASA's Magellan probe is launched toward Venus. The Galileo probe is launched toward Jupiter.

1990 NASA's Magellan reaches Venus and radar-maps its entire surface at a resolution better than 985 feet (300 m). The joint NASA–ESA mission Ulysses is launched to study the solar wind. Its trajectory takes it toward Jupiter where it will get a gravity boost.

1991 NASA's Galileo, on its way to Jupiter, becomes the first probe to pass an asteroid (called Gaspra) and photograph it in close-up. The Japanese launch their solar mission, Yohkoh, which studies the Sun at ultraviolet and X-ray wavelengths.

1992 The Ulysses probe arrives at Jupiter and is swung into a trajectory that passes over the Sun's poles. The Americans launch the probe SAMPEX (Solar, Anomalous and Magnetospheric Particle Explorer) to study cosmic rays, solar flares and the environment of Earth's magnetosphere.

1993 The Galileo probe reaches Ida, another asteroid, on its way to Jupiter. NASA's Mars Observer goes silent just three days before achieving Mars orbit. NASA's Venus probe Magellan is moved to a lower orbit during the first "aerobraking" maneuver to be performed at another planet. From its lower orbit, Magellan makes a map of the gravity field of Venus.

1994 The solar probe Ulysses flies over the Sun's south pole. NASA's Clementine probe orbits and maps the Moon. NASA's Magellan, its important mission complete, finally breaks up in the atmosphere of Venus.

1995 The Galileo probe reaches Jupiter and descends into its atmosphere, where it survives for 58 minutes while taking measurements. Ulysses flies over the Sun's north pole.

1996 Launch of the successful NASA–ESA mission SOHO (the Solar Heliospheric Observatory), to study the Sun in ultraviolet and visible light. The Russians lose contact with Mars 96.

1997 Mars Pathfinder touches down on the Red Planet and deploys the first Martian rover, Sojourner. Cassini–Huygens mission launched.

1998 NASA's Mars Global Surveyor begins mapping the surface of Mars in unprecedented detail.

1999 NASA loses contact with its Mars Climate Orbiter because of an engineering fault.

2000 The probe NEAR–Shoemaker becomes the first to orbit and map an asteroid, called Eros. Afterward, the craft went on to land on the asteroid, guided by mission control. The landing was a bonus, as it was not a scheduled part of the original mission.

2001 Mars Odyssey enters Mars orbit.

2003 31 years after the launch of the Pioneer 10 spacecraft, the Deep Space Network receives the last transmission from the distant probe before losing contact. The craft is 7.6 billion miles (12.2 billion km) from Earth. In May, the Japanese launch the Hayabusa probe (previously known as Muses-C), which will land on an asteroid and return samples. In June, the ESA launches Mars Express, which will carry a British rover to the Red Planet for the first time.

2004 Anticipated arrival of the Cassini–Huygens mission (which was launched in 1997) at Saturn. The Cassini orbiter will study the planet, while Huygens will descend through the atmosphere of its largest moon, Titan, and land on its surface. Also expected is the arrival of the Stardust probe at Comet Wild-2. The aim is to return the first sample of cometary dust to Earth.

2005 Expected arrival of NASA's Deep Impact probe at Comet Tempel. It will fire a projectile and study the resultant crater.

1989 A technician inspects the Galileo probe prior to its launch toward the gas giant, Jupiter.

1997 The Cassini–Huygens mission launches into space on its seven-year journey to Saturn.

2000 The probe NEAR–Shoemaker took this photo of the surface of the asteroid Eros as it orbited it.

Solar and lunar probes

As our nearest neighbor in space, the Moon has been the subject of numerous probes. The Soviet Union was the first nation to successfully study the Moon at close range, with its Luna series. In 1959, Luna 2 and Luna 3 became, respectively, the first projectile ever to hit the Moon and the first to see its farside, while the Americans' Pioneer 4 achieved a flyby later that same year. Since then, with successive probes, and of course using Apollo samples, astronomers have significantly improved their knowledge of the Moon.

The Sun, meanwhile, has also been the subject of intense scrutiny, not surprisingly since it is the nearest star. The 1990s were particularly fruitful for solar astronomy, with NASA, ESA and the Japanese all launching dedicated observatories.

↓ **The NASA–ESA solar probe Ulysses,** seen here in an artist's impression, was the first man-made object to view the Sun by looking straight down on it from the poles. On Earth, we only ever see the Sun perpendicularly to its equator.

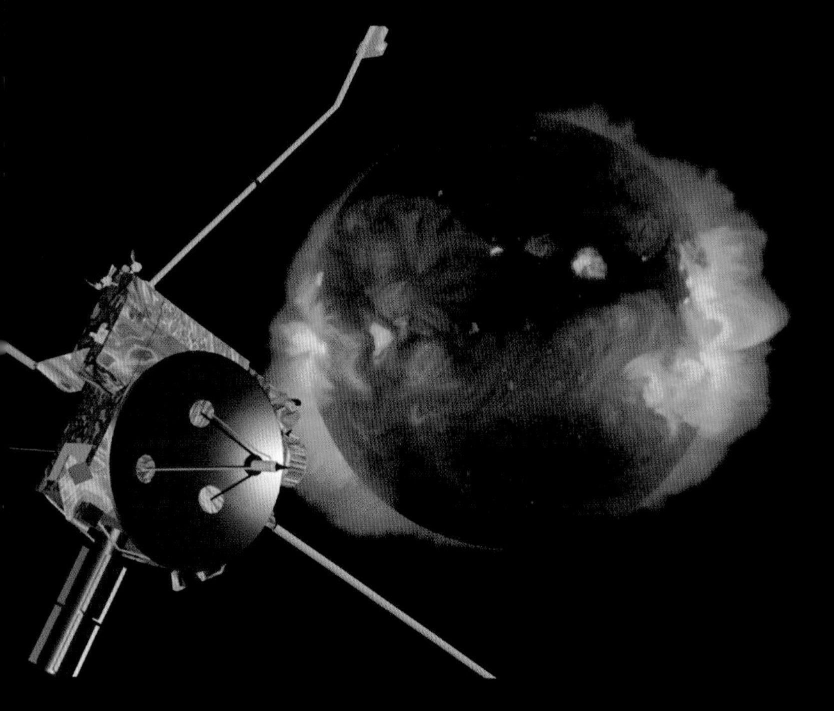

CLEMENTINE MOON PROBE

Launched in January 1994, Clementine was a probe with a primary mission which was military in nature, but which also returned some interesting data from the Moon. The probe was built by the US Naval Research Laboratory, and was designed to test a series of sensors used for the detection and tracking of ballistic missiles. The craft was placed into a lunar orbit, where it mapped the Moon's surface at various wavelengths using four different cameras. One of the successes of the mission was the detection of something very unexpected on the surface of a body as lifeless as the Moon—namely, water. The water was frozen at the bottom of a deep crater near the south lunar pole, permanently shaded from the Sun and so able to remain without evaporating. It is suspected to have been deposited there during a cometary impact. After the Moon, Clementine was put into an orbit around the Sun where it was supposed to encounter an asteroid. But the craft accidentally used up its fuel during a maneuver and the rendezvous was cancelled.

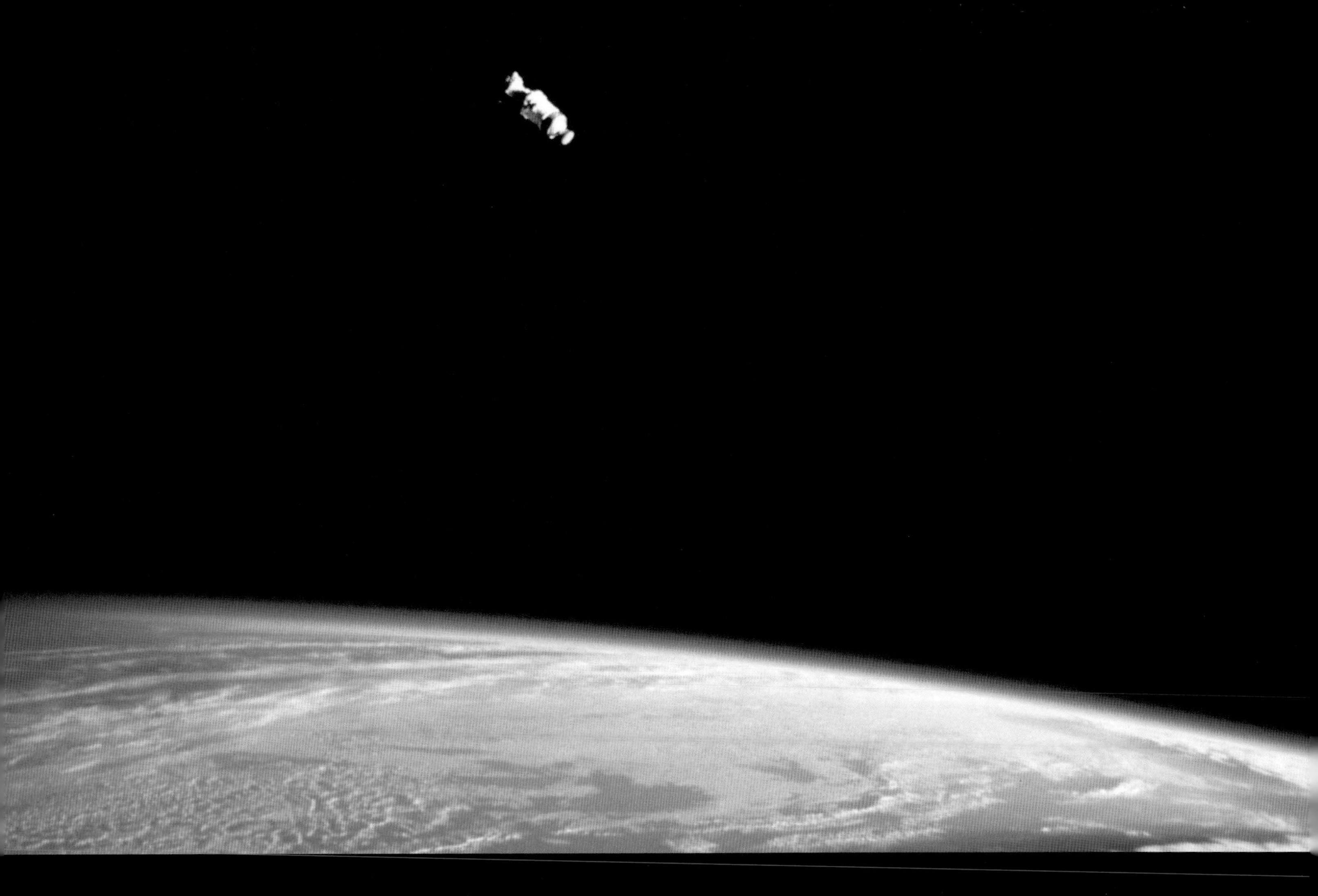

ULYSSES SUN PROBE

In 1990, NASA and ESA, having combined forces, launched a dedicated solar observatory known as Ulysses. Its mission was to study the solar wind—the steady stream of charged subatomic particles that emanates constantly from the Sun's outer atmosphere, or corona. In particular, Ulysses was directed to observe the solar wind coming from the hitherto unseen solar poles. With that aim in mind, in 1992, the spacecraft sailed past Jupiter, and the gravity of that planet hoisted the probe out of the plane of the Solar System and into an orbit that passed over the Sun's poles. Today, Ulysses is still orbiting the Sun. During its more than 12 years of operation it has greatly improved astronomers' understanding of the Sun and its environment.

↑ **Astronauts aboard the space shuttle Discovery** took this photograph of the solar probe Ulysses a short time after it had been deployed, in 1990.

→ **In 1994, the Clementine lunar probe** snapped this false-color topographic map of the western hemisphere of our nearest neighbor in space. The right-hand side (blue) is that part facing Earth, while the other half is the lunar farside, never visible from our planet.

Asteroid and comet probes

As minor components of the Solar System—lacking the scale and grandeur of the planets—asteroids and comets have only recently been the focus of space probes. Many people will remember the first rendezvous of a probe with a comet, that of Giotto with Comet Halley in late 1985. But less well known are the Soviet and Japanese probes that also intercepted the comet. More ambitious encounters with comets are planned, including a landing and a sample–return mission. Several asteroids have now also been in the limelight. For example, the Galileo probe, bound for Jupiter, photographed asteroids Gaspra (1991) and Ida (1993) on its journey, while asteroid Eros became the celebrated subject of the dedicated NEAR–Shoemaker probe in 2000.

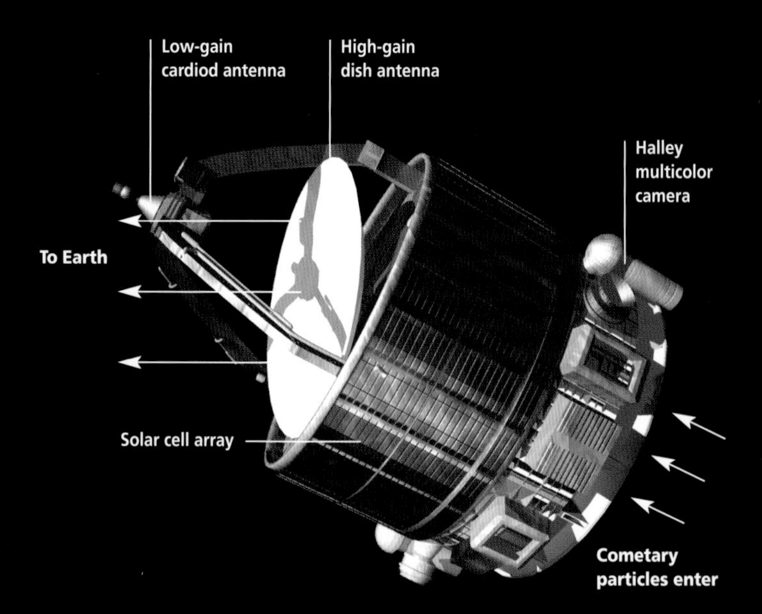

GIOTTO

In the years leading up to 1985, astronomers and interested amateurs across the world eagerly awaited the long-predicted re-arrival of Comet Halley to Earthly skies. But this time around the comet was in for more than just a simple flyby. A battery of five spacecraft—two Soviet, two Japanese and one European—had been designed to intercept the comet as it approached Earth, and to observe for the first time what lies behind the gas and dust that makes comets such spectacular objects in the sky. The best-known probe was the much-publicized Giotto—seen above in an artist's rendering—the first space probe to be launched by the European Space Agency (ESA). Giotto flew to within 370 miles (600 km) of the nucleus, where it analyzed the particles in the gas and dust tails and took many photographs of the comet's heart. Despite several instruments being damaged during the voyage, the craft itself remained intact and went into hibernation. In 1992 it was resurrected and again directed to encounter another comet, known as Grigg–Skjellerup. This mission was not so well publicized, but Giotto flew even closer to this comet than before—skimming the surface just 120 miles (200 km) up.

← **The NEAR–Shoemaker spacecraft** was the first one to orbit an asteroid, specifically Eros, in 2000. This is one of the close-up photographs of the cratered surface. After the mission, the craft was deliberately crashed into the asteroid in an effort to study its surface at as close a range as possible.

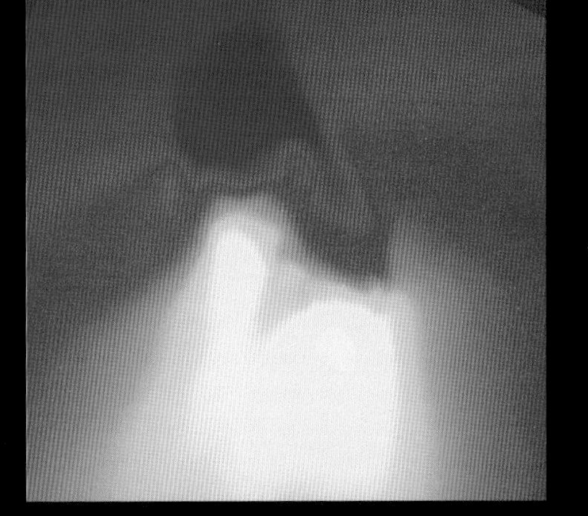

← **In March 1986,** two Soviet craft, Vega 1 and Vega 2, returned more than 1000 photos of Comet Halley, such as the one on the left, as well as navigational data for the subsequent Giotto mission.

↓ **The real star of the Halley encounter** was the first space probe of the European Space Agency (ESA), known as Giotto. In March 1986, Giotto flew within a few hundred miles of the comet's nucleus and took several photos, such as the image below, before its camera was destroyed.

Terrestrial planet probes

As the nearest planets to Earth, the terrestrials (Mercury, Venus and Mars) were the first to attract the attention of space probes. The closest planet, Venus, was the primary target, which the Soviets attempted but failed to reach with Venera 1 in 1961. The following year it was the Americans' Mariner 2 that became the first man-made object to reach another planet. The first successful Mars probe was Mariner 4 in 1965, while almost a decade later, Mercury became the subject of Mariner 10 after it had passed Venus. Two of the terrestrial planets, Mars and Venus, have also been subjected to landings—the first of such were the well-known, and highly informative Viking and Venera probes of the 1970s.

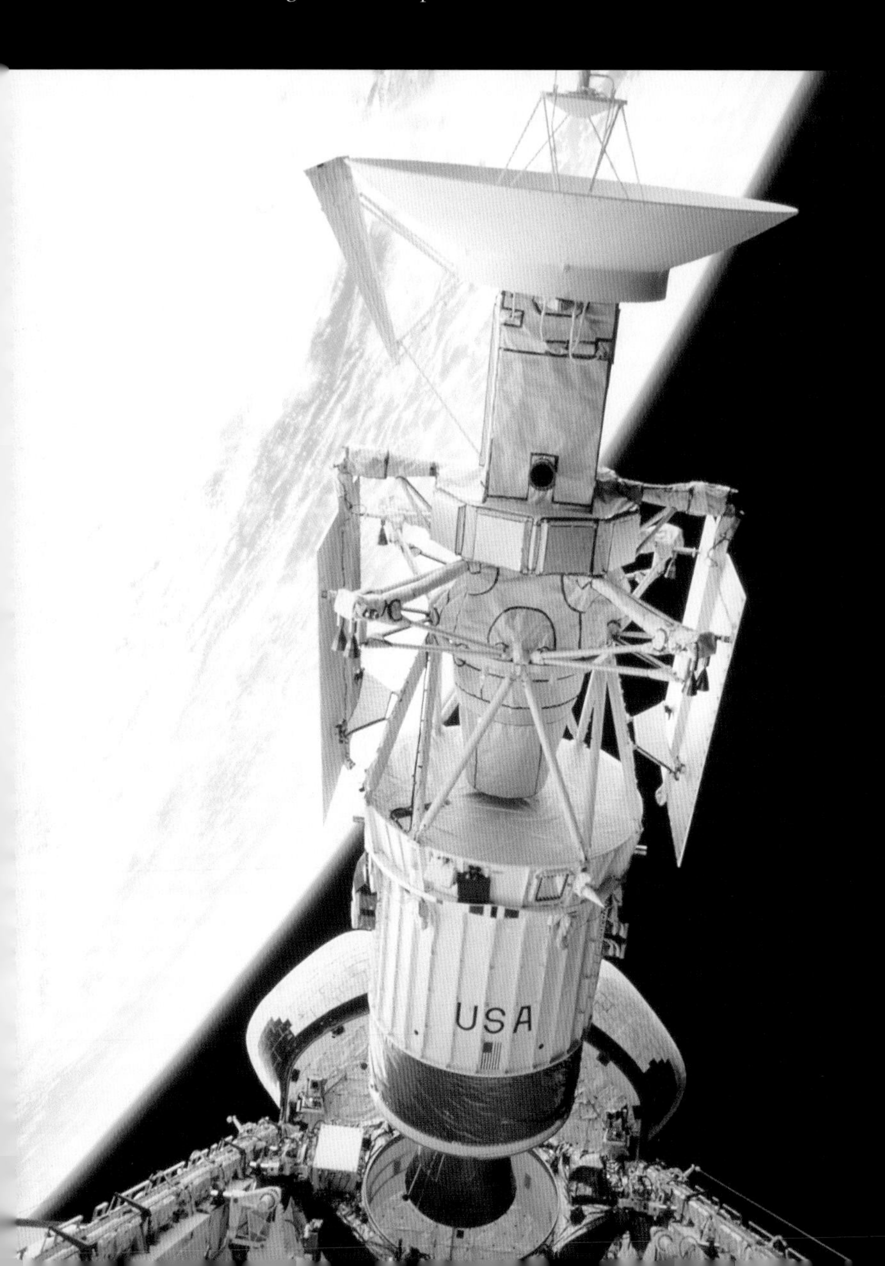

VIKING MISSIONS

In 1975, NASA launched two identical probes to Mars—Viking 1 and 2. These spacecraft were designed with several goals in mind. Upon arrival at Mars, each probe would spend time orbiting the planet and mapping its entire surface at high resolution, studying the two Martian moons, Phobos and Deimos, and also searching for appropriate places where the two landers could set down. The Viking landers, meanwhile—destined to become the first man-made objects to touch another planet—were intended to study the Martian surface, soil, weather and atmosphere, and also to search for any signs of life. Viking 1 (*above*) entered Mars orbit on June 19, 1976, and dispatched its lander the following month, on July 20. Viking 2, meanwhile, arrived at Mars on August 7, 1976, and its lander touched down on September 3. Other probes have been to Mars since, but the Vikings provided perhaps the majority of the information that astronomers have regarding Mars. They showed volcanoes, canyons—even features carved into the Martian landscape via wind or ancient surface-water activity. The results of the search for life, meanwhile, were controversial. The current and official NASA view is that the biology tests proved negative, but a handful of scientists disagree.

← **Magellan, seen in this photograph** taken from the cargo bay of the space shuttle Atlantis, was a highly successful NASA probe that mapped the entire surface of Venus, for the first time, using radar. Venus is covered in clouds so opaque that only radio waves can penetrate to the surface.

← **On July 4, 1997,** the Mars Pathfinder mission touched down on the Red Planet and deployed the first Martian rover. Known as Sojourner, the probe was remotely steered by mission scientists on Earth, and used to analyze the structure and material of several Martian boulders.

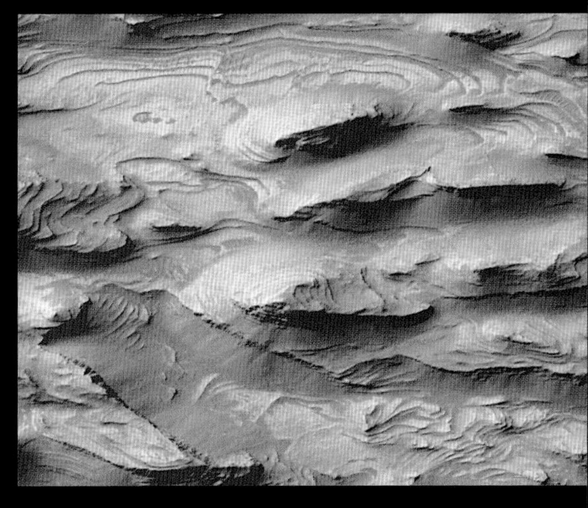

↑ **The Mars Orbiter Camera** on board the Mars Global Surveyor spacecraft took this photo. Some planetary geologists believe that it provides evidence of past surface water on Mars. It shows layers of sedimentary rock, which on Earth are usually deposited by water activity.

← **Mariner 10,** seen here in an artist's impression, was launched in November 1973 on a two-year mission to study Mercury and Venus. Mariner 10 flew by Venus in February 1974, and afterward flew past Mercury three times. During these encounters the probe mapped more than half of Mercury's surface. Mariner 10 was the last of the Mariner probes, and the first time the gravity of a planet (Venus) had been used to gravitationally slingshot a spacecraft to change its course.

Gas planet probes

When NASA's Pioneer 10 probe reached Jupiter in 1973, it became the first man-made object to behold a gas planet in close-up. It was quite an achievement: Jupiter is five times farther from the Sun than Earth. And so began a grand tour of the outer planets. Pioneer 11, the following year, flew even closer to Jupiter than its predecessor and took the first close-up photos of its large (Galilean) satellites. Five years later, in 1979, Pioneer 11 also reached Saturn for the first time. Finally, the remaining two gas planets, Uranus and Neptune, were seen in close-up by the hugely successful Voyager 2 in 1986 and 1989 respectively. More recently, the Cassini–Huygens probe to Saturn was launched in 1997, with its arrival expected in 2004.

↓ **Cassini's path to Saturn:** (1) October 1997, Cassini is launched; (2) April 1998, first Venus flyby; (3) June 1999, second Venus flyby; (4) August 1999, Earth flyby; (5) December 2000, Jupiter flyby; (6) July 2004, Cassini's anticipated arrival at Saturn

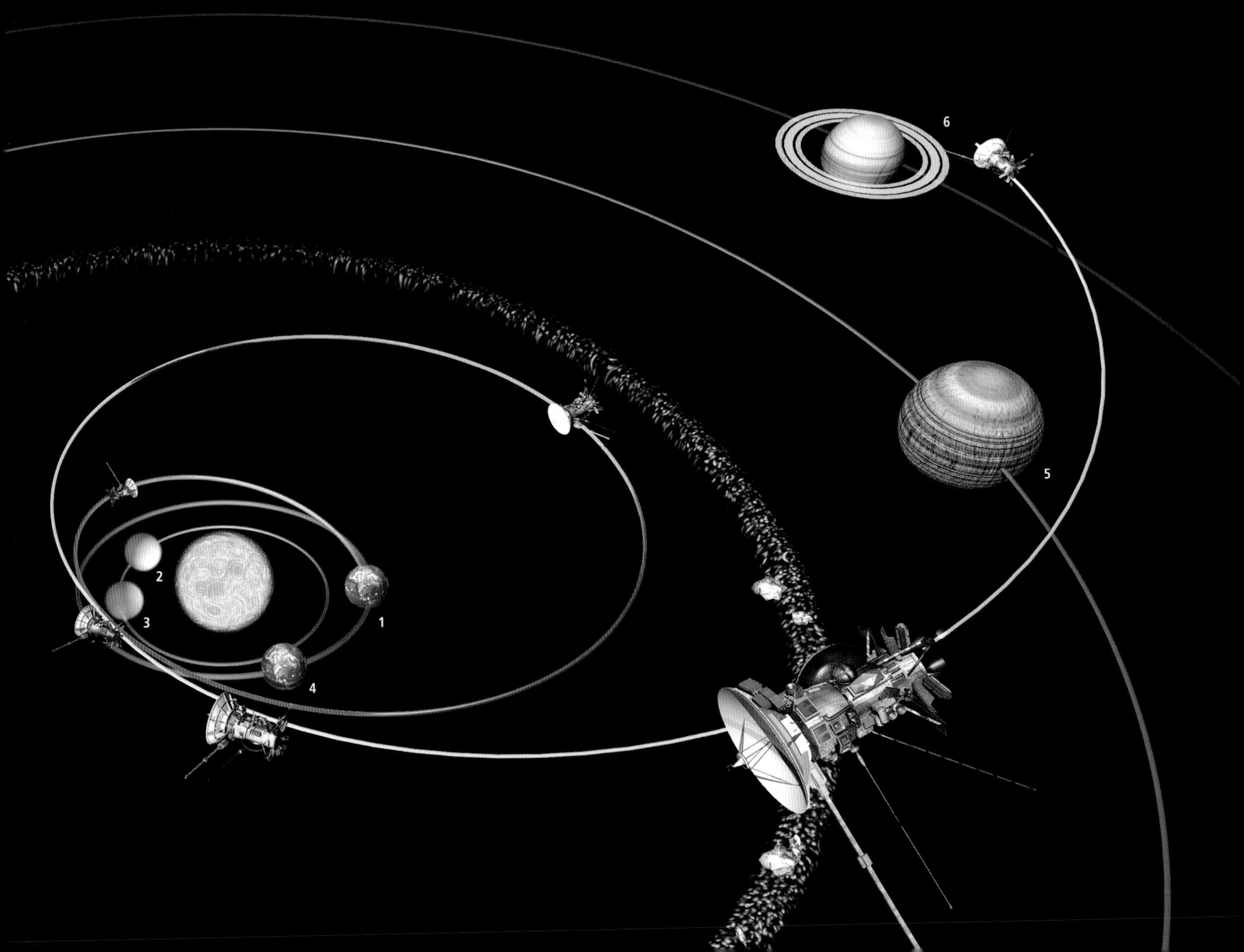

CASSINI–HUYGENS
The Cassini–Huygens mission, launched in 1997, is the first to be sent to Saturn since the Pioneer and Voyager probes of the 1970s and 1980s. As the name suggests, the spacecraft consists of two separate parts, called Cassini and Huygens. Cassini is an orbiter. It is named after the Italian astronomer Giovanni Cassini (1625–1712) who first observed the "division" in Saturn's rings. Upon arrival at Saturn in 2004, Cassini will orbit the ringed gas planet to study its rings, satellites and atmosphere. While there, Cassini will deposit the Huygens probe, which will descend to the surface of Saturn's largest moon, Titan, discovered by the man after whom the Huygens probe is named—Christiaan Huygens (1629–95). Astronomers do not know exactly what Titan will be like. It has a thick nitrogen atmosphere, similar in some ways to Earth's, but it is much colder. The ground may be solid, but some believe that Titan's surface could harbor lakes or seas of liquid methane. As a result of this uncertainty, Huygens—seen here descending to Titan's surface following its release—has been designed to land equally well on land or fluid.

The Solar System

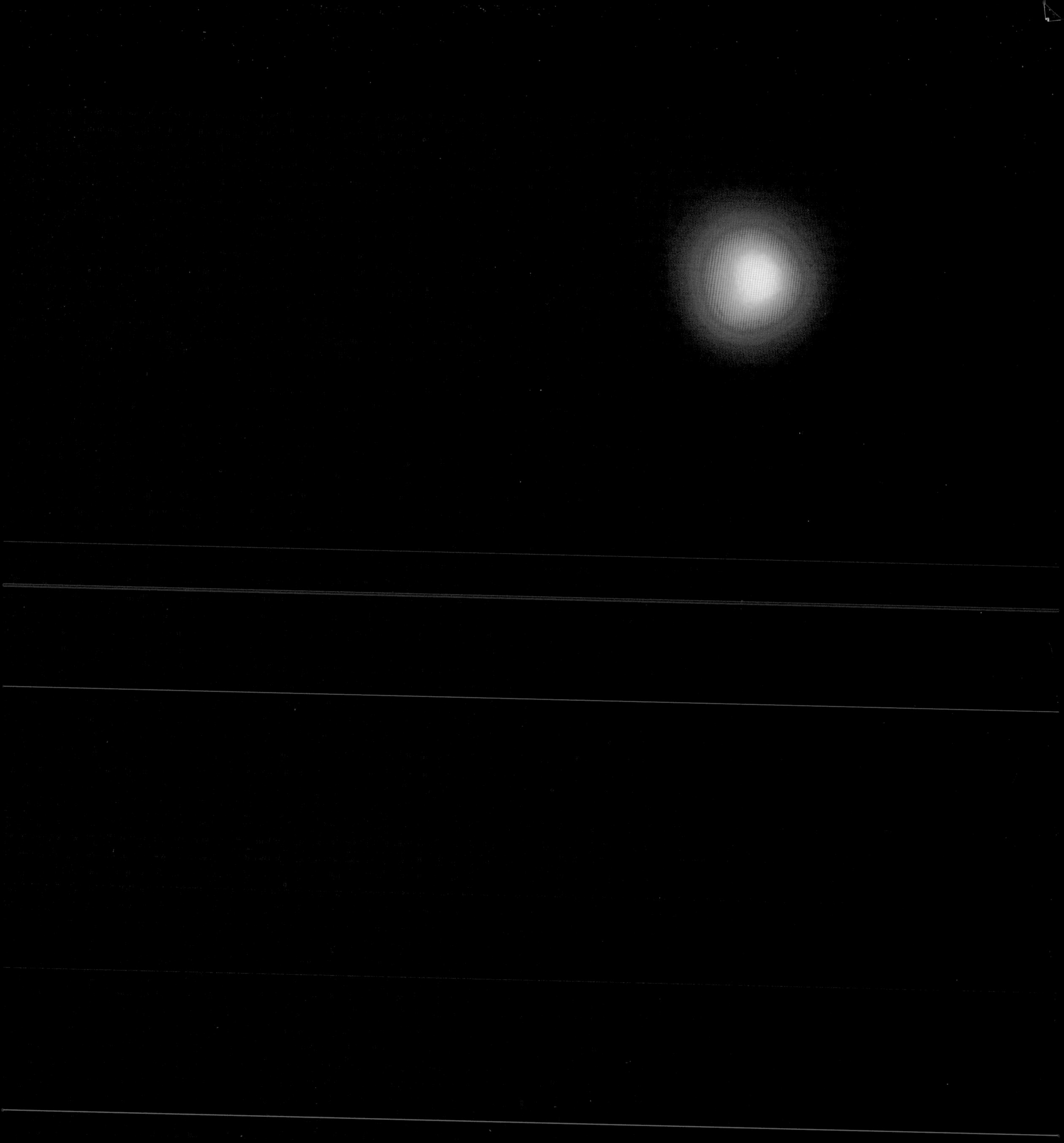

Previous page This is the clearest view we have of the distant planet Pluto and its satellite Charon.

The Solar System is our home in space, a prime piece of celestial real estate some 4600 million years old. This is the planetary system that contains the Sun, the nine known planets—Earth among them—and countless small but equally interesting chunks of debris called comets and asteroids.

Formation of the Solar System

Astronomers have known for some time how the Solar System formed. The theory is that the planets and the Sun condensed from a vast cloud of spinning gas and dust—the Solar Nebula. The idea was conceptualized as early as the eighteenth century. German philosopher Immanuel Kant first proposed the theory, although it did not garner attention until, half a century later, Frenchman Pierre-Simon, Marquis de Laplace, reintroduced it.

According to current understanding, the first step in creating the Solar System was the demise of another star. Massive shockwaves from a supernova blast, ending the life of that star, caused a nearby cloud of gas and dust to collapse under its own gravity (1). After one or two million years, the collapsing cloud (the Solar Nebula) had flattened out by virtue of its rotation, and a primitive protosun had begun to form at the center of the disk (2). Over the next million years, grains of carbon, rock, ice and other materials lumped together in the disk in a process known as accretion. As more of them stuck together, the planets gradually took shape (3), with the planets being fully formed after 10 to 100 million years (4).

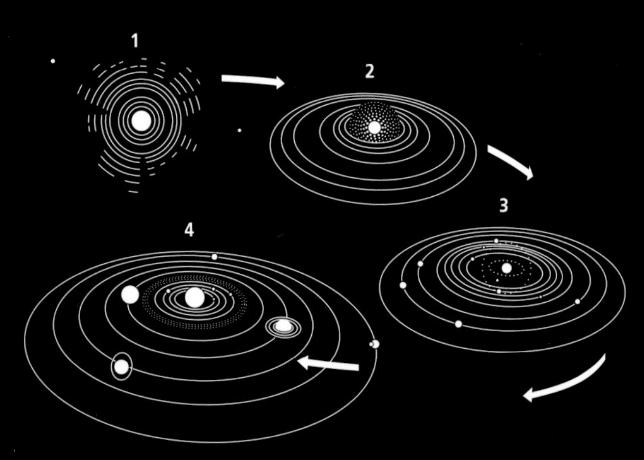

The modern Solar System

The modern Solar System is comprised of several different zones, which reflect how the Solar System was born. The innermost zone is that of the terrestrial planets: Mercury, Venus, Earth and Mars. These bodies are comparatively small, rocky and metallic. When the planets were forming, these heavy materials collected near the Sun, drawn-in by its gravity, which is why the inner planets are hard and dense. The next zone is the asteroid belt, a region between the orbits of Mars and Jupiter where millions of tumbling boulders of rock and iron can be found. These are leftovers from the planet's formation. Beyond the asteroids is the realm of the giant planets. These worlds—Jupiter, Saturn, Uranus and Neptune—are made mostly of lightweight gases and liquids such as hydrogen and helium. The giants formed in a region around the Sun where these elements were abundant. They could not condense closer-in as it was too hot, so the giants formed far from the Sun. Finally, beyond the giant planets is where comets are found. They occasionally intrude into the inner Solar System, of course, but most of them exist far from the Sun in two vast and separate reservoirs called the Kuiper Belt and the Oort Cloud.

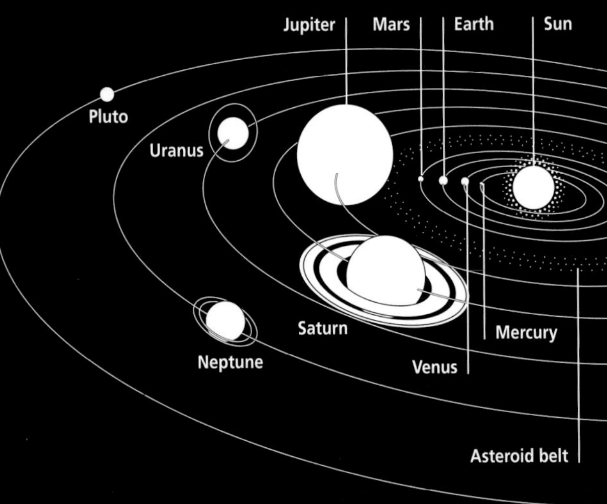

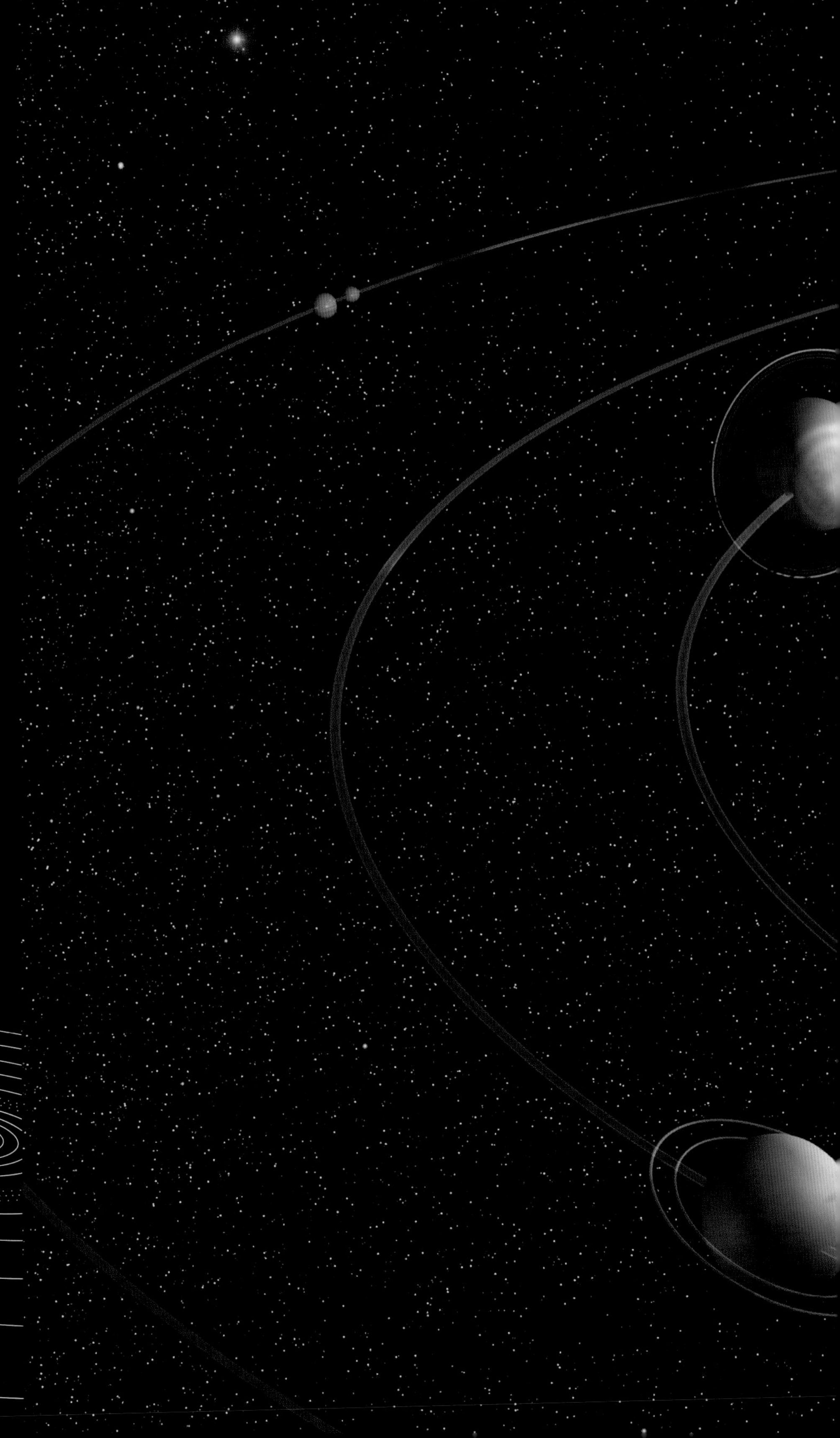

Future of the Solar System

The Sun and the planets were born together some 4600 million years ago. An unimaginably long lifetime by human standards—but still a mortal one. One day, the Sun will begin to die and with it, the Solar System.

At present, the Sun is on the "main sequence," a stage where is converts hydrogen into helium. Astronomers estimate that the Sun has consumed about 35 to 45 percent of its original supply of fuel, thus it has enough to last another 5000 to 8500 million years, depending on which model is correct. Once the hydrogen starts to run out, the Sun will become a red giant—a monstrous and exceptionally luminous star, well over one hundred times its present diameter, and thousands of times brighter (1). It will engulf the orbits of Mercury, Venus, and possibly Earth, gobbling up the scorched and barren planets in the process. Then, over the next several hundred million years, the red giant will pulsate, contracting and expanding several times, throwing off a shell of gas with each pulsation and so reducing its overall mass (2). Finally, the jettisoned envelope will shine briefly as a planetary nebula (3), while the exposed core of the Sun, a white dwarf, will slowly fade over billions of years (4).

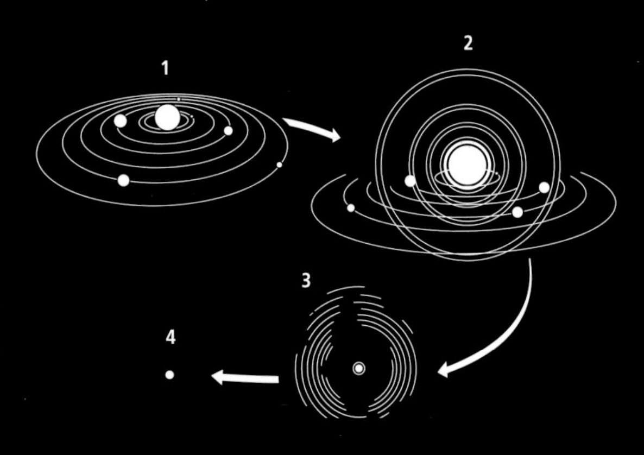

The planets

Aside from the Sun, the main components of the Solar System are the planets, nine of them in total. There have been speculations about a tenth planet, but this remains doubtful. The combined material of all the planets amasses to about 447 Earths, which is still only about 0.13 percent of the mass of the Sun. Of all the planets, Jupiter is by far the most massive, being heavier than all the others put together.

All of the planets except Mercury and Venus have moons, and with the exception of Pluto fall into two classes. Four of them—Mercury, Venus, Earth and Mars—are rocky and are known as the terrestrial worlds. The larger four planets—Jupiter, Saturn, Uranus and Neptune—are called gas giants, being composed of mainly hydrogen and helium, and are similar in composition to the Sun. The gas giants all have rings, the most famous of which are those of Saturn. All of the planets orbit the Sun in elliptical paths—a shape like a slightly squashed circle. However, with the exception of Mercury, Mars and Pluto, most of the orbits are very close to circular.

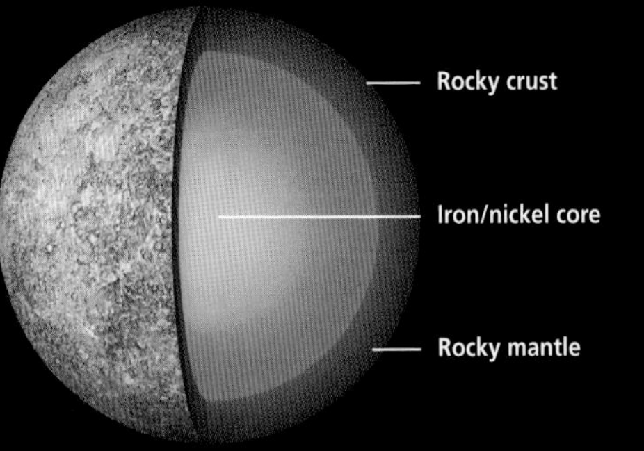

Mercury shows an ancient Moonlike face that is covered with craters and basins. Its unusually large iron and nickel core may have been the result of a massive collision early in its history that removed part of the planet's rocky mantle.

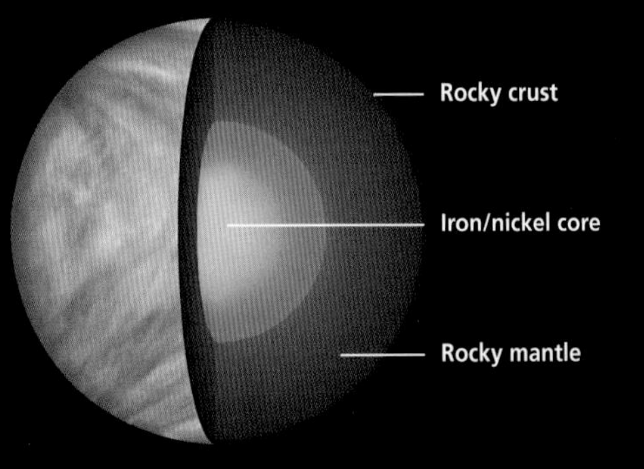

Venus is cloaked in a layer of white sulfuric acid clouds, but the radar spacecraft that have mapped its rocky surface have revealed mountain ranges, volcanoes and lava flows. Approximately 60 percent of the surface is a lava plain.

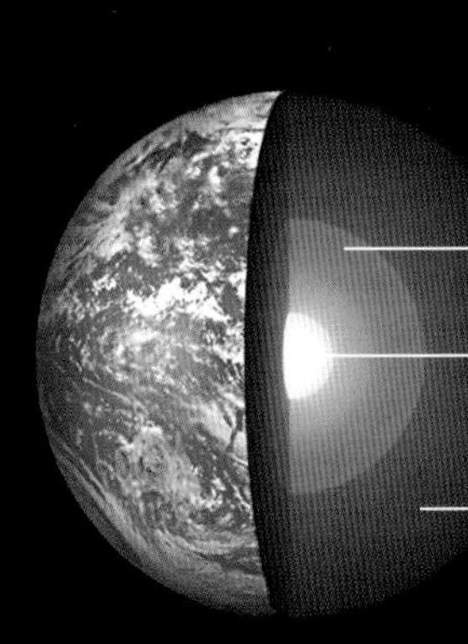

Earth was so intensely hot when it formed that it melted, causing the heaviest materials, such as iron and nickel, to sink to the center, forming the core, while lighter materials rose toward the surface and separated into the mantle and the crust.

DISTANCE FROM THE SUN

The distances given below are averages. Because the orbit of a planet is elliptical, rather than circular, its distance from the Sun varies—Mars, for example, travels up to 155 million miles (249 million km) from the Sun and travels as close as 129 million miles (207 million km). Pluto's orbit is so elliptical that at times it is actually closer to the Sun than Neptune.

Sun

Mercury 36 million miles (58 million km)

Venus 67 million miles (108 million km)

Earth 93 million miles (150 million km)

Mars 142 million miles (228 million km)

Asteroid Belt

Jupiter 483 million miles (778 million km)

Saturn 890 million miles (1432 million km)

Uranus 1784 milli (2871 million km)

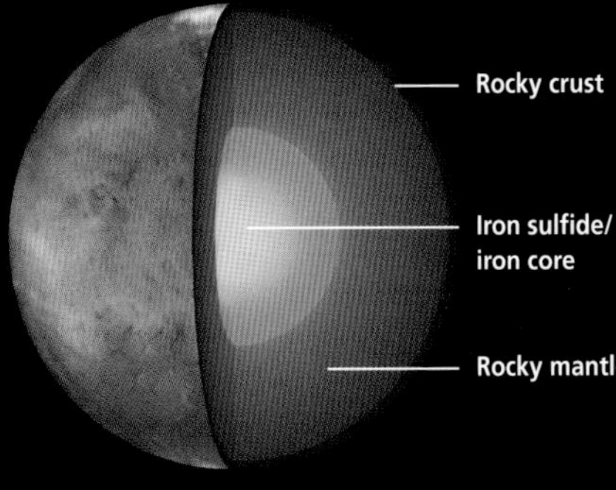

Mars is a frigid desert world that has two distinct types of terrain, each occupying about half of the surface. In the south are the older highlands with many craters, while in the north a plain relatively free of craters lies a few miles higher.

Gaseous hydrogen
Liquid hydrogen
Rocky core
Metallic hydrogen

Jupiter is a giant ball of gas, almost entirely hydrogen and helium, and it lacks a solid surface. The hydrogen and helium are gaseous in the upper layers, but behave like a liquid and then like a liquid metal as the depth increases. The center is dense rock.

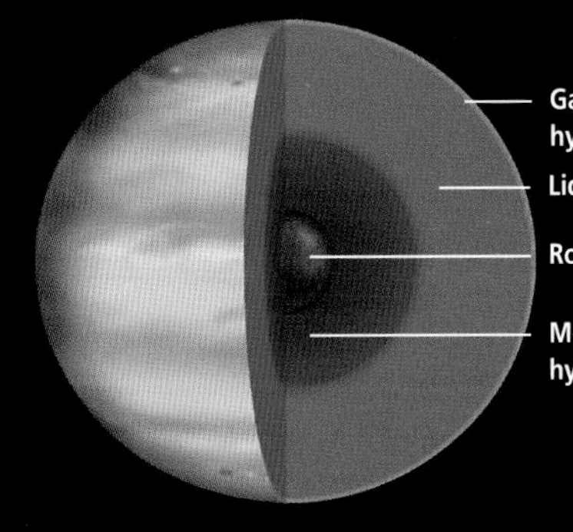

Saturn has no solid surface and, like Jupiter, is almost all hydrogen and helium, with a small, dense core. Although it is almost as big as Jupiter, it has only 30 percent as much mass and would float if it were placed in water.

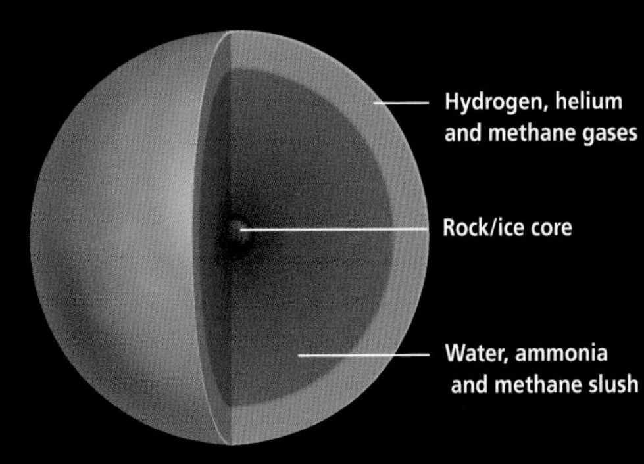

Uranus has a deep upper atmosphere of hydrogen, helium and methane. Beneath this is a hot, slushy mixture of water, methane and ammonia, with a rocky core. Unlike the other gas giants, Uranus does not seem to have an internal heat source.

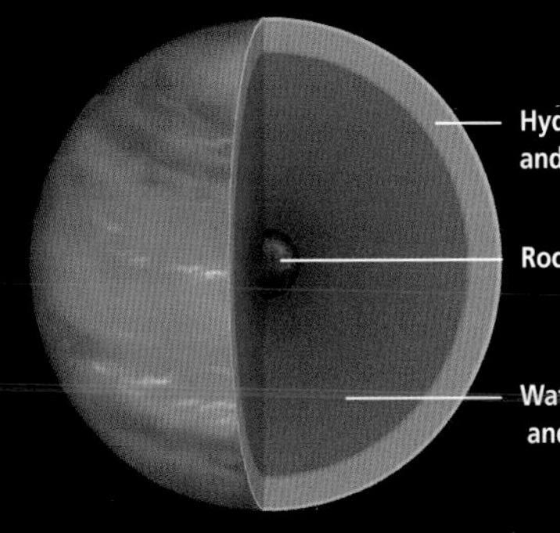

Neptune is colored blue by the methane in its gaseous surface. It has a deep "ocean" of water, ammonia and methane and a hot rocky core. The interior bubbles up almost twice as much heat as the surface receives from the Sun.

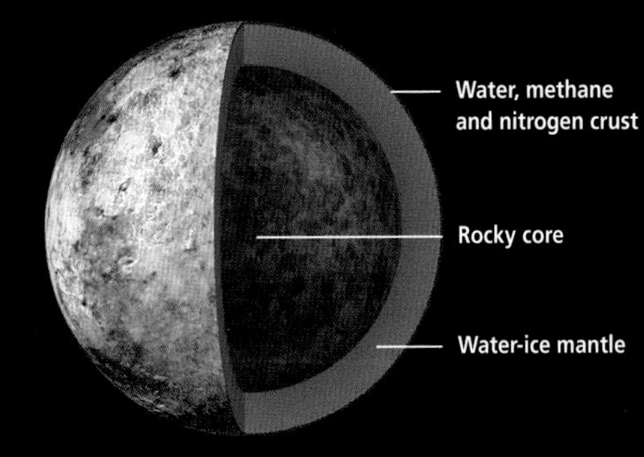

Pluto has a transient, cold atmosphere of nitrogen and methane, and a thin crust of water, nitrogen and methane ice. No probe has yet studied the planet, but scientists speculate that it has a mantle of water ice and a large rocky core.

Neptune 2795 million miles (4498 million km)

Pluto 3675 million miles (5914 million km)

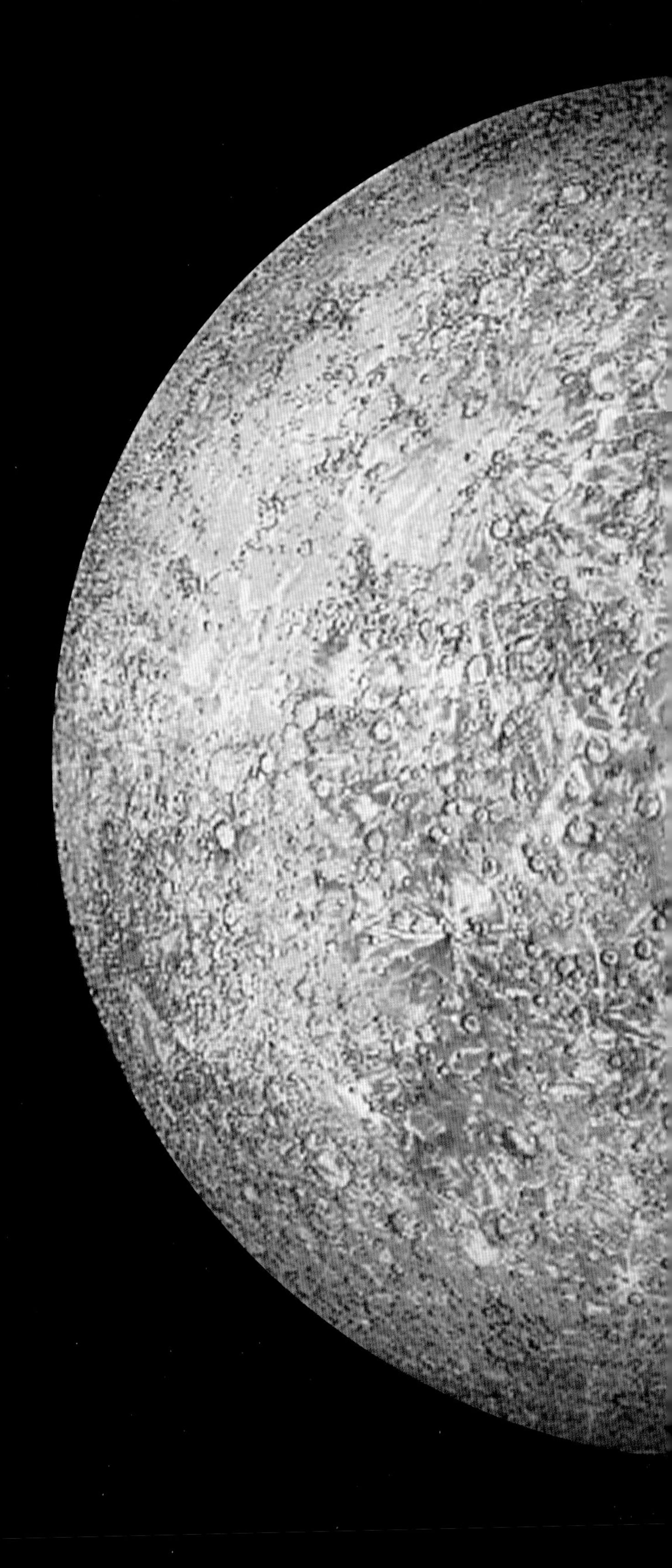

Mercury

At face value, Mercury resembles the Moon—an endless sea of craters upon craters, dominated by the large impact scar called Caloris. The first planet from the Sun and the second smallest after Pluto, Mercury is a scorched, airless ball of iron covered with a thick crust of rock. Its large iron content suggests that it originally had a much thicker rocky mantle, which was blown off in a collision with another protoplanet when the Solar System was still forming. Mercury has no satellites.

ATMOSPHERE

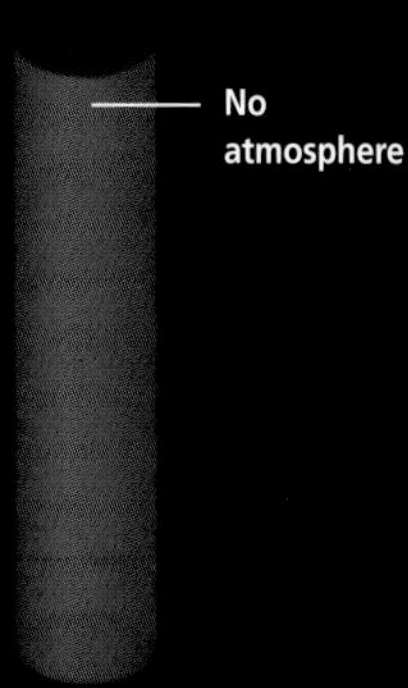

PLANET STATISTICS

Origin of name *Mercurius*, messenger of the Roman gods
Discovered Known since antiquity
Diameter 3031 miles (4878 km), 38.2% of Earth's
Mass 0.055 x Earth
Volume 0.06 x Earth
Farthest from Sun 43.4 million miles (69.8 million km)
Closest to Sun 28.6 million miles (46.0 million km)
Mean surface temperature 800°F (430°C)
Sunlight strength 450–1040% of Earth's
Apparent magnitude -2.0 to +3.0
Surface gravity 0.38 gee (38% of Earth's)
Magnetic field strength 0.003 gauss (1% of Earth's)
Number of satellites 0

↓ **A Mariner 10 view of the surface** of Mercury near the Caloris Basin—an enormous impact structure measuring some 800 miles (1300 km) across. The impact sent powerful shock waves through the planet's nickel–iron core.

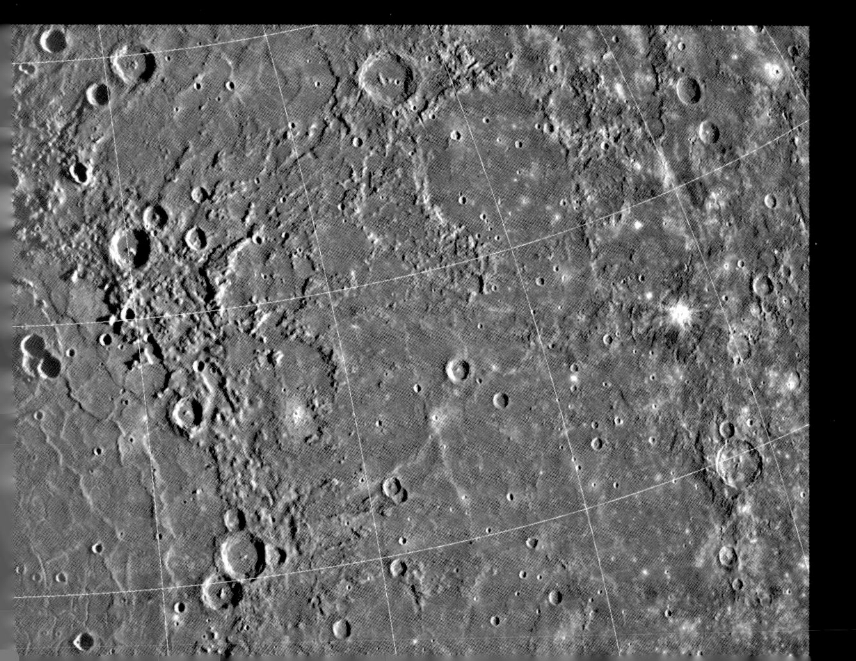

THE LAYERS OF MERCURY

Mercury is so dense for its size that it must have a large iron core. Astronomers suspect the planet was originally larger with more extensive outer layers—before a collision with a rogue protoplanet, billions of years ago, blasted them into space.

Crust The crust is the thin outermost layer that includes the surface. Mercury's surface is heavily cratered and therefore old. Cracks in the crust, called scarps, tell us that Mercury contracted in the past as it cooled.

Mantle Below the crust is a solid rocky (silicate) mantle about 350 miles (550 km) thick, occupying the outermost 25 percent of the planet.

Core Mercury has a global magnetic field, albeit small, so its large nickel–iron core could be partially liquid, with circulating currents there generating the magnetic field as the planet spins. The core extends from the center to 75 percent of the way to the planet's surface. No other terrestrial planet has such a large iron content.

SIZE COMPARISON

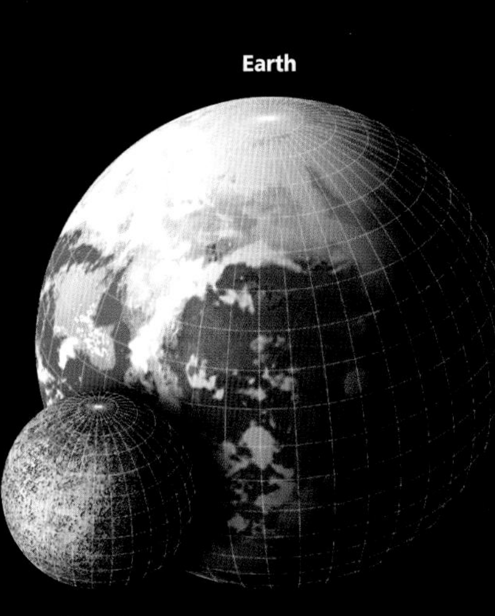

ORBIT STATISTICS

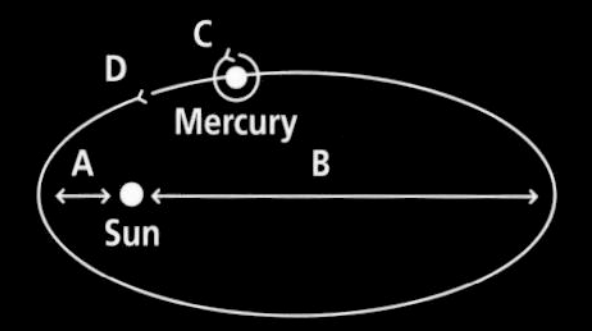

A 28.6 mil. miles (46.0 mil. km)
B 43.4 mil. miles (69.8 mil. km)
C rotates in 58.65 Earth days
D orbits in 87.98 Earth days

Axial inclination 0°
Angle of orbit to ecliptic 7°

Mercury: features

Comparatively little is known about the planet Mercury. Only a single probe—Mariner 10 in 1974 and 1975—has ever been sent there, mapping about half of its surface. It is battered, cracked and ancient. The rest of its landscape, we know nothing about at all. Only distant Pluto is more mysterious.

Mercury's orbit is the second most eccentric, after Pluto. It swings rapidly around the Sun, completing one revolution in about three Earth months. When farthest from the Sun, Mercury is one and a half times farther out than when at its closest approach. So from the surface of Mercury, the diameter of the Sun as seen in the sky varies a great deal, from twice the apparent diameter as seen from Earth to more than three times that. Mercury also has the greatest range of surface temperature. With the Sun high overhead the surface roasts at around 800°F (430°C), almost as hot as Venus, but it is more than 1080°F (600°C) cooler in the opposite hemisphere, where the Sun is well out of sight.

MAPPED HEMISPHERE

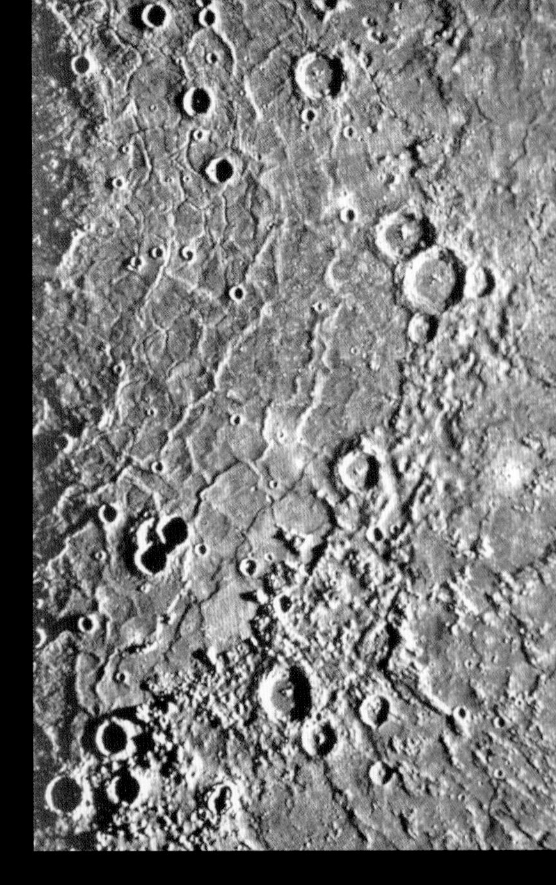

THE SURFACE OF MERCURY

Mercury has a truly ancient surface. The planet lacks an atmosphere and has never had water on its surface, so the only erosion comes from the impact of meteorites or the occasional comet. Because the rate of impacts in the Solar System fell off dramatically 3800 million years ago, most of Mercury's surface has not changed since then. Now, Mercury is an orbiting fossil. As expected, then, Mercury appears somewhat like the Moon, peppered with impact craters of all sizes (*above*), with the largest, Caloris, large enough to contain Texas. There are some regions that are relatively smooth, though, called smooth plains, found near Caloris and toward the north pole. These are probably volcanic in origin, layers of lava laid down billions of years ago, burying some older craters. The other significant surface features are the scarps—cracks in the crust that form cliffs up to 2½ miles (4 km) high.

→ **Mercury's orbit around the Sun** is quite elongated, although the ellipticity in this diagram has been greatly exaggerated for clarity. The orbit also precesses. That is, the entire orbit appears to rotate slowly with time, while Mercury's distance from the Sun (at aphelion and perihelion) remains fixed.

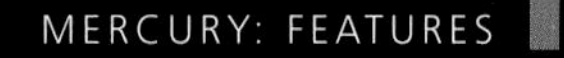

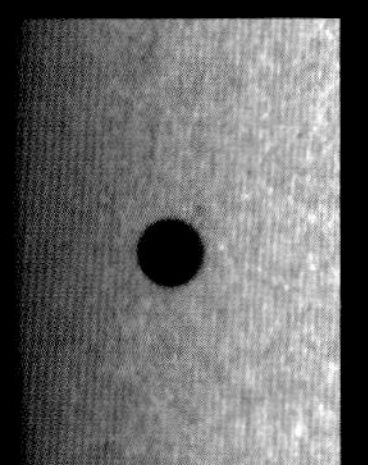

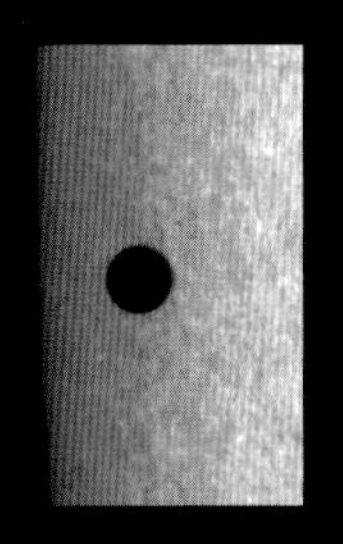

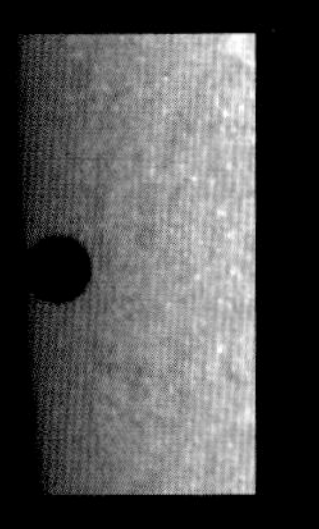

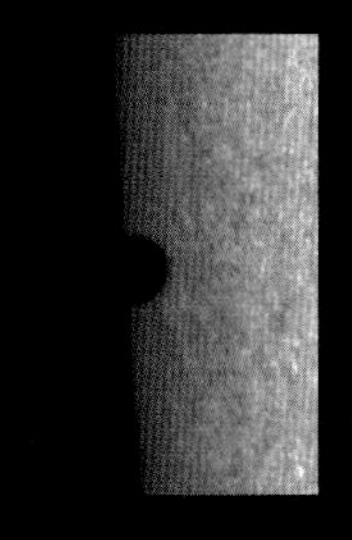

← **Occasionally, as seen from Earth**, Mercury passes in front of the Sun in its orbit—an event known as a transit. This series of images (from left to right) follows the progress of one such occurrence on May 7, 2003.

MERCURY'S ORBIT

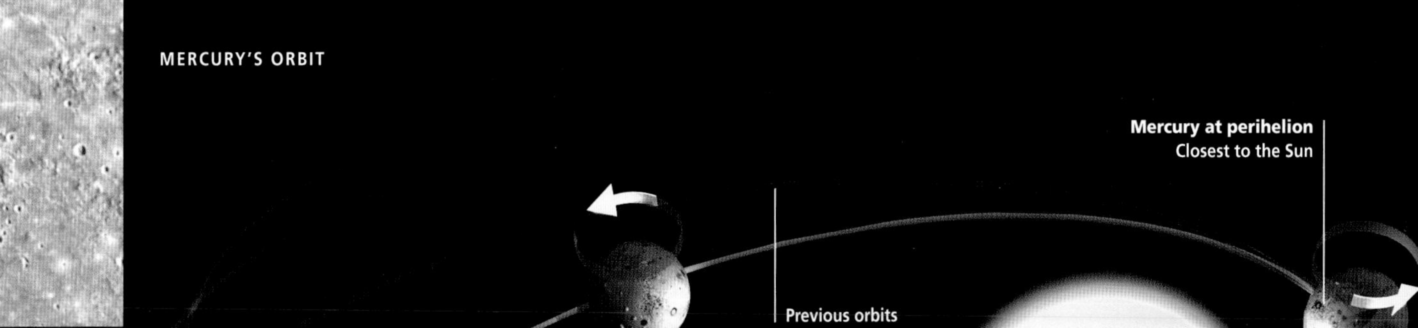

Venus

Venus has been called Earth's twin because of its similar size and mass. But our nearest planetary neighbor is anything but Earthlike. The surface is permanently hidden behind an atmosphere of carbon dioxide choked with clouds of sulfuric acid. This noxious sky pushes down on the surface with a pressure equivalent to that at the bottom of a lake nearly 3000 feet (900 m) deep. The atmosphere is a heat trap, too. Despite being farther from the Sun than Mercury, Venus is hotter. Both lead and tin would melt there.

ATMOSPHERE

Other 0.8%
Nitrogen 3.2%
Carbon dioxide 96%

PLANET STATISTICS

Origin of name *Venus*, Roman goddess of love and beauty
Discovered Known since antiquity
Diameter 7521 miles (12,104 km), 94.8% of Earth's
Mass 0.95 x Earth
Volume 0.86 x Earth
Farthest from Sun 67.7 million miles (108.9 million km)
Closest to Sun 66.8 million miles (107.5 million km)
Mean surface temperature 900°F (480°C)
Sunlight strength 190% of Earth's
Apparent magnitude -4.0 to -4.6
Surface gravity 0.9 gee (90% of Earth's)
Magnetic field strength <0.00002 gauss (<0.007% of Earth's)
Number of satellites 0

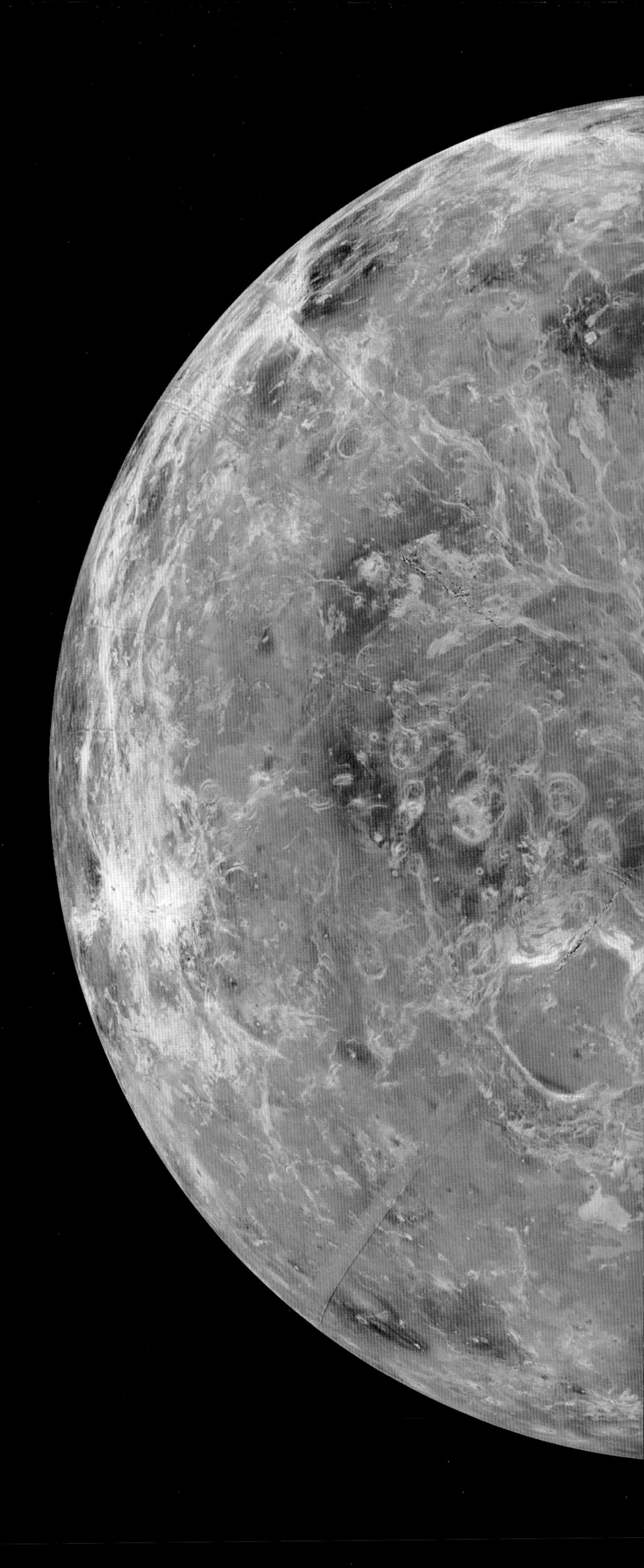

↓ **This false-color radar photograph** of the surface of Venus reveals a region near a now-extinct volcano. In other regions, Venus may still be volcanically active.

THE LAYERS OF VENUS

Venus, similar to Earth in size and density, is not unlike Earth on the inside, either. Its various layers are comparable to Earth's and their sizes are in a similar proportion. One difference is that Venus does not have a two-component core.

Crust The Venusian crust is comparatively thin, extending to a depth of 30 miles (50 km). The surface is remarkably flat, with only 10 percent of the planet having a relief greater than 7 miles (10 km). The thin crust is one of the reasons why Venus was geologically active in the recent past.

Mantle Below the crust is a rocky mantle, again solid, extending almost from the surface to a depth of about 2000 miles (3000 km).

Core Venus has a fairly large metal core, probably composed of the metals nickel and iron, which takes up about half of its radius. The planet has at most a very weak magnetic field, so its core is most likely solid.

SIZE COMPARISON

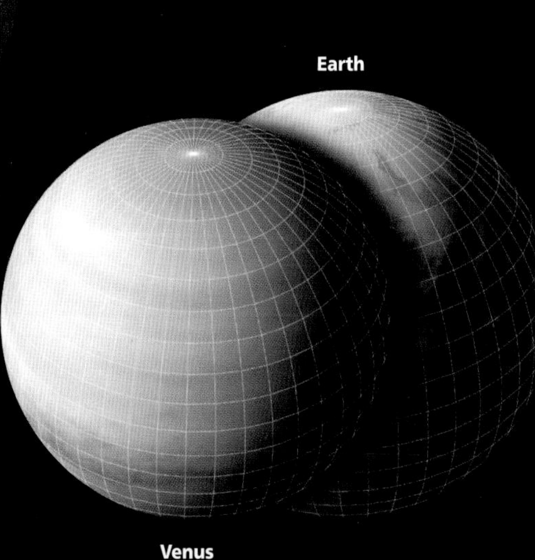

ORBIT STATISTICS

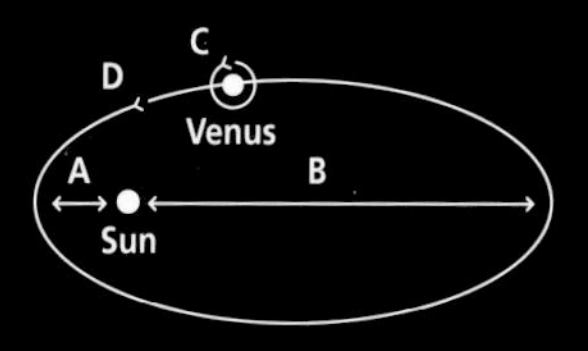

A 66.8 mil. miles (107.5 mil. km)
B 67.7 mil. miles (108.9 mil. km)
C rotates in 243.01 Earth days
D orbits in 224.7 Earth days

Axial inclination 177.4°
Angle of orbit to ecliptic 3.4°

Venus: surface

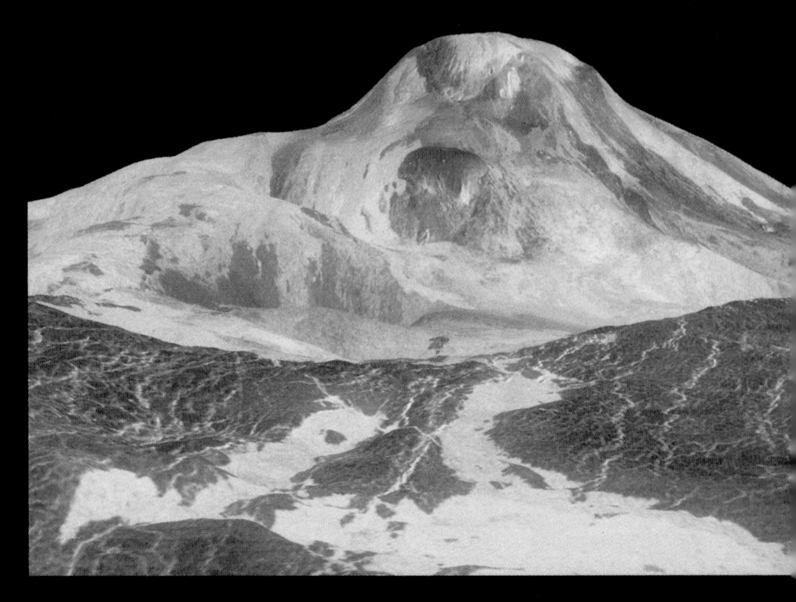

Astronomers had little idea what lay beneath Venus' thick atmosphere until, in 1989, NASA's Magellan probe mapped its surface using cloud-penetrating radar. There is little surface relief—only 20 percent of the planet has local height variations greater than 1½ miles (2 km). The remaining surface is comprised of vast lowland planes. The biggest surprise for astronomers, though, was that Venus is geologically young—the oldest features are only 500 million years old. The reason is that Venus was recently, and may still be, volcanically active. Photos show conclusively how volcanic structures dominate the landscape. One type of volcanic feature that Venus has are the so-called pancake domes. These resemble blisters, and were formed when gelatinous lava leaked onto the surface, cooled, spread out and cracked.

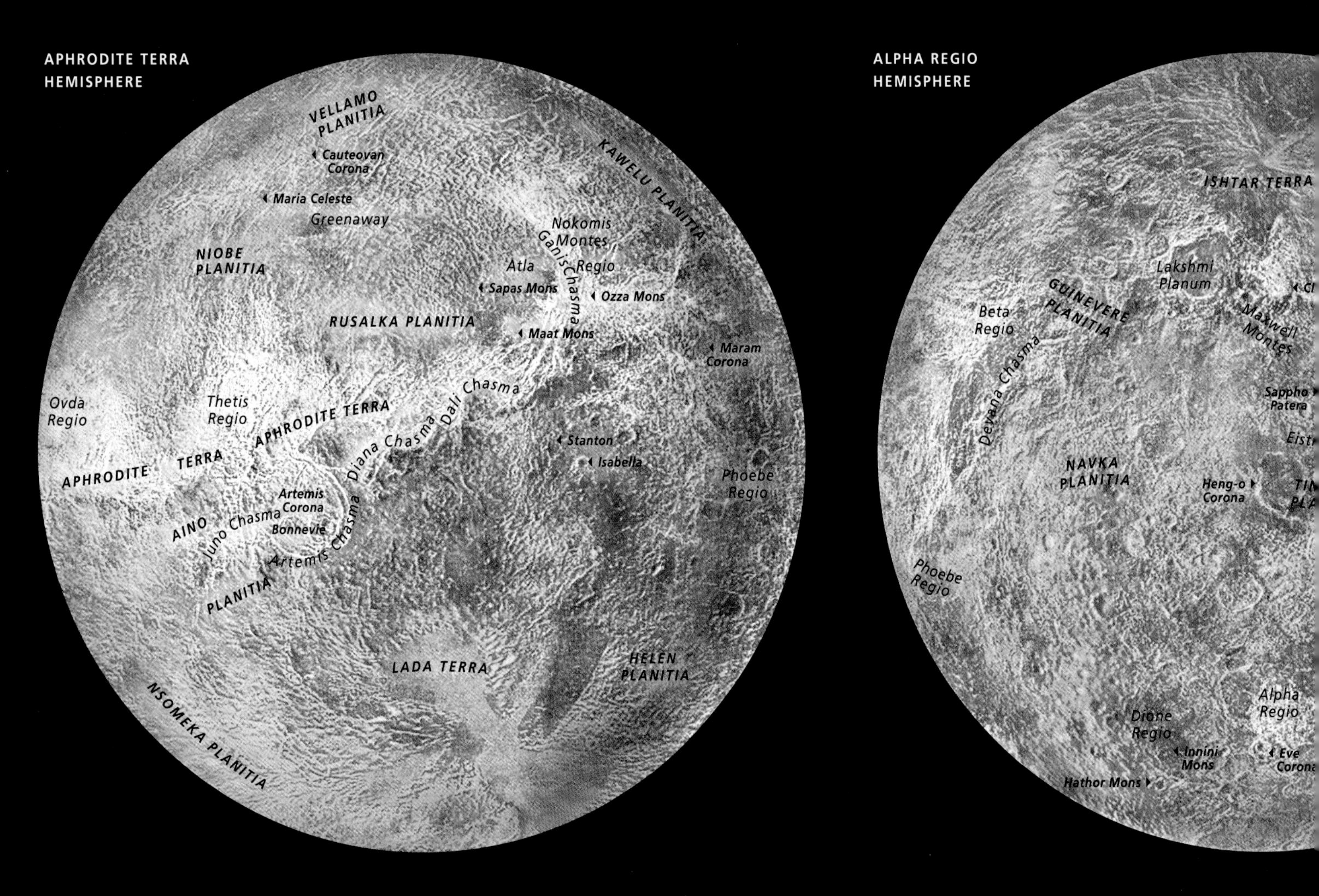

APHRODITE TERRA HEMISPHERE

ALPHA REGIO HEMISPHERE

← **The peak of Maat Mons** is seen in this three-dimensional simulated view, constructed from data collected from the Magellan probe. Maat Mons, jutting up 5 miles (8 km), is one of the highest features found on Venus. Lava flows hundreds of miles across the fractured plains, as seen in the foreground.

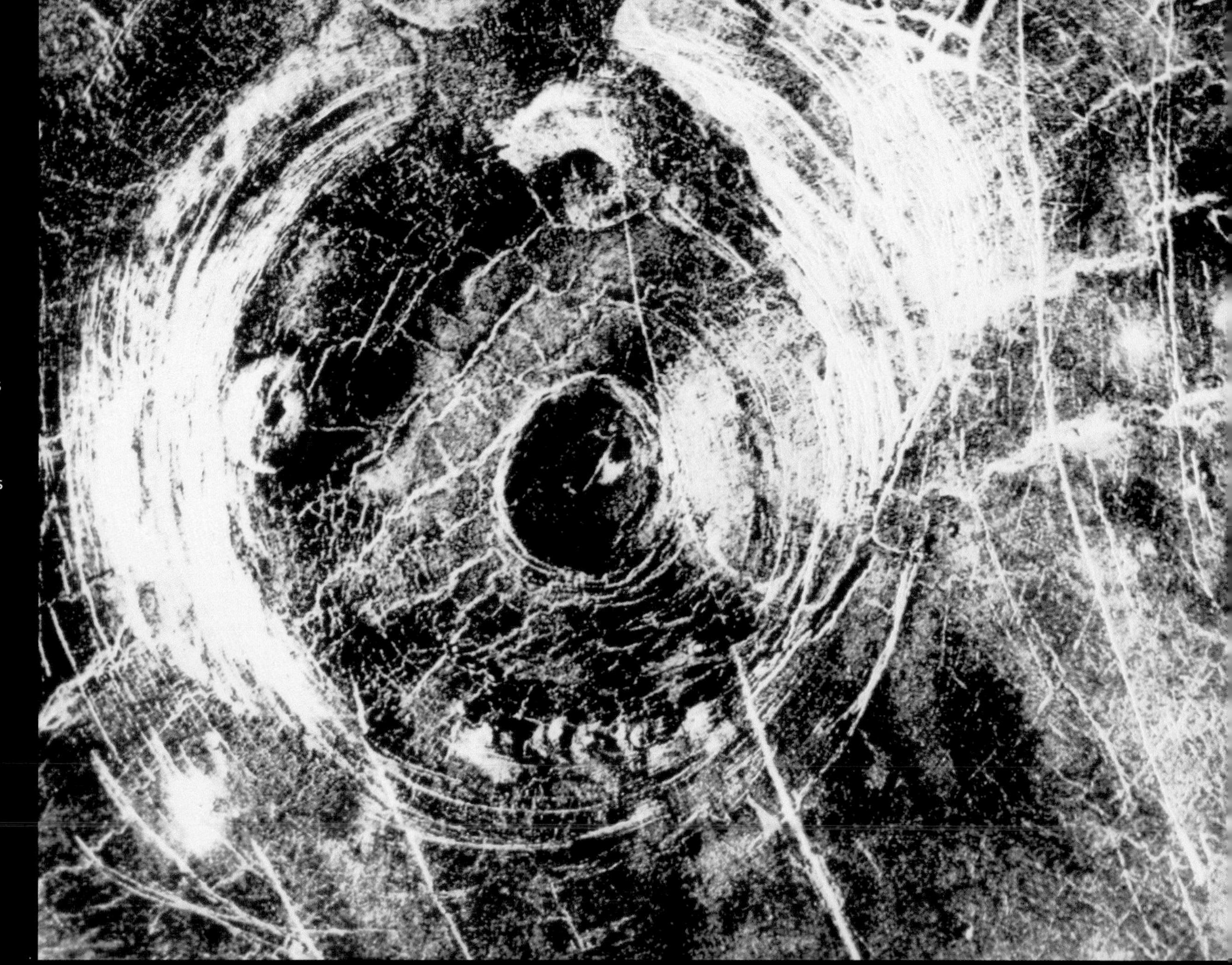

↑ **Magellan snapped this image** of the Aine Corona. The corona (the large circular structure) is 125 miles (200 km) across and is thought to have formed following intense volcanic activity.

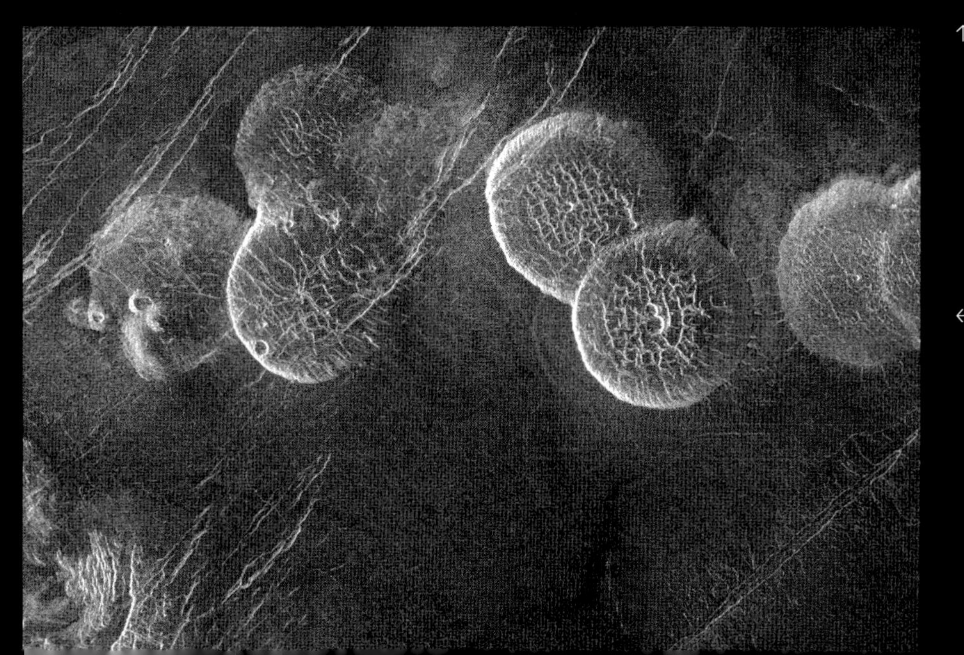

← **The pancake domes** are found in the Alpha Regio region of Venus. Each of the domelike hills measures some 16 miles (25 km) across and were likely formed by the successive heating and cooling of lava.

Earth

Third planet from the Sun and largest of the so-called terrestrial worlds, Earth is the only planet in the Solar System capable of sustaining liquid surface water. Fully 70 percent of the surface of our planet is covered in it, many miles deep in places. The atmosphere is primarily nitrogen and oxygen, and the climate (among other things) has been conducive to the appearance of widespread life. Like some other Solar System bodies, Earth is still volcanically active.

ATMOSPHERE

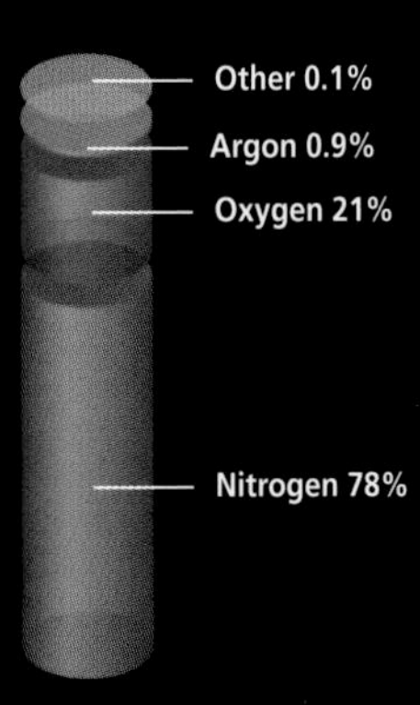

PLANET STATISTICS

Origin of name *Nerthus*, an ancient Germanic goddess
Diameter 7926 miles (12,756 km)
Mass 1.314 x 10^{25} pounds (5.973 x 10^{24} kg)
Volume 0.3 trillion cubic miles (1.1 trillion km^3)
Farthest from Sun 94.5 million miles (152.1 million km)
Closest to Sun 91.4 million miles (147.1 million km)
Mean surface temperature 72°F (22°C)
Highest peak 29,035 feet (8850 m), Mount Everest
Lowest trough -36,200 feet (-11,033 m), Marianus Trench
Water coverage 71%
Surface gravity 1.0 gee
Magnetic field strength 0.31 gauss
Number of satellites 1

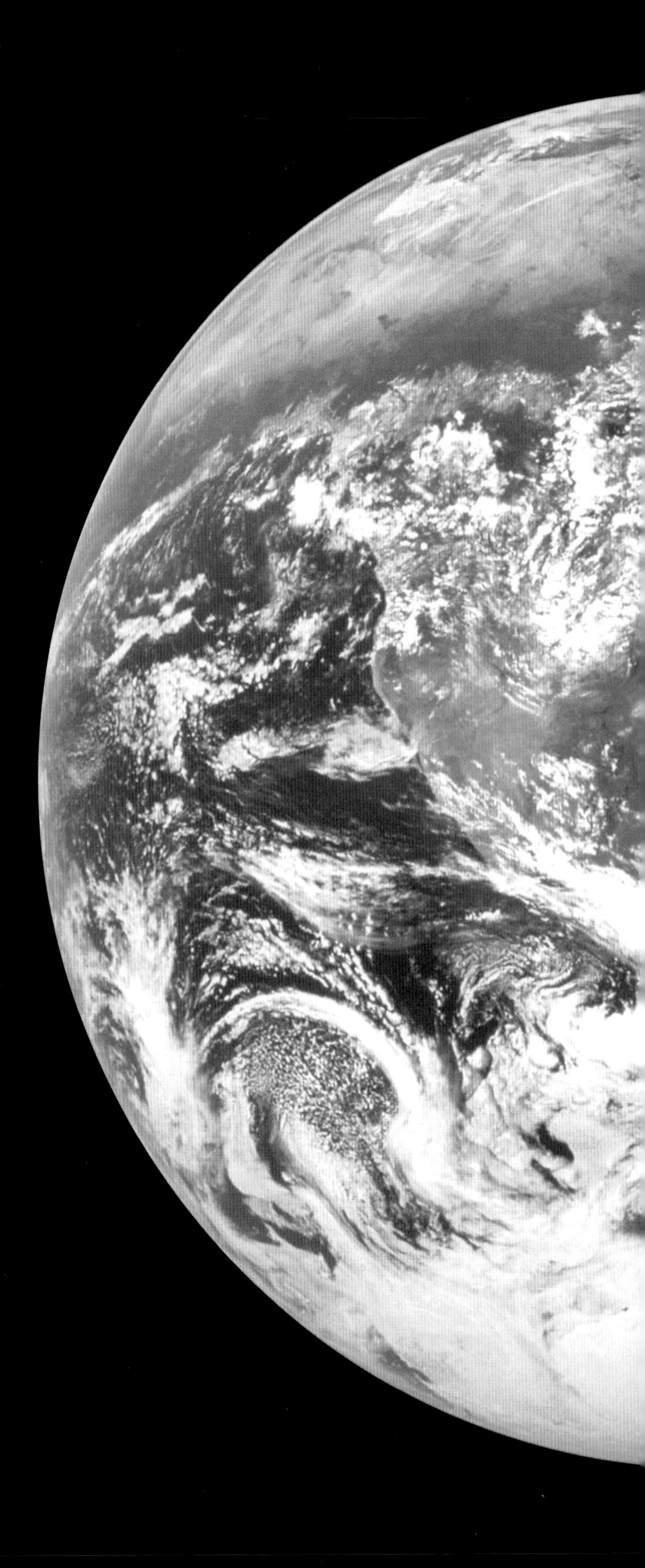

↓ **The Spanish Canary Islands** off the west coast of Africa peep from beneath the clouds in this satellite photo of July 2002. Here the wind is blowing north to south (from top to bottom in the image), producing the turbulent cloud patterns.

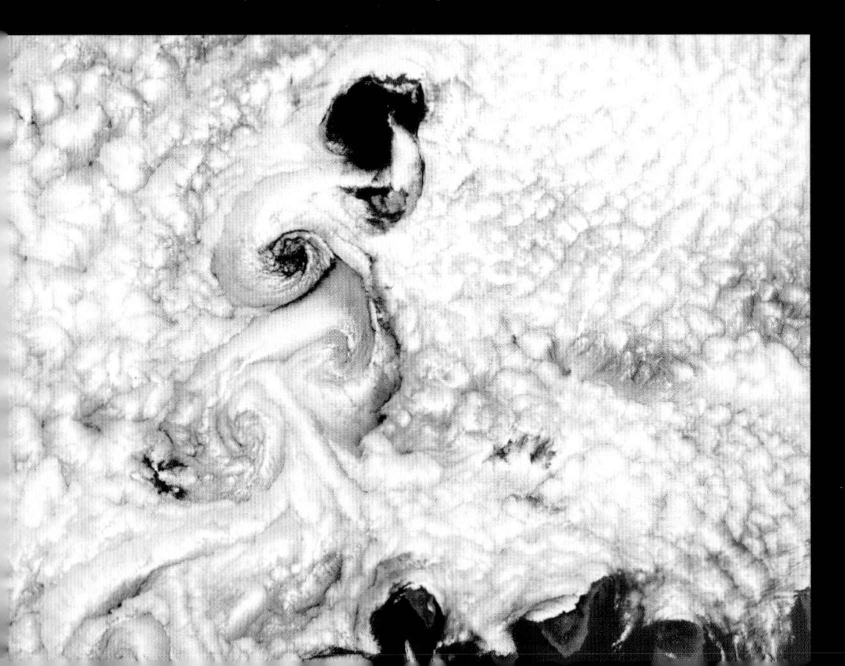

THE LAYERS OF EARTH

Earth is the only terrestrial planet whose interior has four components—the addition being a clearly defined inner core. As a result of seismologica measurements, our planet's interior is known much more accurately tha that of the other planets.

Crust Earth's crust varies in thickness from about 5 miles (8 km) under the oceans to 45 miles (70 km) beneath the continents. It is composed of igneous rocks—those derived from volcanism.

Mantle The mantle, pliable at the top but still solid, extends to a dept of about 1800 miles (2900 km) and is composed primarily of oxygen, magnesium, silicon and iron—a substance known as olivine.

Solid outer core Below the mantle is the outer edge of the core, made of nickel–iron. Its radius is about 2200 miles (3500 km), covering more than half of the planet's interior.

Liquid inner core The inner core is liquid, about 750 miles (1200 kr in radius. The temperature there may reach 11,000°F (6000°C). No other terrestrial planet has this two-component core structure.

SIZE COMPARISON

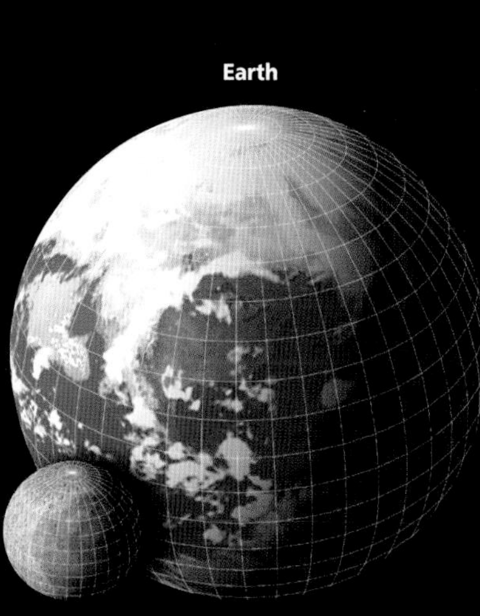

ORBIT STATISTICS

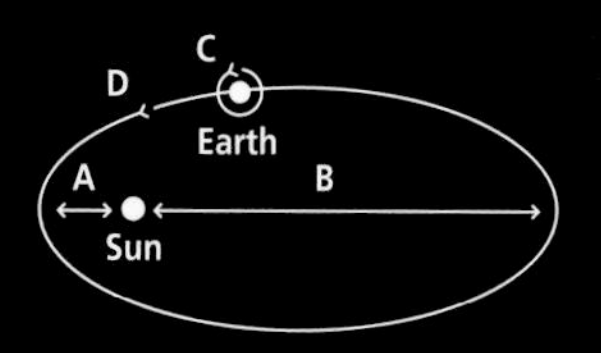

A 91.4 mil. miles (147.1 mil. km)
B 94.5 mil. miles (152.1 mil. km)
C rotates in 23 hr 56 min 4.1 sec
D orbits in 365.36 Earth days

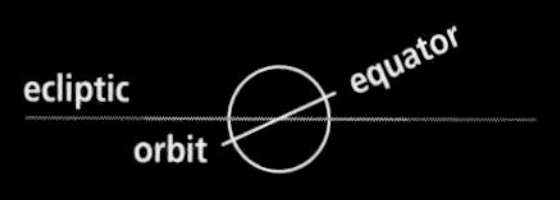

Axial inclination 23.5°
Angle of orbit to ecliptic 0°

arth: atmosphere

ırth's atmosphere is medium in density—100 times thicker than Mars', ıt 100 times thinner than Venus'. In contrast to these two planets, hose atmospheres are filled with carbon dioxide, Earth's contains very tle of this noxious gas. Instead, much of our planet's carbon dioxide dissolved in the oceans and kept out of harm's way. The primary mponent of our atmosphere, 78 percent by volume, is nitrogen. xygen comprises 21 percent, argon 0.9 percent, and various other bstances including water vapor complete the cocktail. The water vapor most apparent when it condenses to form clouds. These can cover much as a quarter of the surface at any given time. Otherwise the mosphere is transparent. Curiously, Saturn's moon Titan has an even enser nitrogen atmosphere. But it is exceedingly cold and inhospitable.

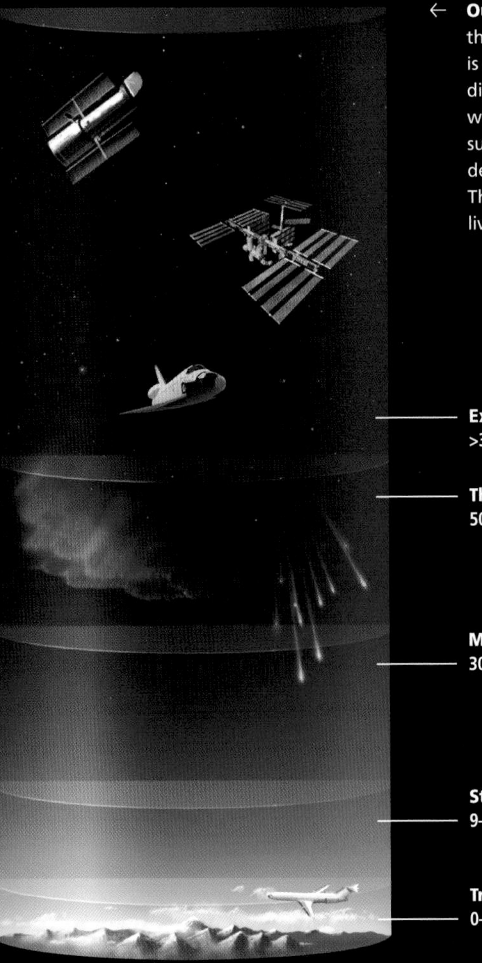

← **Our atmosphere,** like those of all the planets, is divided into several distinct regions, each with different properties such as temperature, density and pressure. The region that we all live in is the troposphere.

Exosphere
>310 miles (>500 km)

Thermosphere
50–310 miles (80–500 km)

Mesosphere
30–50 miles (50–80 km)

Stratosphere
9–30 miles (15–50 km)

Troposphere (Sea level)
0–9 miles (0–15 km)

← **The aurorae as seen on Earth** are caused by particles from the Sun interacting with Earth's magnetic field in the thermosphere region of our atmosphere—more than 50 miles (80 km) above the poles.

↓ **The Sun shines through** Earth's atmosphere in this view taken from the perspective of space.

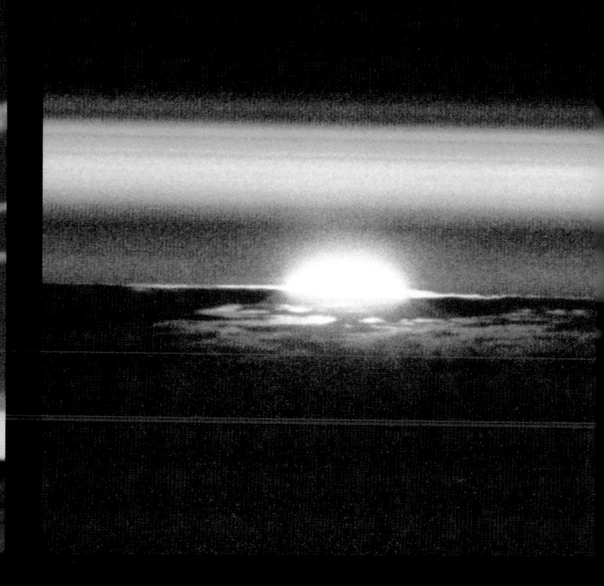

← **As seen from the orbiting space shuttle Columbia** in late 2002, the Moon starts its slow descent behind Earth and shines through our atmosphere. The Moon is actually a crescent, illuminated by the Sun from below, but the whole Moon face is visible since Earth is also reflecting sunlight onto it.

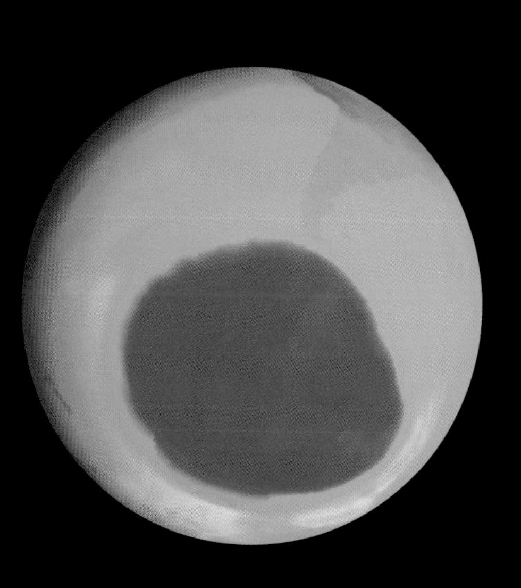

OZONE LAYER

One of the molecules in our atmosphere is called ozone. It is a form of oxygen in which three atoms are glued together. Ozone is a precious commodity, because it absorbs the harmful ultraviolet radiation that emanates from the Sun, and thus keeps us all from getting skin cancer—mostly. However, our atmospheric ozone is at risk. Compounds called CFCs (chlorofluorocarbons)—which are used in aerosols and then find their way into our atmosphere—and other substances cause a chain reaction that destroys ozone and renders us more vulnerable to the Sun's ultraviolet onslaught. This false-color photo (*left*) taken over Antarctica shows the ozone concentration —blue for the weakest and red for the highest. Evidently there is a hole (dark blue) in the layer that changes with the seasons, being at its worst in the late winter and early spring in the south. Luckily, there are signs that the hole overall is gradually becoming smaller with time.

Earth: geography

Earth is a unique planet, and not merely because it is the only world we know is capable of supporting life. No other planet in the Solar System has a geology or climate quite like Earth's. One thing that makes Earth different is its crust. It is divided into several different plates that are constantly on the move across the surface, building mountains as they go. All the other terrestrial worlds are single-plate planets. Earth also has two-thirds of its surface covered in water. Together, these features ensure that the planet's surface is never constant. Meteorites do impact from time to time, but by and large their craters are eradicated very quickly on astronomical timescales. By contrast, the surface of the Moon has not changed significantly in billions of years.

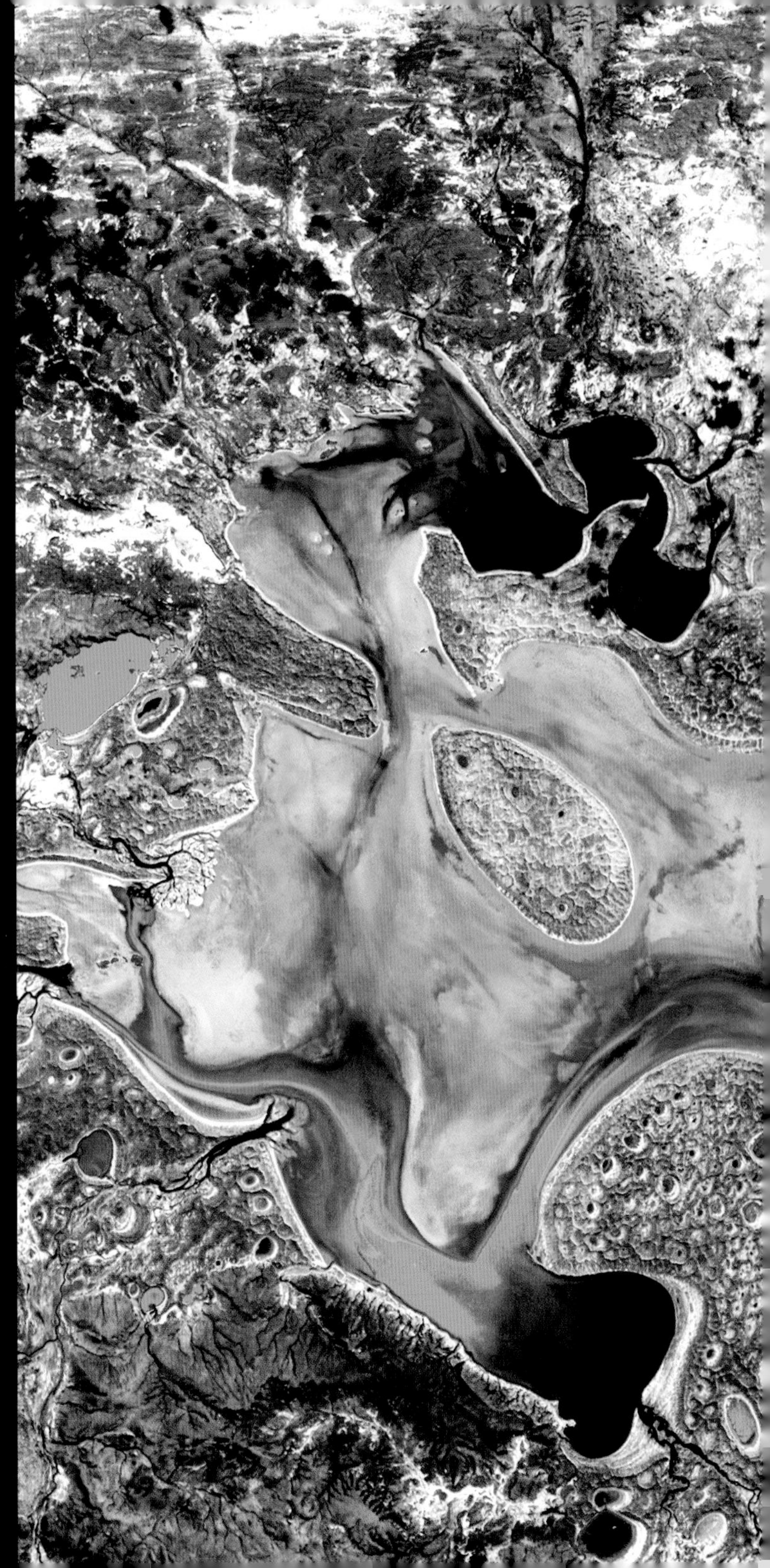

COMPARING EARTH

Earth's surface has a moderate degree of relief. From the bottom of the deepest trench to the highest peak there are about 12½ miles (20 km). Venus, slightly smaller than Earth, has a similar relief of about 8½ miles (13.7 km). But Mars, although only half the size of Earth, has much higher peaks and deeper canyons. Its highest volcano, Olympus Mons, towers 16 miles (26 km) above the local Martian mean level. The horizontal scale in this diagram is greatly reduced for clarity.

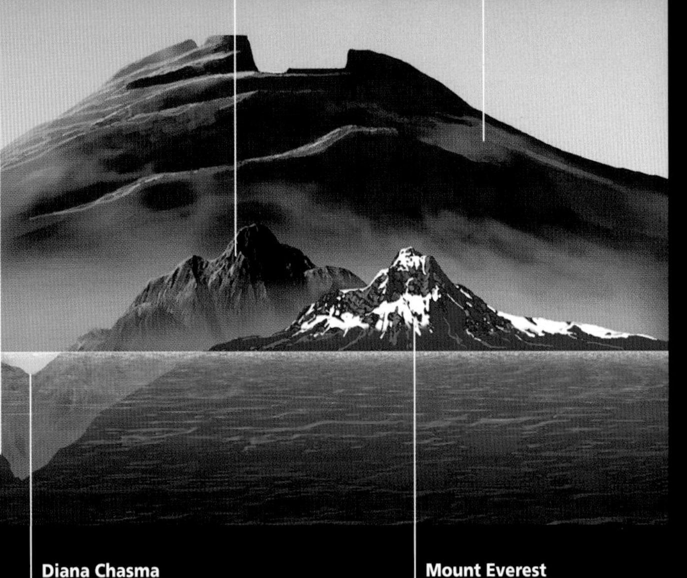

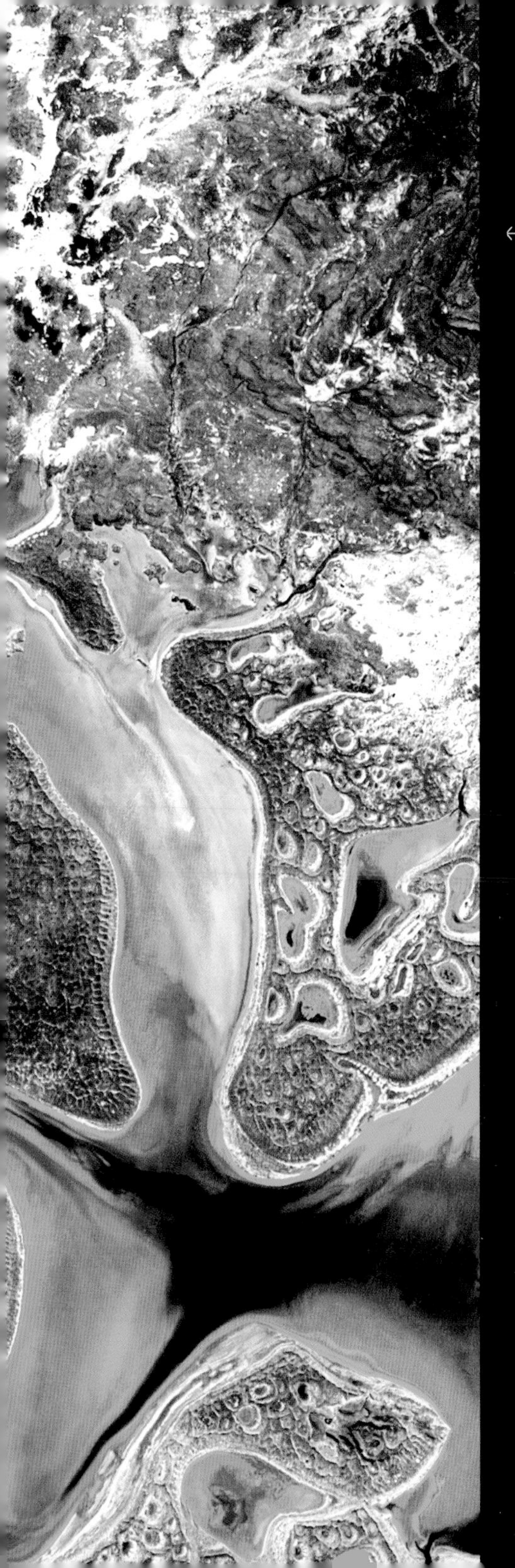

← **This colorful image shows** a Landsat-satellite view of Lake Carnegie in Western Australia. This is an ephemeral lake—it contains water only during periods of heavy rainfall. At other times of the year the "lake" is reduced to a vast patch of cracked mud.

→ **The Mississippi is the longest** river in North America, with a total length of 2350 miles (3780 km) from its source in Lake Itasca, Minnesota, to the river delta (*right*) in the Gulf of Mexico. A delta is the name for the land formation that results when a river drains into a larger body of water.

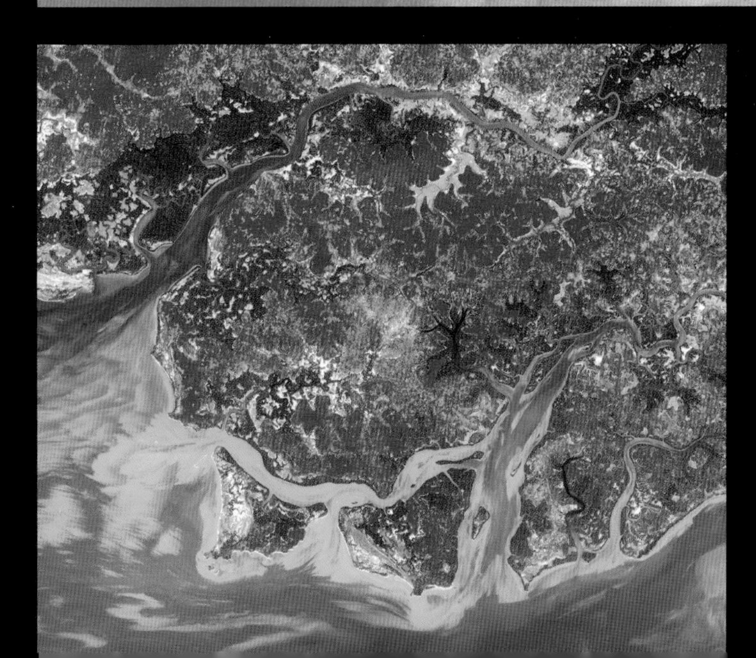

→ **Guinea-Bissau,** the small western-African country, reveals bizarre patterns along its coastline in this aerial photograph. They are the result of silt—tiny particles of sediment—draining into the Atlantic Ocean, and carried by various rivers such as the Geba

The second-smallest terrestrial world is Mars, fourth from the Sun. Mars is about half the size of Earth. It has a thin atmosphere of carbon dioxide, polar caps of dry ice, and an active weather system. Although now thought to be volcanically dead, Mars is covered in ancient volcanoes, including the largest in the whole Solar System, Olympus Mons, while impact craters are also common. There are even indications that Mars may once have had oceans, or at least seas, of liquid water. But this issue is still in debate.

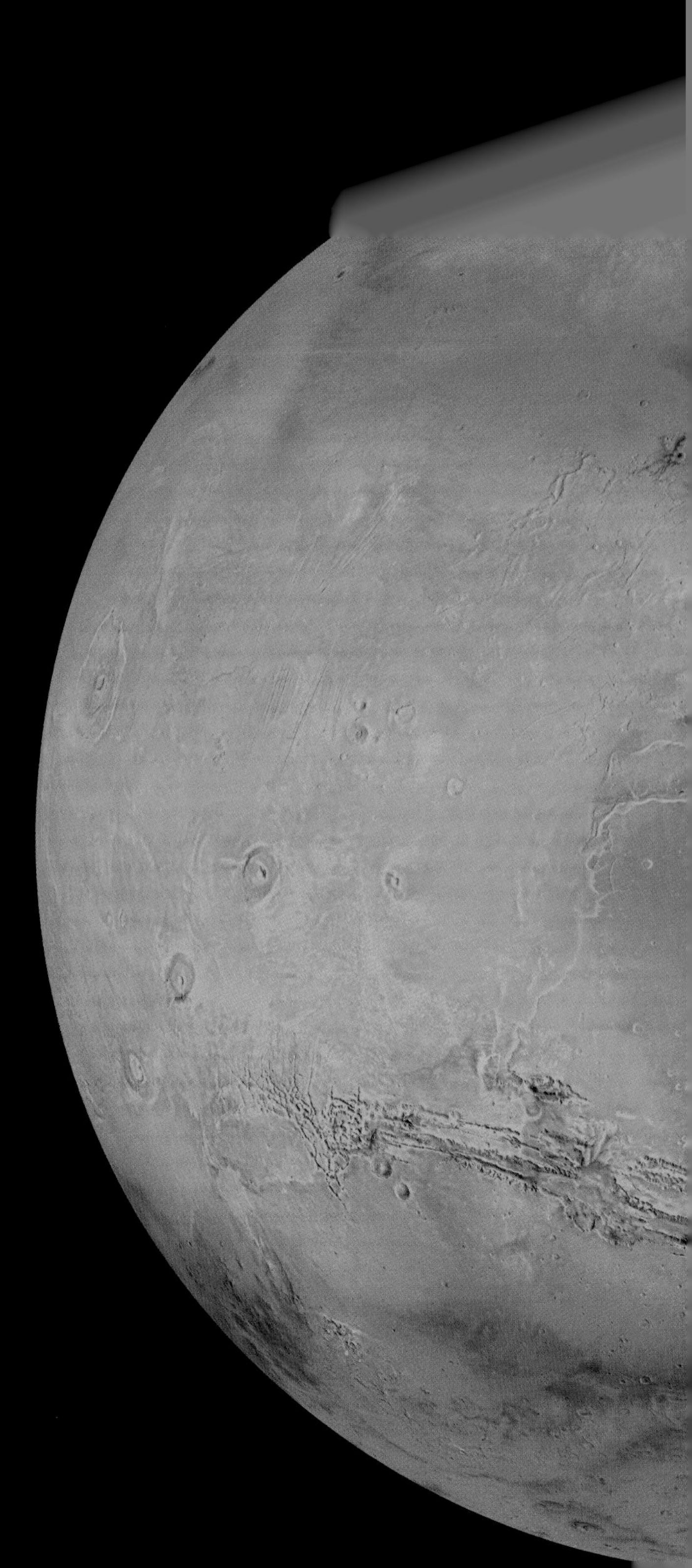

ATMOSPHERE

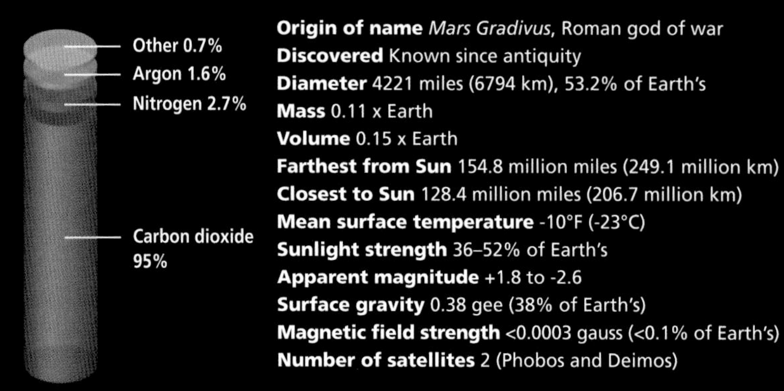

PLANET STATISTICS

Origin of name *Mars Gradivus*, Roman god of war
Discovered Known since antiquity
Diameter 4221 miles (6794 km), 53.2% of Earth's
Mass 0.11 x Earth
Volume 0.15 x Earth
Farthest from Sun 154.8 million miles (249.1 million km)
Closest to Sun 128.4 million miles (206.7 million km)
Mean surface temperature -10°F (-23°C)
Sunlight strength 36–52% of Earth's
Apparent magnitude +1.8 to -2.6
Surface gravity 0.38 gee (38% of Earth's)
Magnetic field strength <0.0003 gauss (<0.1% of Earth's)
Number of satellites 2 (Phobos and Deimos)

↓ **Sunrise on Mars, as captured by the Sojourner rover** during the American Pathfinder mission of July 1997. Mars has long been the center of planetary study.

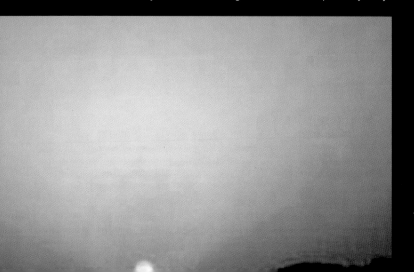

THE LAYERS OF MARS

A cross-section of the interior of Mars looks somewhat like that of Venus. However, there are differences in composition between the layers of Mars and Venus, with Mars having a lower overall density and therefore containing lighter minerals.

Crust The Martian crust is thought to be quite thick, because it is difficult to see how a thin crust would be able to support such massive volcanoes. Olympus Mons, for example, is the largest volcano in the known Solar System. The crust may extend to a depth of about 75 miles (120 km).

Mantle The thick mantle is similar in density to either Earth's or Venus'. It is probably composed of a mixture of substances including olivine (an oxygen–iron–magnesium compound found throughout the terrestrial worlds), iron-oxide and even some water.

Core Mars' core occupies no more than 30–50 percent of its interior (shown here with a core at the upper end of that scale) and may be partially molten. It probably has a lower density than Earth's core, so may be made of iron and iron sulfide, not the heavier nickel–iron of Earth's core.

SIZE COMPARISON

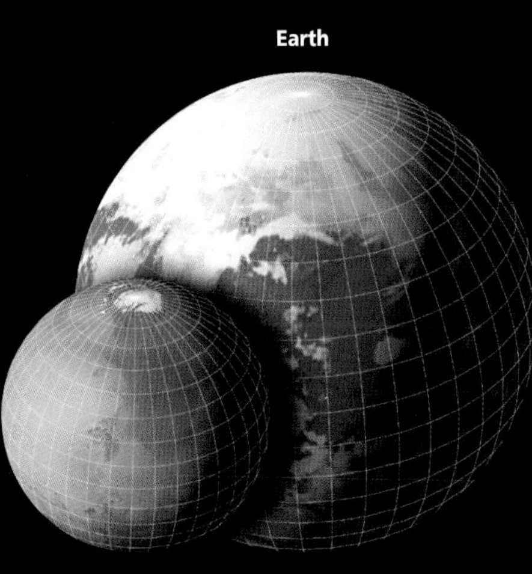

ORBIT STATISTICS

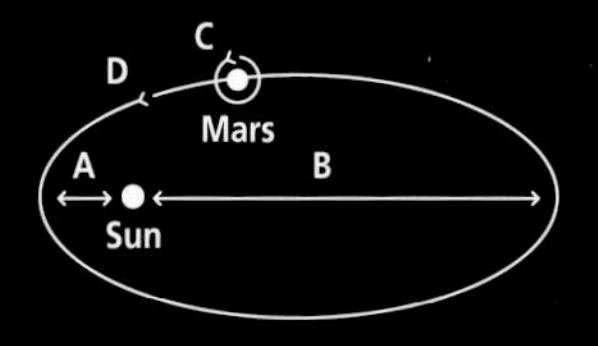

A 128.4 mil. miles (206.7 mil. km)
B 154.8 mil. miles (249.1 mil. km)
C rotates in 24 h 37 min 23 sec
D orbits in 686.98 Earth days

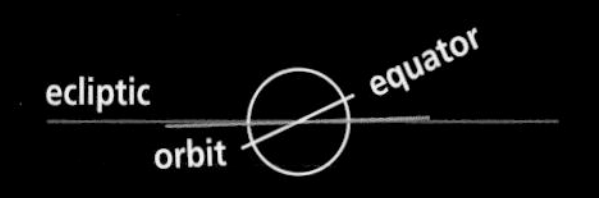

Axial inclination 23.98°
Angle of orbit to ecliptic 1.85°

Mars: surface

Since antiquity, Mars has been known as the Red Planet, and for good reason. The surface is strewn with large rocks and a dusting of fine red soil. The red color comes from particles of oxidized iron. In effect, Mars is a rusty planet. The terrain is divided into two quite different regions. The south is very heavily cratered. But in the north it is the volcanoes (now probably extinct) that dominate. This region is geologically the youngest, lava from the volcanoes having smothered the craters underneath before solidifying. The other striking Martian features are the canyons, which some believe were carved long ago by ancient waterways. Indeed, some so-called gullies on Mars seem to have been carved by liquid water very recently. However, the jury is still out, as these might equally be the result of the action of liquid carbon dioxide.

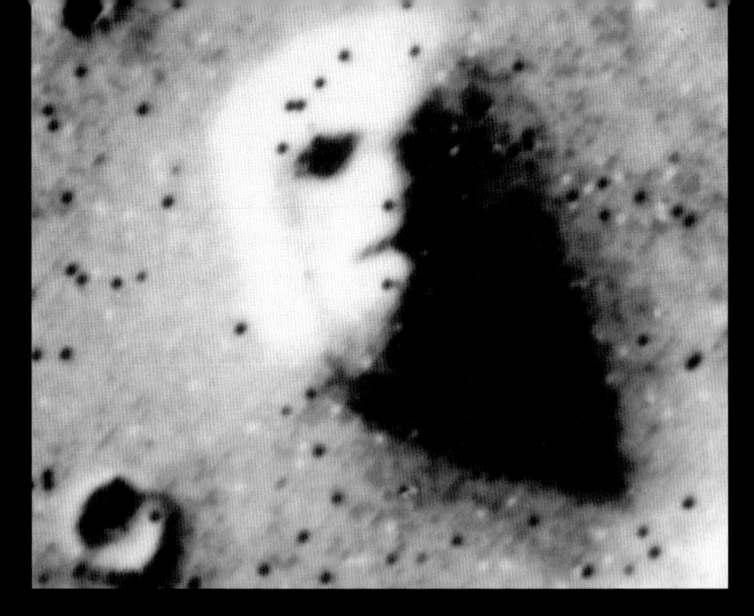

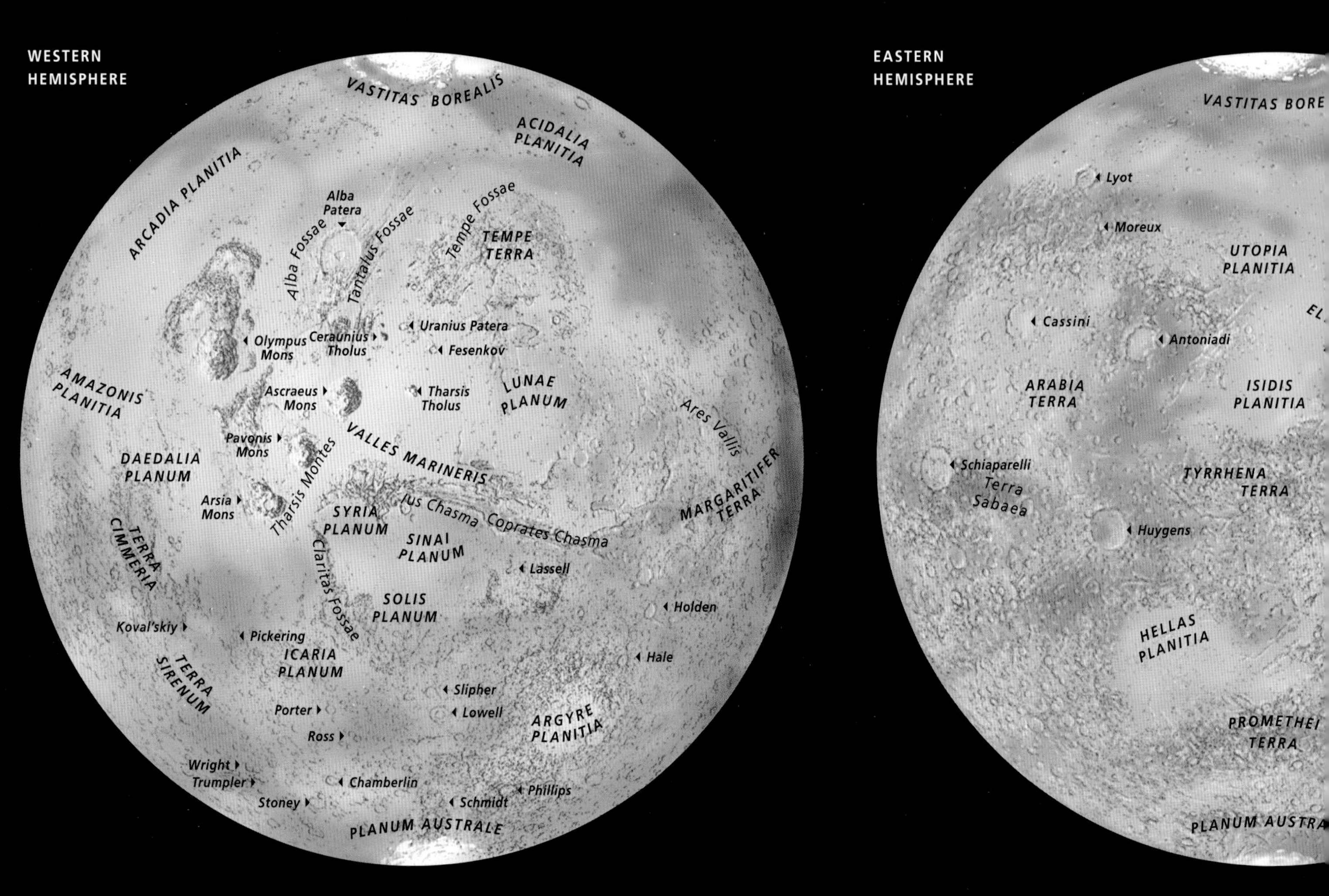

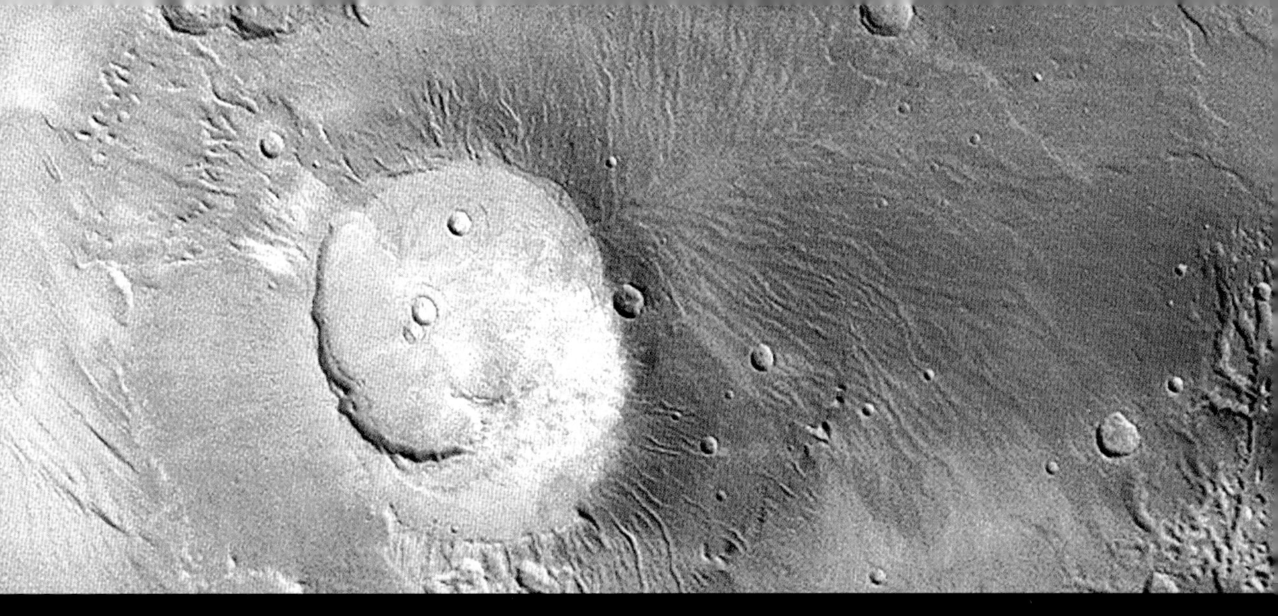

← **In 1976, the Viking orbiters** returned this famous image of a rock formation on Mars that resembles a human face. It is in fact known as the "Face on Mars." But more recent photos taken by the Mars Global Surveyor craft under different lighting conditions show the "Face" for what it really is—a trick of the light.

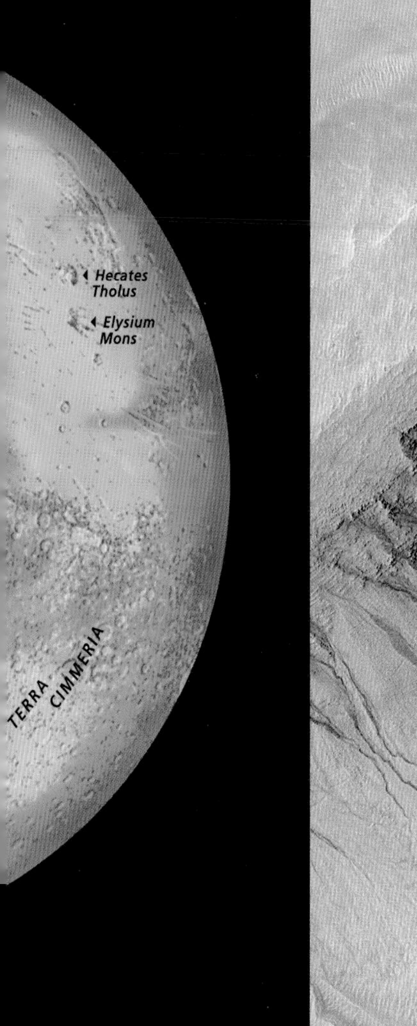

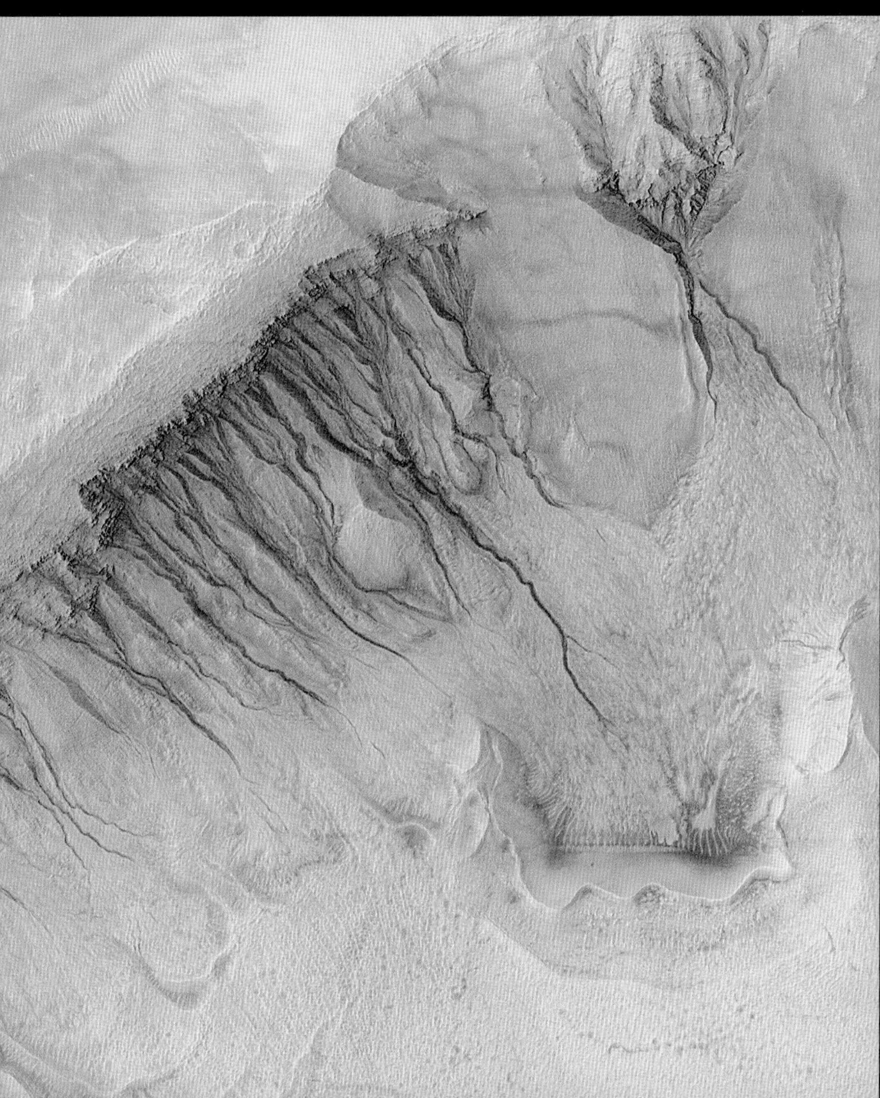

↑ **Apollinaris Patera,** one of numerous volcanoes on Mars, is seen center-left of this image with a patch of cloud clinging to its summit. Like many Martian volcanoes, this is a giant. Its crater measures some 50 miles (80 km) across, while the whole structure juts 3 miles (5 km) above the surrounding local level.

← **There is much evidence** of recent water flow on Mars, and this is perhaps one of the most striking images. It reveals apparent vast gullies, carved by water (or possible liquid carbon dioxide, if the surface conditions are just right) flowing down the side of the Newton Basin—a large crater in the Terra Sirenum region of Mars. Whether this landscape was caused by water or carbon dioxide, these are geologically very recent formations—perhaps a mere million years old.

Mars: satellites

Mars' satellites, Phobos and Deimos, are among the smallest in the Solar System. They are not native to Mars, but are in fact asteroids that were captured into Martian orbit. Like asteroids, both are heavily cratered and irregularly shaped. Both worlds have very dark, sooty surfaces that reflect only 2 percent of incident sunlight. Phobos is the closest to Mars and the largest, at 16¼ miles (26.2 km) across its longest axis, orbiting Mars three times for every rotation of the planet. Deimos, meanwhile, is about half the size of Phobos, and hangs in the sky over three times farther above the planet. If you viewed them from the surface of Mars, Phobos would appear about half the size of our own Moon, while Deimos would be just one-fifth the size of Phobos—little more than a starlike speck.

MARTIAN SATELLITES IN ORBIT

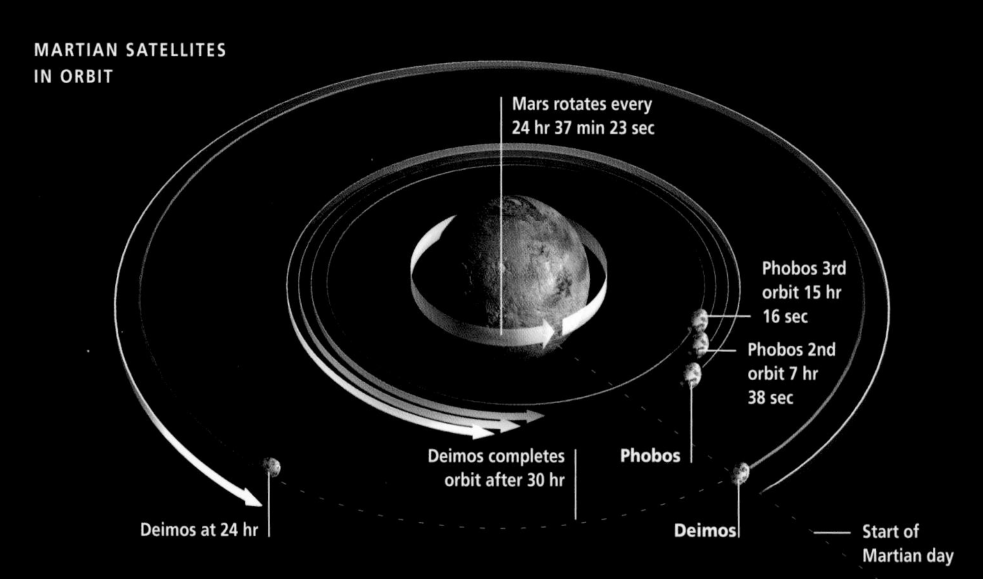

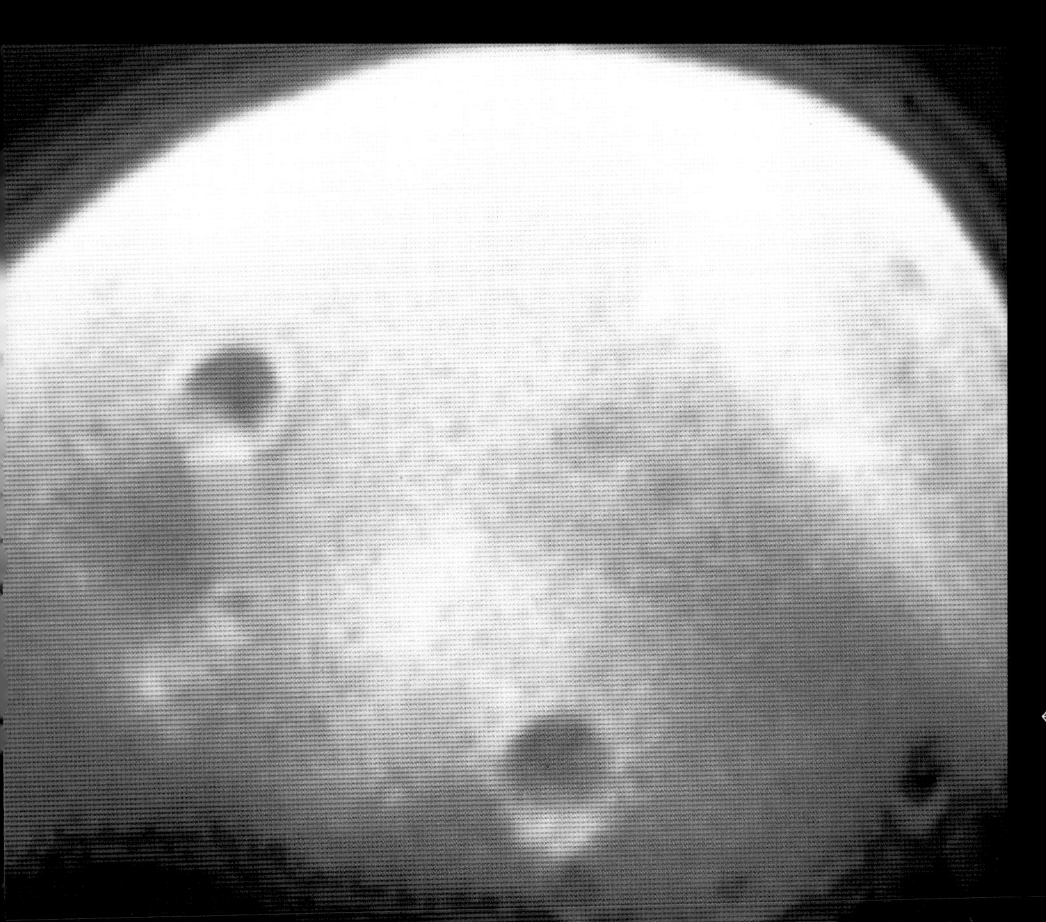

SATELLITES IN ORBIT

Martian satellites, Phobos and Deimos orbit Mars at a distance of 3700 miles (6000 km) and 12,500 miles (20,000 km) respectively above the red sands. Phobos is so close to Mars that it completes more than three orbits for every Martian day. So from the surface of Mars, Phobos rises in the west and sets in the east, overtaking the Martian rotation. The proximity and orbit mean that Phobos is doomed. In about 40 million years it will fall to the surface of Mars to create a new impact crater. Deimos orbits farther out and its orbit, by contrast, is stable.

SATELLITE DATA

Satellite	Orbit period	Distance from Mars
Phobos	0.319 day	5830 miles (9380 km)
Deimos	1.791 days	14,580 miles (23,460 km)

← **This is the smaller of the two Martian moons, Deimos,** as photographed by the Viking 1 Orbiter in 1977. Being so small, it is irregularly shaped, and its surface is covered in ancient impact scars.

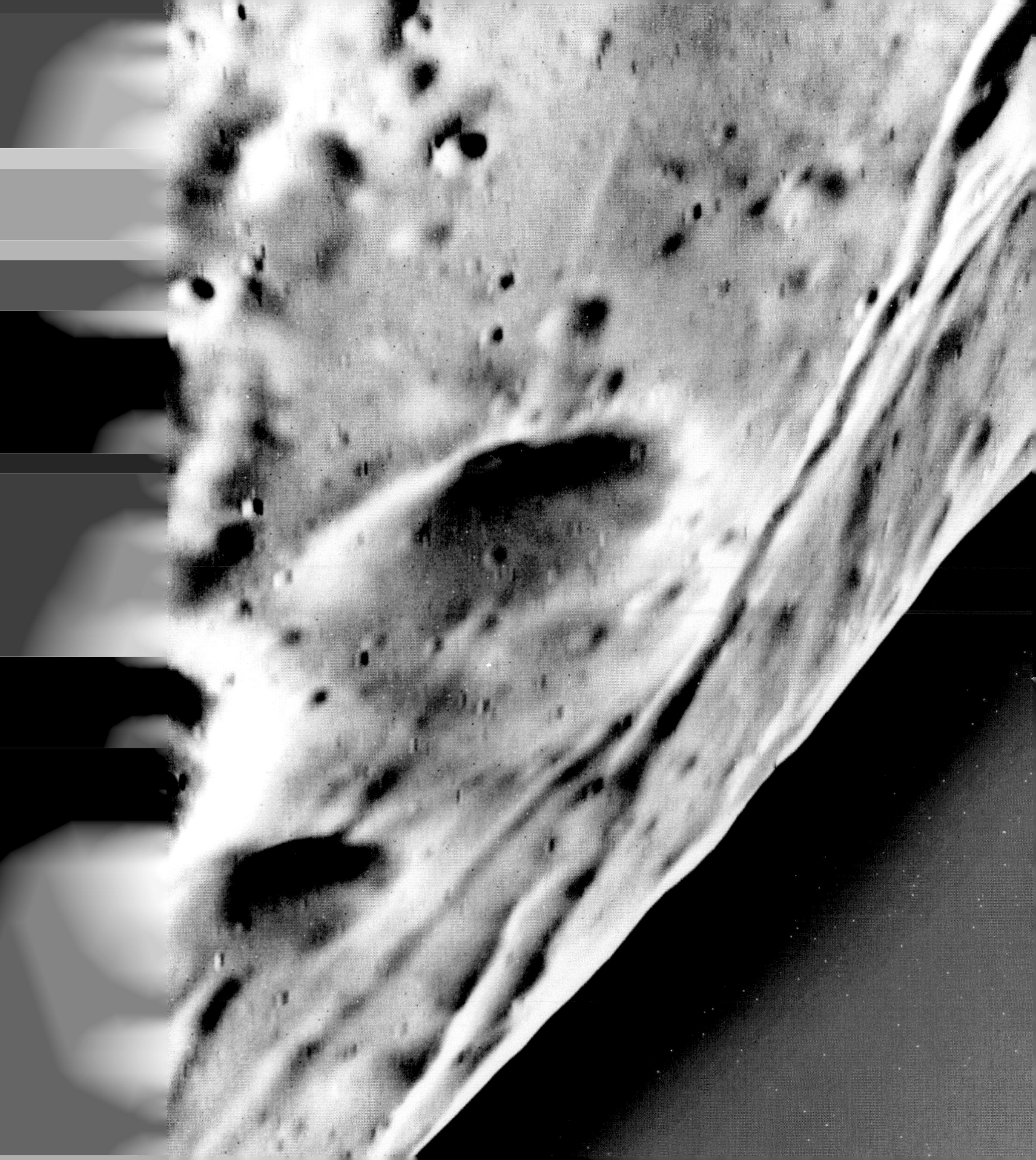

Jupiter

Jupiter is easily the Solar System's largest world—it is big enough to contain all the other planets and more. This is the first of the gas giants—worlds composed mainly of fluids rather than rock and metal, but possibly having rocky or icy cores. Jupiter is distinctively banded, but it is perhaps most famous for its Great Red Spot—an enormous hurricane-like storm, larger than Earth. The planet has several dozen satellites—our knowledge of the exact number increases all the time. Four of these are among the largest known.

ATMOSPHERE

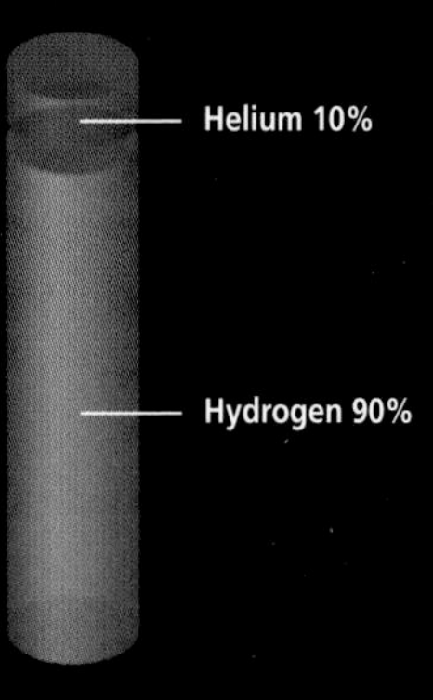

PLANET STATISTICS

Origin of name *Jove*, Roman god of heaven and Earth
Discovered Known since antiquity
Diameter 89,405 miles (143,884 km), 1127.9% of Earth's
Mass 317.7 x Earth
Volume 1323 x Earth
Farthest from Sun 506.9 million miles (815.7 million km)
Closest to Sun 460.4 million miles (740.9 million km)
Mean surface temperature -240°F (-150°C)
Sunlight strength 3–4% of Earth's
Apparent magnitude -1.2 to -2.5
Surface gravity 2.3 gee (230% of Earth's)
Magnetic field strength 4.28 gauss (1380% of Earth's)
Number of known satellites 61

↓ **The Hubble Space Telescope snapped** multiple comet impacts on Jupiter in 1994—seen in this image as the prominent dark spots marking the surface.

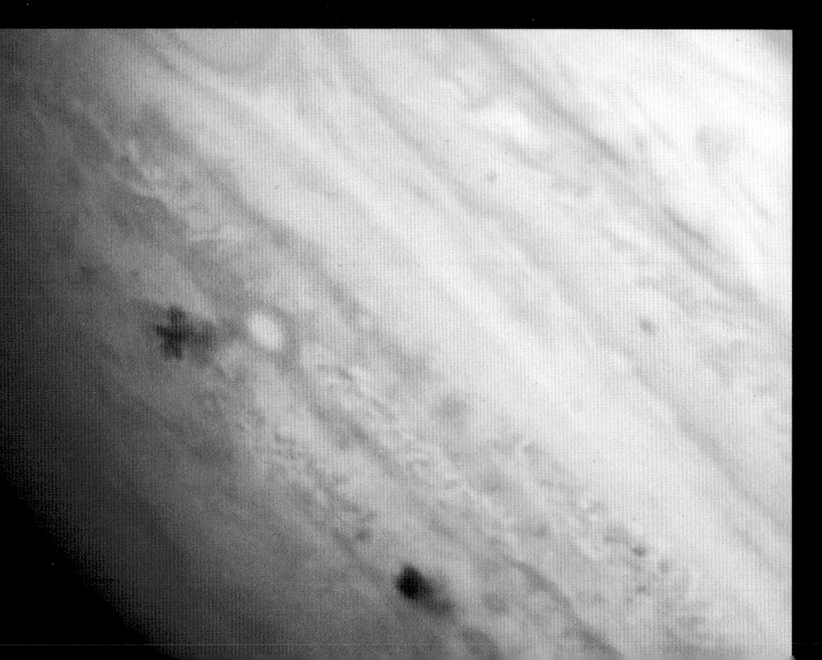

THE LAYERS OF JUPITER

Jupiter, being a gas giant, is nothing at all like the terrestrial planets on the inside, as one might expect. Its vast magnetic field strongly indicates the presence of a substance known as liquid metallic hydrogen.

Atmosphere Jupiter has no solid surface. Instead, its atmosphere—which is mainly molecular hydrogen and some helium—blends seamlessly into the fluid layer beneath, the outer mantle.

Outer mantle This region is mostly hydrogen, in keeping with the rest of the planet. It is under so much pressure that it has become a liquid, known as liquid hydrogen.

Inner mantle The inner mantle, which takes up most of Jupiter's interior, is even denser than the outer mantle. The pressure has converted the hydrogen into a metallic fluid called liquid metallic hydrogen. Like a metal, it is very efficient at conducting electricity.

Core The core is about 10–15 times as massive as the entire Earth and composed of rock and ice, possibly differentiated—that is, with rock in the center and ice farther out.

SIZE COMPARISON

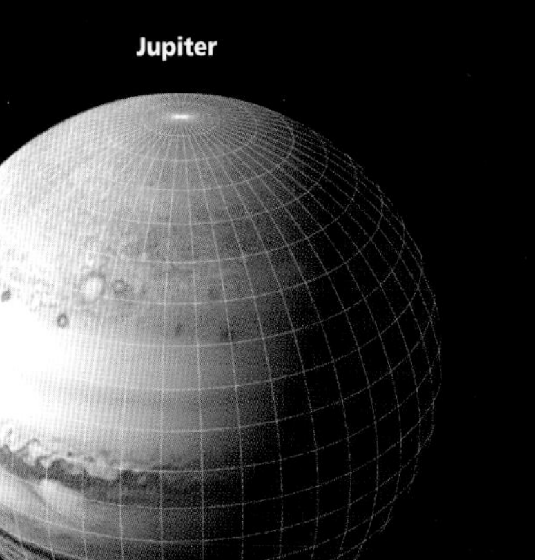

ORBIT STATISTICS

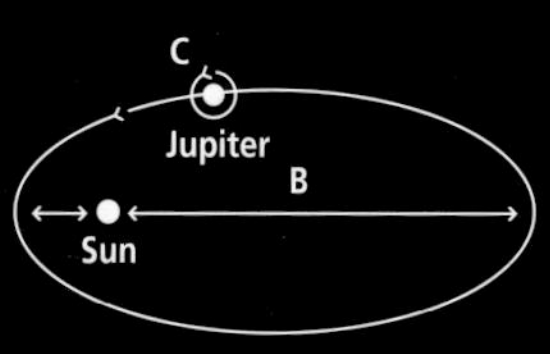

A 460.4 mil. miles (740.9 mil. km)
B 506.9 mil. miles (815.7 mil. km)
C rotates in 9 h 50 min 30 sec
D orbits in 11.86 Earth years

Axial inclination 3.1°
Angle of orbit to ecliptic 1.3°

Jupiter: features

Jupiter is not only the largest planet in the Solar System, it is also arguably one of the most attractive. This world has no solid surface, since it is almost entirely fluid. Instead, the images we see come from the outermost edge of its animated atmosphere, a thick shroud of hydrogen and helium. Its most obvious features are its bands, visible from Earth in even a modest telescope. The dark regions are called belts and the brighter regions, at a higher altitude and so catching the sunlight, are called zones. Both of these features are stretched around the planet by its rapid rotation. The other striking Jovian characteristic is its so-called Great Red Spot (*right*, in detail). It's a vast, hurricane-like storm three times larger than the entire planet Earth, and has existed for centuries. Its vibrant color comes from the presence of certain molecules, most notably phosphine. Like Saturn, Jupiter also hosts a system of rings. But they are very dark and transparent, and only visible from within Jupiter's shadow, shielded from the glare of the Sun. And unlike the rings of the other planets, those of Jupiter contain only microscopic fragments.

MAP OF JUPITER

THE RINGS OF JUPITER
The rings of Jupiter are extremely sparse, especially in comparison to Saturn. In fact they are not at all visible from Earth, and even from the Jupiter system itself you would only spot them clearly if you were within the shadow of the planet. Jupiter's rings consist of three components—the Halo ring, the Main ring and the Gossamer ring. Unlike the rings of all the other planets, those of Jupiter are made of tiny fragments of rock comparable in size to particles of smoke. The pressure of the Sun's radiation, and also the strong magnetic forces at work within the Jovian environment, regularly remove some of these fleeting particles, since they are very lightweight. However, the ring system is most likely maintained through the regular action of meteorite impacts on the surfaces of the Jovian satellites. Dust kicked-up during these impacts is thrown into orbit around Jupiter and eventually repopulates the rings.

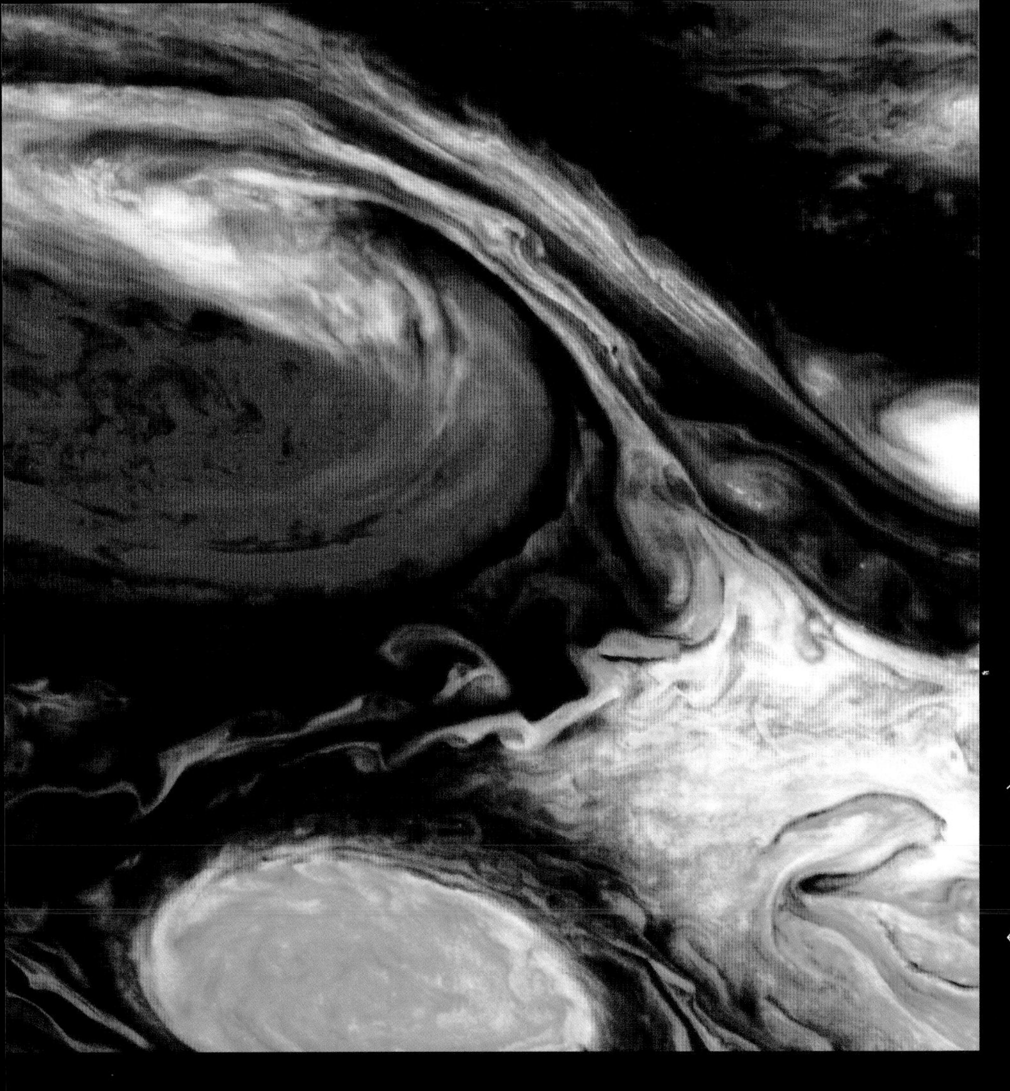

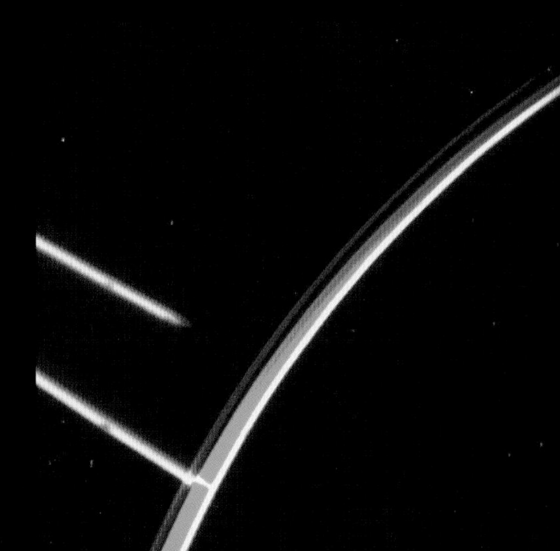

↑ **This photo reveals Jupiter's sparse ring system** (seen in orange), as viewed by Voyager 2. The apparent gap in the top portion of the ring is caused by the massive shadow of Jupiter, which is falling on it.

← **The Great Red Spot** is a giant Jovian storm system that has existed as far back as telescopic records go. It scoots around the planet, dragged by the high-speed rotation, gobbling up smaller storms that it passes. Around three Earths could fit into this system.

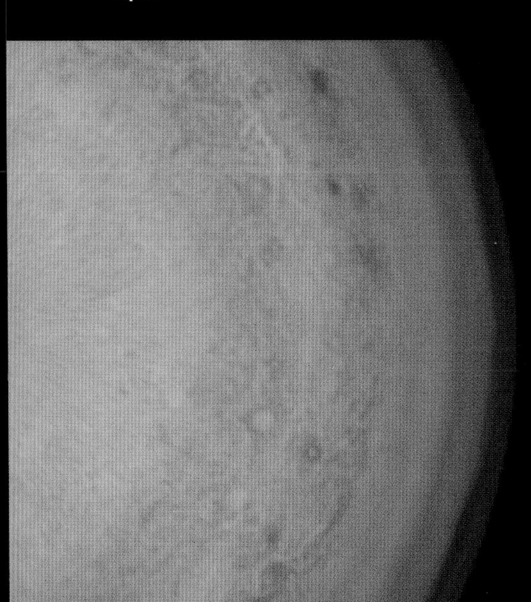

Jupiter: satellites

Jupiter excels in more than just its size: it also has the highest count of natural satellites. To date, the planet has an incredible 61 moons. These include the four so-called Galilean satellites discovered in 1609, and 57 much smaller worldlets exposed more recently, of which only 34 currently have names. The new additions are little more than captured asteroids, like Mars' Phobos and Deimos. But the Galilean satellites, named after Galileo, are significantly more substantial. They are all in the top seven of the Solar System's largest satellites, and one of them (Ganymede) is even larger than a planet, Mercury. These satellites formed in orbit around Jupiter when the planet itself was still taking shape, and each is unique. Io is the closest to Jupiter, rocky and volcanic; Europa is rocky with a cracked surface of frozen water; Ganymede is similar to Europa, but with a much older surface; and Callisto's icy surface is the most battered in the known Solar System.

↓ **Europa's surface, seen here in false color,** resembles a vast, fractured ice rink. Its surface is a shell of water-ice that may hide a liquid ocean beneath.

SATELLITES IN ORBIT
Jupiter has 61 known moons, although the total may measure in the hundreds. This diagram shows the orbits of the four largest, so-called Galilean satellites: Io, Europa, Ganymede and Callisto. Io is so close-in that its massive parent nearby flexes its interior and renders the tiny world volcanic. These worlds are probably primordial—that is, they formed in orbit around Jupiter while the planet itself was growing.

SATELLITE DATA

Satellite	Orbit period	Distance from Jupiter
Io	1.77 days	262,000 miles (422,000 km)
Europa	3.55 days	417,000 miles (671,000 km)
Ganymede	7.15 days	665,000 miles (1.07 mil. km)
Callisto	16.69 days	1.17 mil. miles (1.88 mil. km)

→ **Io passes in front of Jupiter** and casts a shadow on its parent world, in this photo snapped by the Galileo probe in December 2000.

↓ **Each of the Galilean moons is different.** Io is the most volcanic world in the known Solar System, and is predominantly made of rock. Europa has a scored and cracked surface of solid water-ice wrapped around a rocky interior. Ganymede and Callisto have ancient icy surfaces—scarred during eons of cratering. Callisto has more craters per unit surface area than any other known astronomical body.

IO

EUROPA

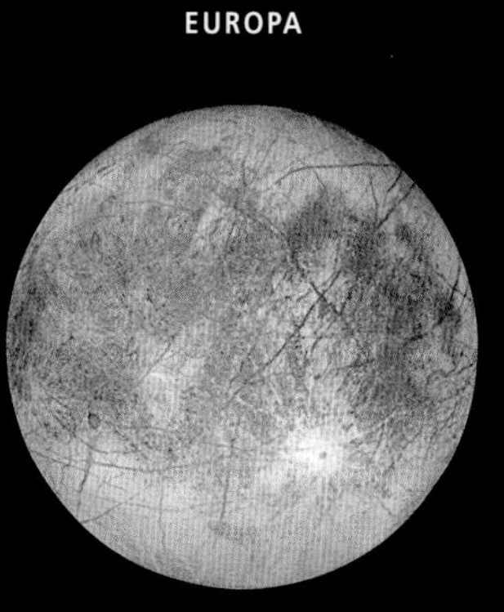

CALLISTO

GANYMEDE

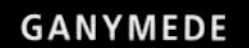

Saturn

The second-largest planet is Saturn, a kind of pale Jupiter with a similar composition. Saturn, like all gas planets, has a system of rings, but those of Saturn are without doubt the finest, easily visible from Earth even with a small telescope. They are composed of countless icy blocks tumbling around the planet, each like an independent satellite and some as large as a house. Another interesting feature about Saturn is its shape. Because of its rapid rotation—once in 10 hours, 13 minutes and 59 seconds—it is noticeably flattened at the poles.

ATMOSPHERE

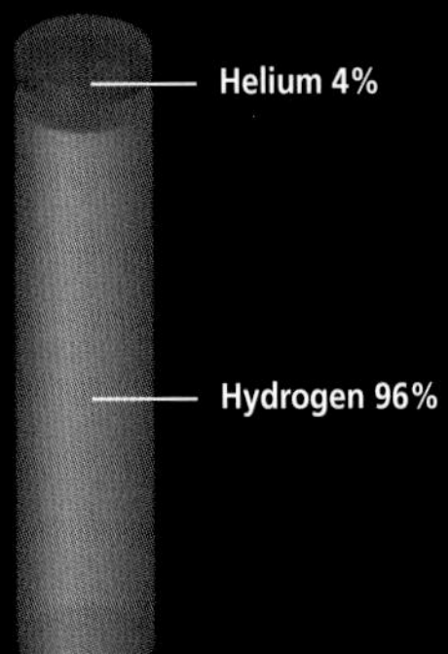

Helium 4%

Hydrogen 96%

PLANET STATISTICS

Origin of name *Saturnus*, Roman god of agriculture
Discovered Known since antiquity
Diameter 74,898 miles (120,536 km), 944.9% of Earth's
Mass 95.2 x Earth
Volume 752 x Earth
Farthest from Sun 933.9 million miles (1503 million km)
Closest to Sun 837.6 million miles (1348 million km)
Mean surface temperature -110°F (-80°C)
Sunlight strength 1% of Earth's
Apparent magnitude +0.6 to +1.5
Surface gravity 1.16 gee (116% of Earth's)
Magnetic field strength 0.22 gauss (71% of Earth's)
Number of known satellites 31

THE LAYERS OF SATURN

The interior of Saturn is very similar to that of Jupiter, with exactly the same materials but with the various layers having slightly different proportions. Saturn is such a lightweight world for its size that a solid plastic duck of the same density would float on water.

Atmosphere Saturn's atmosphere of hydrogen and helium is relatively pale because, being farther from the Sun than Jupiter, its clouds condense lower down. They are thus relatively obscured by layers of haze above them.

Outer mantle As in Jupiter, this region is composed of liquid hydrogen. However, the comparatively low density of Saturn compared to Jupiter means that this zone must occupy a smaller fraction of the planet.

Inner mantle Again, as with Jupiter, this zone is composed of the highly conducting liquid metallic hydrogen. Because of Saturn's lower mass, the internal pressures are lower, so the amount of liquid metallic hydrogen is smaller.

Core Saturn may have a solid core, like Jupiter, composed of rock and ice. As in Jupiter, the rock and ice may be differentiated, or they could be evenly mixed.

SIZE COMPARISON

Earth

ORBIT STATISTICS

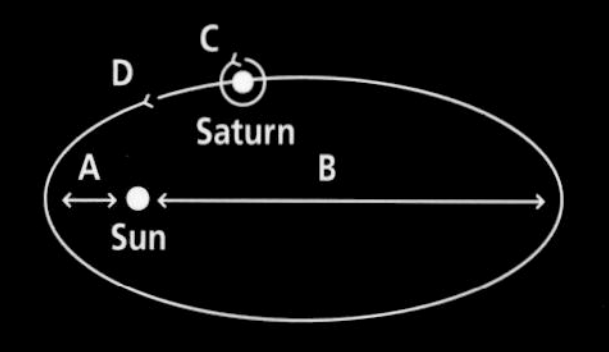

A 837.6 mil. miles (1348 mil. km)
B 933.9 mil. miles (1503 mil. km)
C rotates in 10 hr 13 min 59 sec
D orbits in 29.41 Earth years

ecliptic
equator
orbit

Axial inclination 27°
Angle of orbit to ecliptic 2.48°

Saturn: rings

All of the giant worlds have rings, but the famous ornaments of Saturn are in a class of their own. Like all ring systems, Saturn's are composed of countless boulders of various sizes, each like a mini-satellite on its own independent orbit. But Saturn's rings are particularly conspicuous because the ring particles are bright—each is covered in, or formed completely from, highly reflective ice. Saturn's rings are extremely flat. Though they measure more than 170,000 miles (280,000 km) across—well over half the distance between Earth and the Moon—the majority of the rings are just a few feet to a few hundred feet deep. Studies shows that ring systems ought not last longer than 100 million years or so. Thus, either the rings of Saturn were created comparatively recently from the break-up of a small encroaching comet, or events such as these regularly replenish the planet's rings.

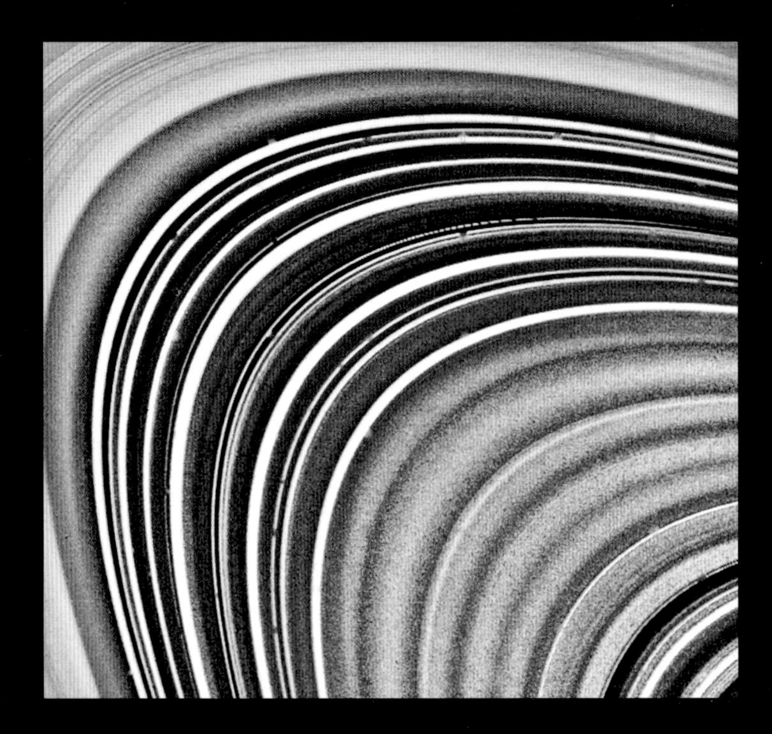

↑ → **Planetary scientists were suprised** when they got their first close-up photos of the Saturnian rings, provided by the Voyager space probes in 1980 and 1981. They expected the rings to be smooth, their particles evenly distributed. Instead, the gravity of tiny moons embedded in the rings (not visible in these photos) herds the particles together to form hundreds of discrete ringlets. The colors in these images have been enhanced to more clearly emphasize the rings' structure.

THE RINGS OF SATURN

Brighter than those of any other planet, the rings of Saturn are made up of particles of ice that on average reflect about 60 percent of incident sunlight. They are divided up into seven major regions, named for the first seven letters of the alphabet, but only A, B and C are readily visible from Earth. Between A and B there is the celebrated Cassini Division, named for the astronomer who first noticed this gap in the system. The outermost edge of the E ring spans some 600,000 miles (960,000 km)—more than twice the distance from Earth to the Moon—but the bulk of the rings measure 170,000 miles (280,000 km) across.

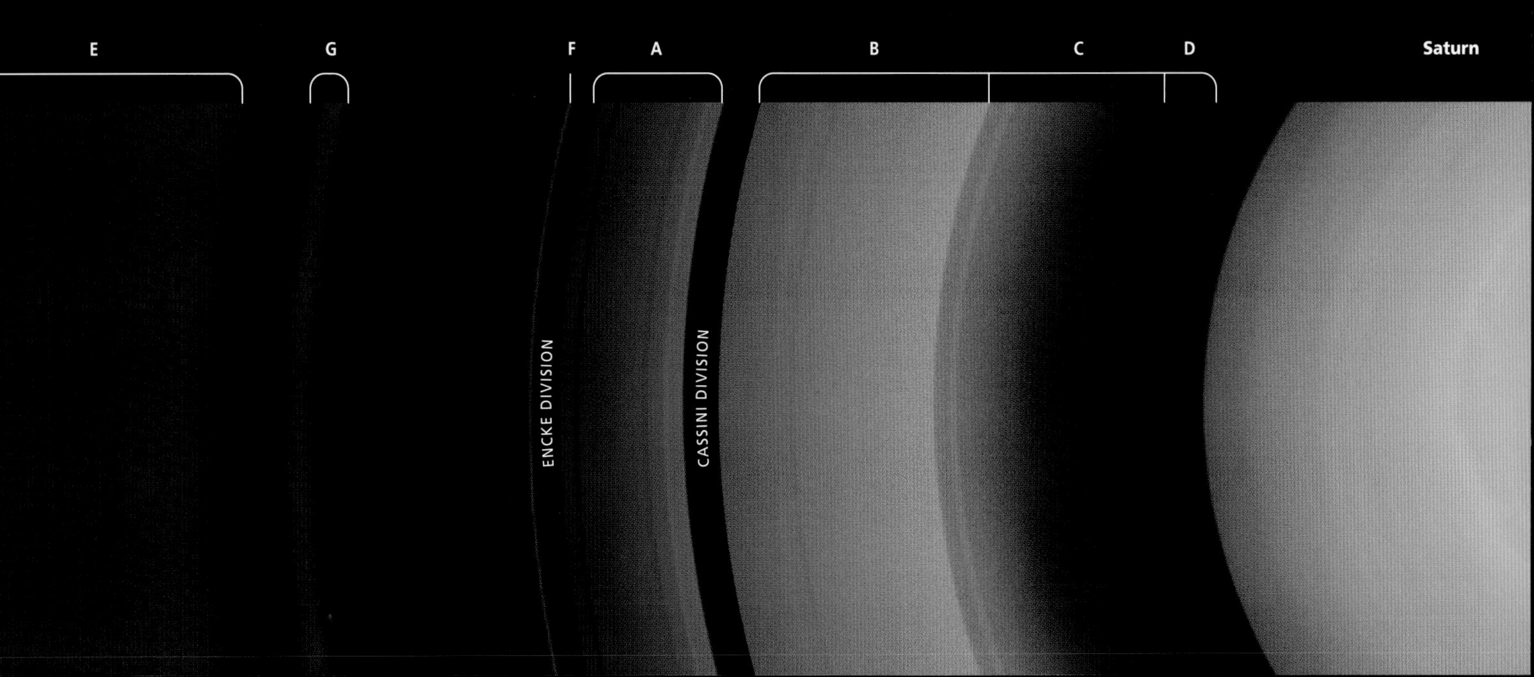

Saturn: satellites

Being the second-largest planet with the second-greatest gravitational field, Saturn—like Jupiter—has netted a fair number of asteroids and enslaved them as its satellites. Currently, Saturn has 31 known moons, one of which, discovered in 2003, has yet to be named. Again, as with Jupiter, Saturn has a number of sizable satellites, not just captured asteroids. But with the exception of Titan, they are all nonetheless substantially smaller than our own Moon and the Galilean satellites of Jupiter. Titan, though, is special, with a fitting name. At 3200 miles (5150 km) across, this is the second-largest moon in the Solar System, easily dwarfing all its Saturnian neighbors. Even more interestingly, Titan has a thick nitrogen atmosphere, the only satellite with a substantial sky. But being so far from the Sun, Titan is freezing and hostile.

↓ **The surface of Titan, Saturn's largest moon,** is not visible from space. Its surface, like that of Venus, is permanently hidden beneath its thick atmosphere.

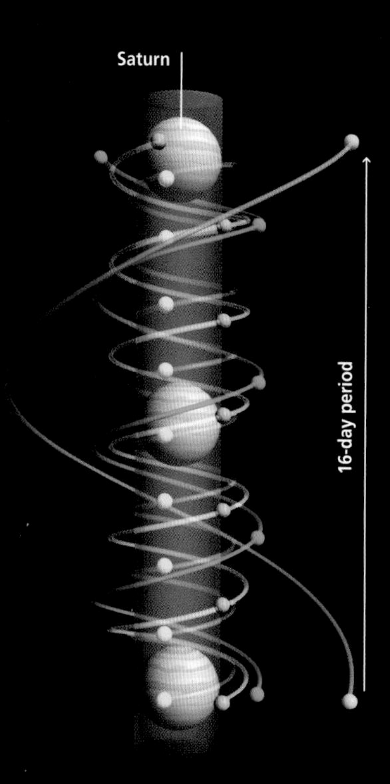

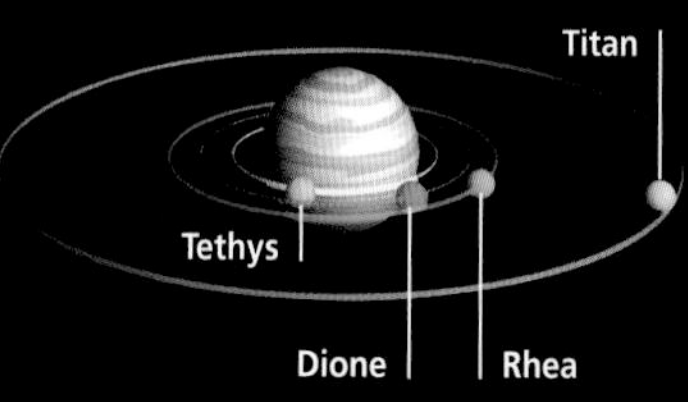

SATELLITES IN ORBIT

Like Jupiter, Saturn is a mini planetary system all in itself, commanding an impressive array of satellites. This illustration shows four of the most important, including Titan. But the picture is by no means complete, as there are many other worlds both inside the orbit of Tethys and beyond that of Titan.

SATELLITE DATA

Satellite	Orbit period	Distance from Saturn
Tethys	1.88 days	183,000 miles (295,000 km)
Dione	2.74 days	235,000 miles (377,000 km)
Rhea	4.52 days	327,000 miles (527,000 km)
Titan	15.95 days	759,000 miles (1.22 mil. km)

↑ **This is a Voyager 2 photograph of the night side of Titan,** taken from a distance of 563,000 miles (907,000 km). The atmosphere scatters sunlight and renders the atmosphere visible around the moon's limb.

← **The thick, hazy atmosphere of Titan** shines blue in this Voyager 1 photo from 1980. Titan's atmosphere is primarily nitrogen, like Earth's, but it is far, far colder.

↓ **This composition shows four of Saturn's largest moons,** measuring from 320 miles (500 km) across for Enceladus to 940 miles (1520 km) across for Rhea. All are much smaller than Titan, which is larger than Mercury.

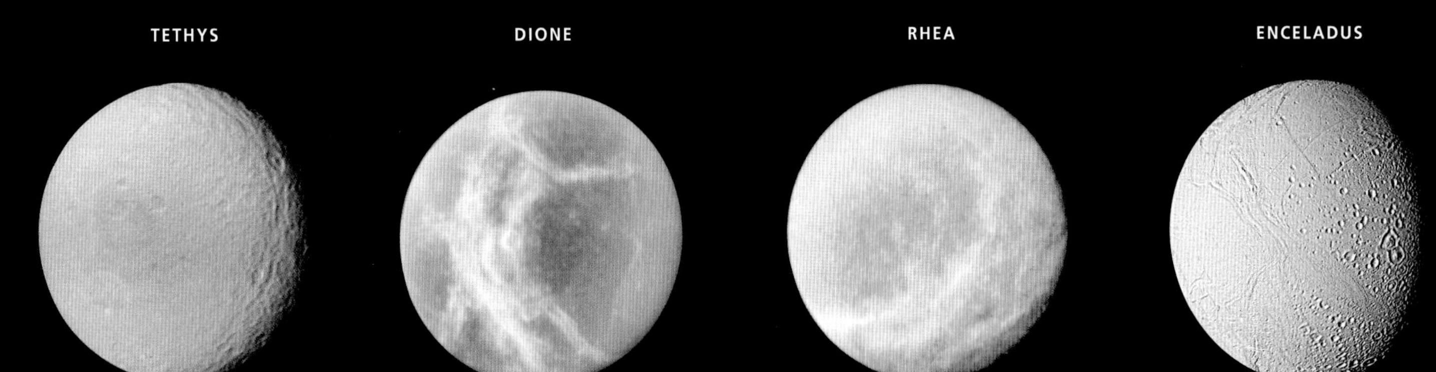

Uranus

Uranus is the third-largest planet, just slightly larger than Neptune and four times the size of Earth. This is a featureless blue–green ball, the color dictated by a high methane content in the atmosphere. There are none of the distinctive bands or storms that mark Jupiter and Saturn, and the rings—while clearly defined in close-up photos—are dark and narrow. Curiously, Uranus is tipped over by nearly 100 degrees. It could be that, long ago in its history, it suffered a glancing collision with another massive protoplanet.

ATMOSPHERE

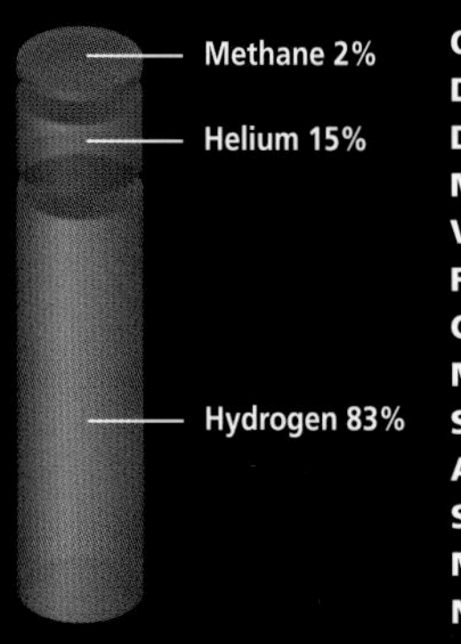

PLANET STATISTICS

Origin of name *Ouranos*, Greek primeval god of the sky
Discovered March 1781, by William Herschel
Diameter 31,763 miles (51,118 km), 400.7% of Earth's
Mass 14.5 x Earth
Volume 64 x Earth
Farthest from Sun 1866 million miles (3003 million km)
Closest to Sun 1702 million miles (2739 million km)
Mean surface temperature -355°F (-215°C)
Sunlight strength 0.2–0.3% of Earth's
Apparent magnitude +5.5 to +5.9
Surface gravity 1.17 gee (117% of Earth's)
Magnetic field strength 0.23 gauss (75% of Earth's)
Number of known satellites 27

↓ **Hubble snapped this infrared image of Uranus,** its surrounding rings and several satellites. The colors show more detail than is apparent in the optical band.

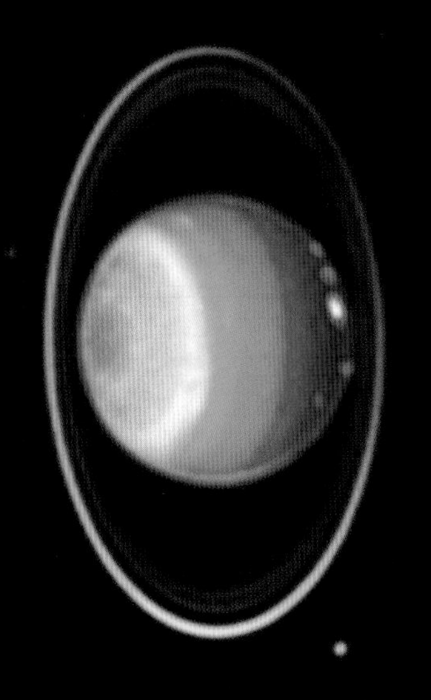

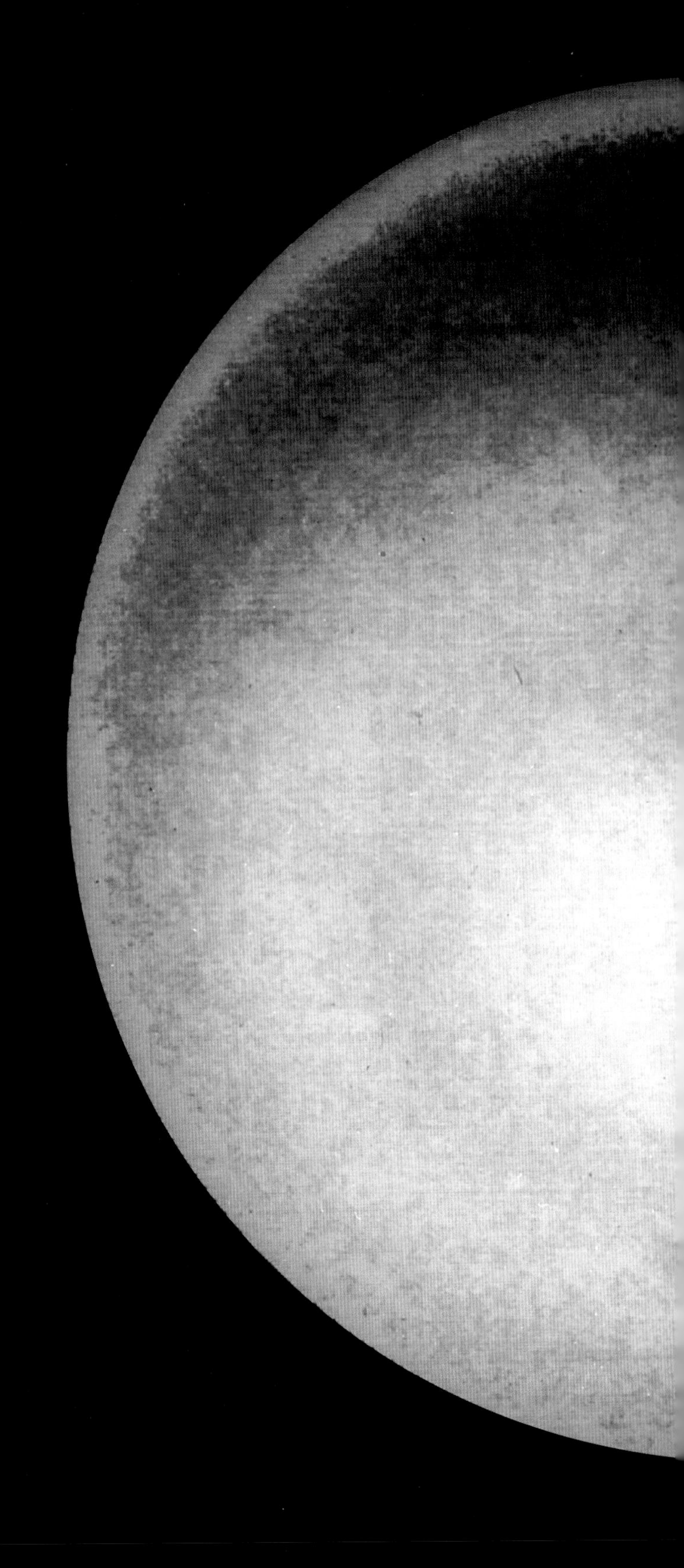

THE LAYERS OF URANUS

Uranus is dense compared to Jupiter and Saturn, so its interior must contain heavier compounds than just hydrogen—probably an assortment of various ices. For this reason, Uranus has been called an "ice giant," to distinguish it from the gas giants Jupiter and Saturn.

Atmosphere Uranus' atmosphere is a bland, practically featureless green sea of hydrogen and helium. The planet has none of the bands and colors that typify Jupiter and Saturn. This is because it is so cold, being fully twice as far from the Sun as Saturn, that its clouds reside deep in the atmosphere.

Mantle The atmosphere thickens until it mixes smoothly with the layer beneath it, the mantle. This is where most pure hydrogen resides.

Outer core Uranus has a large core divided into two distinct regions. The outer core is a slushy ball of mixed ices of water, ammonia and methane. It contains much of the planet's mass. In fact, "ices" is a little misleading, because under the crushing pressures in the planet's interior these fluids will be very hot.

Inner core Being farther from the Sun than Jupiter and Saturn, Uranus' solid core—if it has one—will probably contain a larger proportion of ices compared to rock.

SIZE COMPARISON

Uranus

Earth

ORBIT STATISTICS

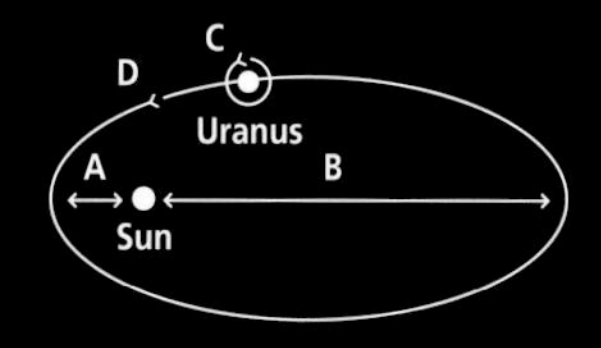

A 1702 mil. miles (2739 mil. km)
B 1866 mil. miles (3003 mil. km)
C rotates in 17 hr 14 min
D orbits in 84.04 Earth years

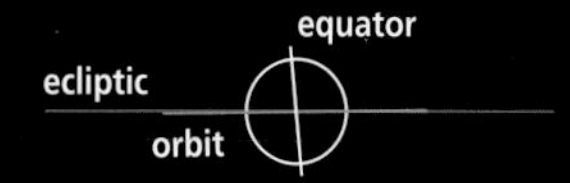

Axial inclination 97.9°
Angle of orbit to ecliptic 0.77°

Uranus: features

Because of its strange axial inclination—tipped over almost at right angles to the plane in which it orbits—Uranus has an 84-year trip around the Sun that brings with it some interesting seasonal variations. Each of the poles faces the Sun for more than 20 years, during the summer and winter solstices. At these times, the hemisphere facing away from the Sun in plunged into a profound darkness for two decades while the other basks in unyielding solar energy—although the strength of the sunlight here is less than one percent as strong as on Earth. Between the solstices the day–night pattern is more normal, dictated by the planet's spin.

THE SLOW ORBIT OF URANUS

↓ **Uranus takes 84.4 Earth years** to complete an orbit, so each pole receives a long period of constant sunlight followed by a long period of darkness. Currently Uranus is nearing its autumn equinox.

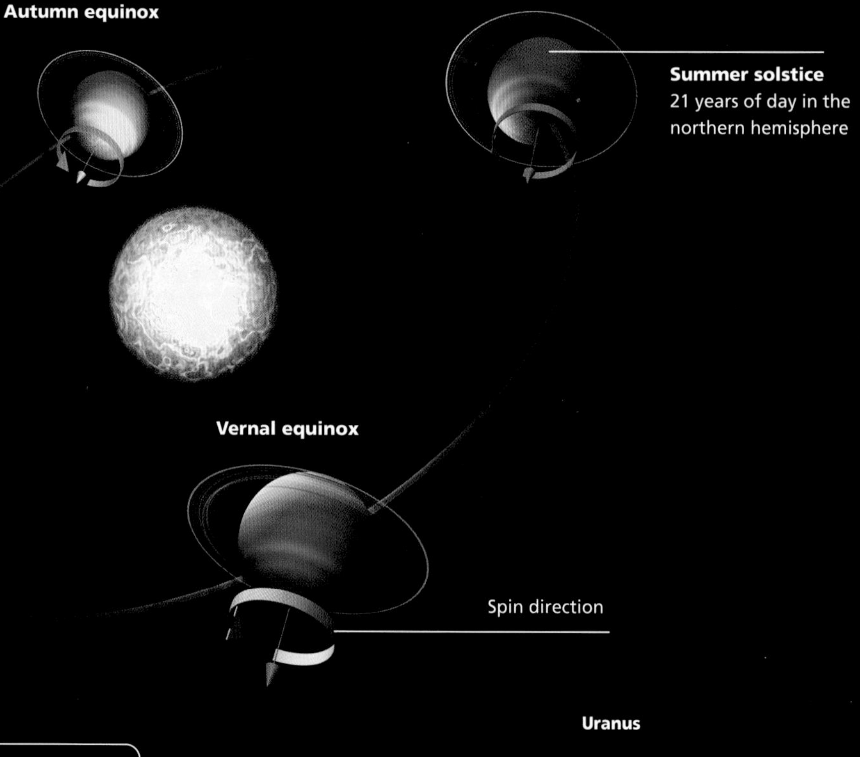

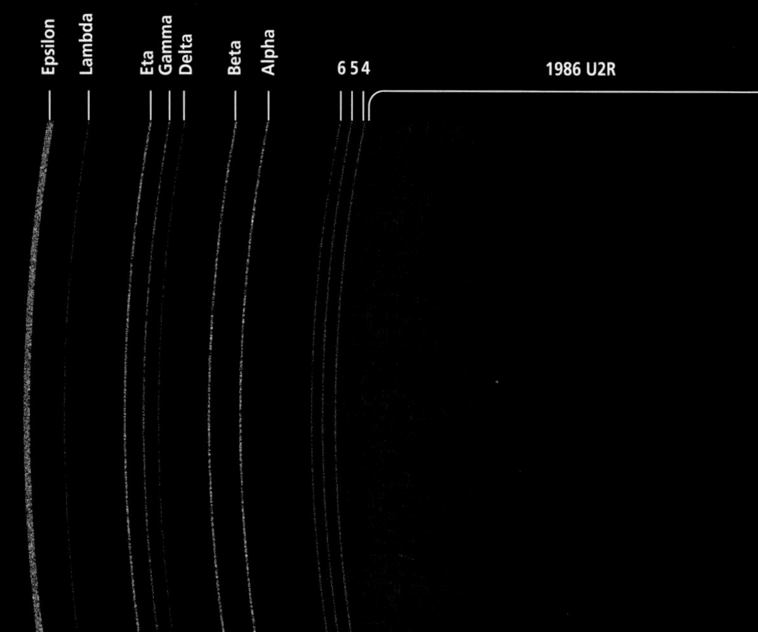

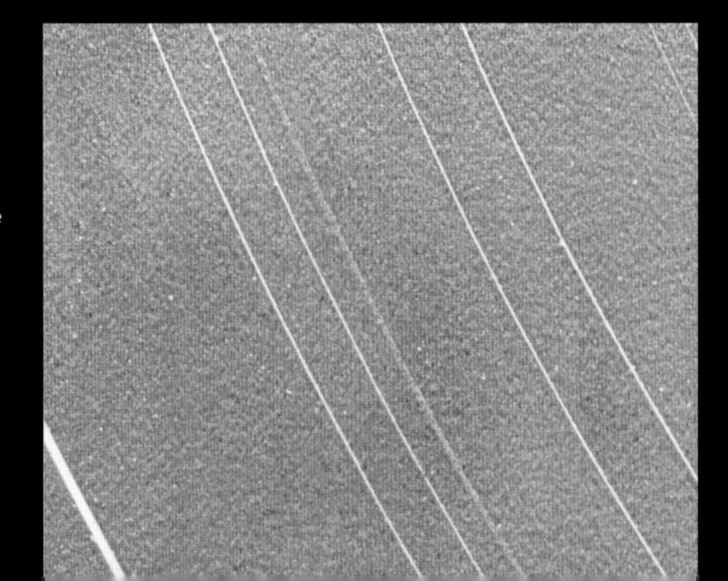

THE RINGS OF URANUS

The rings of Uranus were discovered in 1977 from Earth, when the planet passed close to a star on the sky. As the star and planet drew closer together, astronomers noticed that the light from the star dipped a few times before it vanished behind the planet, then again the same number of times after the star reappeared on the other side of Uranus. The dipping was caused by the star temporarily passing behind a series of previously unknown, discrete rings. Photographed up-close by the Voyager 2 probe, Uranus' 11 rings are very dark compared to Saturn's—their constituent particles covered in a layer of sooty grime. Seven of the rings are named for letters of the Greek alphabet. The one most recently found, in 1986 by Voyager 2, is 1986 U2R.

↑ ← **These photos, both obtained in 1986** by the Voyager 2 probe during its Uranus flyby, reveal the extremely narrow, discrete nature of its rings. Both images show all but the broad and diffuse innermost ring, 1986 U2R. Epsilon is the brightest (and outermost) ring. The colors indicate the different compositions in the particles that make up the rings.

Uranus: Satellites

Uranus has a total of 27 known moons. The largest five are the so-called classical satellites—Miranda, Ariel, Umbriel, Titania and Oberon—discovered between 1787 and 1948. Voyager 2 found 10 more satellites in 1986 when it flew past Uranus. Since then, others have been found in ground-based searches. All except the largest five measure just a few tens of miles across, and are more than likely to be captured asteroids. However, the classical moons themselves are not particularly large. The biggest, Titania, is 980 miles (1580 km) in diameter, a little less than half the size of our own Moon and therefore far smaller than, for example, the largest moons of Saturn, Jupiter and Neptune. All of the classical moons are heavily cratered and, as expected this far from the Sun, they are rich in ice. The smallest, Miranda, is indeed an oddball, and its patchwork geology suggests that it once broke up and later reformed.

↓ **The surface of Ariel,** seen here in a Voyager 2 photo, is riddled with craters and cracks—evidence perhaps of stretching caused by the powerful tides of Uranus.

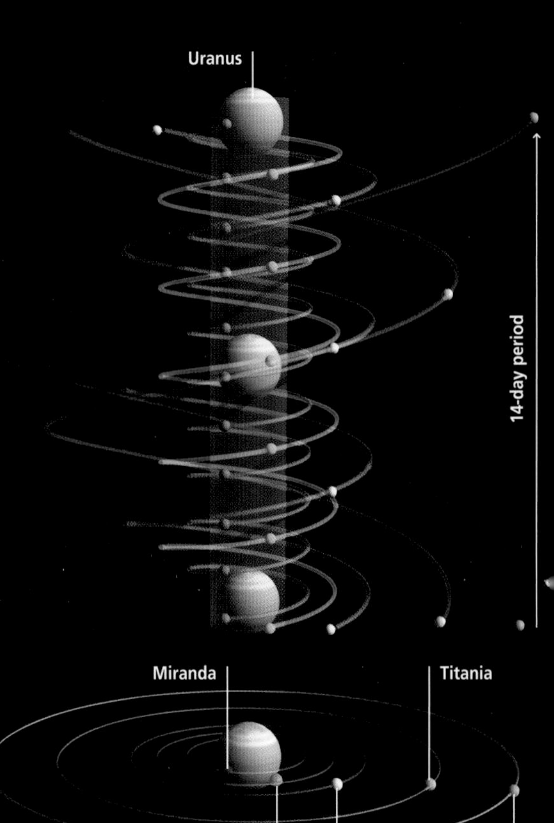

SATELLITES IN ORBIT
This diagram shows the orbits of the five largest, so-called classical moons of Uranus—from the center outward, they are Miranda, Ariel, Umbriel, Titania and Oberon. In total, 14 days are shown, corresponding to just over one orbit of Oberon and 10 orbits for Miranda. The other 22 Uranian satellites are all much smaller than Miranda, and are found both inside Miranda's orbit and outside that of Oberon.

SATELLITE DATA

Satellite	Orbit period	Distance from Uranus
Miranda	1.41 days	80,000 miles (129,000 km)
Ariel	2.52 days	118,000 miles (191,000 km)
Umbriel	4.14 days	165,000 miles (266,000 km)
Titania	8.71 days	271,000 miles (436,000 km)
Oberon	13.46 days	363,000 miles (584,000 km)

← **Miranda's surface is extremely bizarre.** Its patchwork of different geologies makes it look as though it has been pulled apart and then stuck back together again—and in fact, this is precisely what astronomers think has happened. They suggest that the moon was once shattered by a collision, and later reformed in orbit around Uranus. The cliffs seen at the bottom of this image jut out several miles above Miranda's surface.

↓ **The largest four moons of Uranus** are seen in this composition. The largest is Titania (980 miles / 1580 km across) and the smallest is Ariel. At 720 miles (1160 km) across, Ariel is not quite large enough to be spherical. Miranda (not shown) is less than half the size of Ariel.

ARIEL

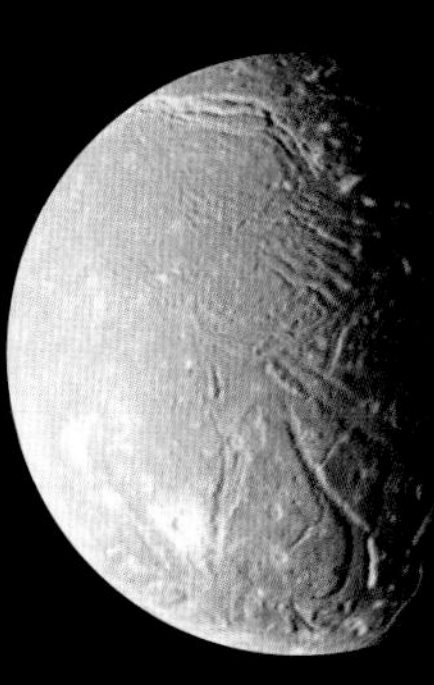

UMBRIEL

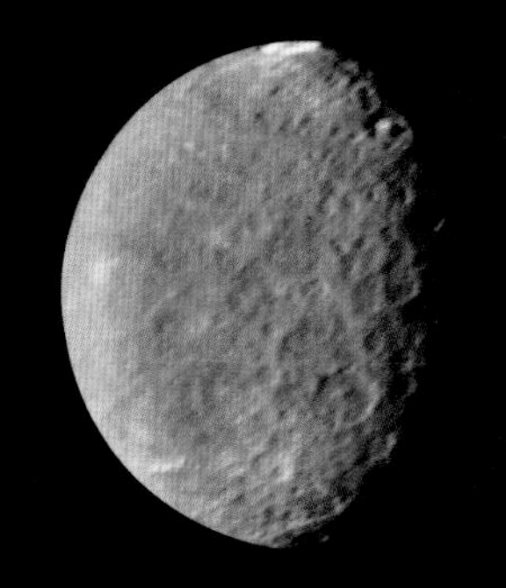

OBERON

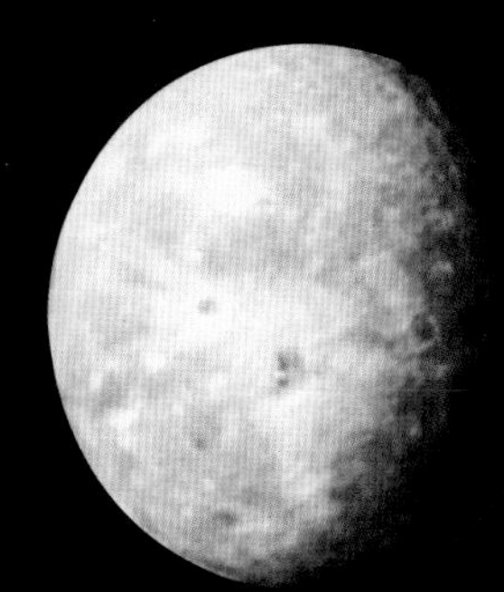

TITANIA

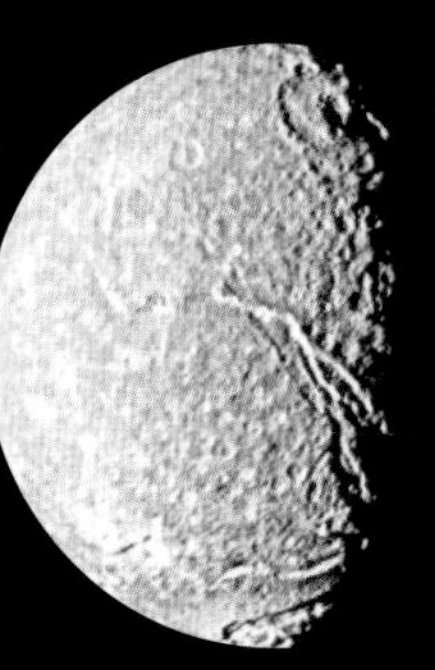

Neptune

Neptune is very similar in mass to Uranus, and in its appearance too. However, Neptune has a little more detail visible in its atmosphere. There are a few atmospheric bands, some wispy cirrus clouds made of methane crystals, and even occasional storm systems. The most famous storm, discovered in 1989, was the Great Dark Spot, but this has since dissipated. Like the other giant planets, Neptune has a ring system (although they often seem to appear merely as ring arcs) and a small army of satellites.

ATMOSPHERE

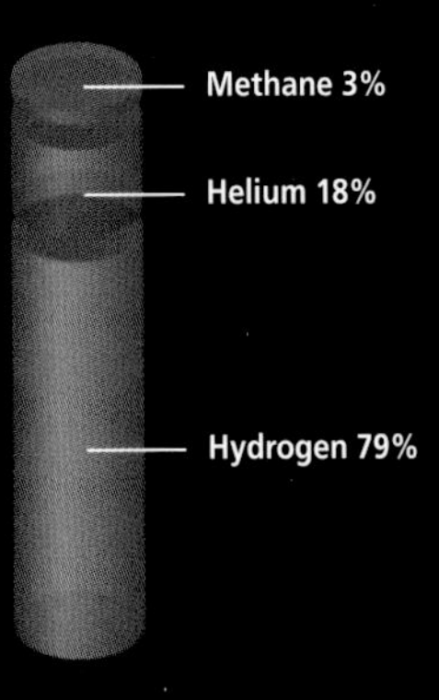

PLANET STATISTICS

Origin of name *Neptunus*, Roman god of water
Discovered September 1846, by Johann Galle
Diameter 30,778 miles (49,532 km), 388.3% of Earth's
Mass 17.1 x Earth
Volume 54 x Earth
Farthest from Sun 2825 million miles (4546 million km)
Closest to Sun 2769 million miles (4456 million km)
Mean surface temperature -265°F (-220°C)
Sunlight strength 0.1% of Earth's
Apparent magnitude +7.9
Surface gravity 1.77 gee (177% of Earth's)
Magnetic field strength 0.14 gauss (45% of Earth's)
Number of known satellites 13

↓ **This Voyager 2 close-up of Neptune** shows wispy clouds of methane crystals scooting around high in the planet's atmosphere.

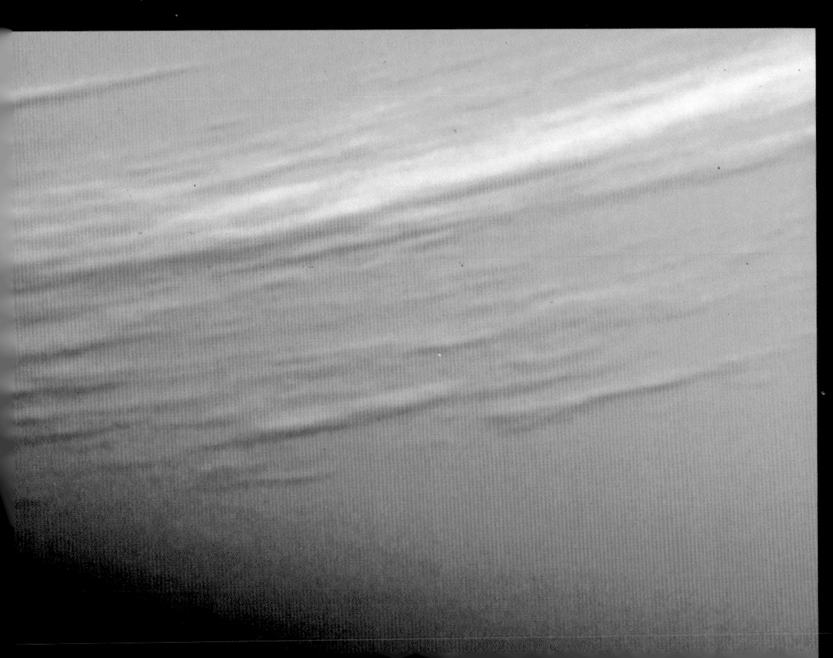

THE LAYERS OF NEPTUNE

Neptune, like Uranus, is an ice giant, rich in compounds such as water-ice, methane and ammonia. The two planets have very similar interiors with practically identical materials, but they differ in proportion compared with each other.

Atmosphere Neptune's deep blue, hydrogen-rich atmosphere is surprisingly colorful and patterned, considering how far it is from the Sun—twice as far as Uranus, which has no visible clouds at all. The clouds and bands on Neptune are probably driven by the planet's internal heat source, something that is lacking on Uranus.

Mantle As with Uranus, a soupy layer of hydrogen and other gases comprise Neptune's mantle. It may be thinner than that of Uranus.

Outer core Again, there is a slushy outer core, a hot chemical soup composed of water, ammonia, methane and various other substances in liquid form.

Inner core The inner core may be slightly larger than Uranus', and is again a sphere of rock and/or ice, possibly with a majority of ices.

SIZE COMPARISON

Neptune

ORBIT STATISTICS

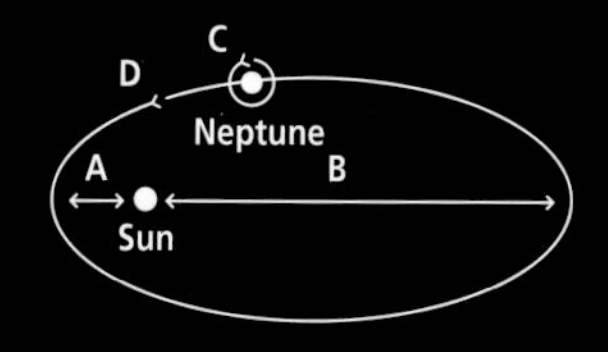

A 2769 mil. miles (4456 mil. km)
B 2825 mil. miles (4546 mil. km)
C rotates in 16 hr 6 min 6 sec
D orbits in 164.8 Earth years

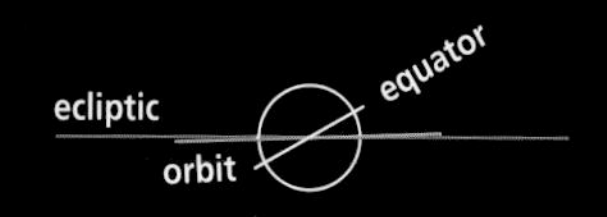

Axial inclination 29.6°
Angle of orbit to ecliptic 1.77°

Neptune: rings

It came as no real surprise to astronomers when, in 1989, the Voyager 2 probe confirmed the presence of Neptune's rings. Five are known in total. The most well defined rings are those called Adams and Le Verrier. They were named after the two scientists who in the mid-1800s—before Neptune's discovery—independently calculated where Neptune ought to be on the sky, based on the position of the planet Uranus. Extending just 30 miles (50 km) and 70 miles (110 km) in radial extent, these rings are very narrow. The innermost ring is called Galle. It is named in honor of the German astronomer who, in 1846, using Adams' and Le Verrier's calculations, actually discovered Neptune close to its predicted position. The widest ring, meanwhile, is Lassell, extending for 2500 miles (4000 km). All of Neptune's rings are extremely dark. They likely consist of boulder-sized fragments of ice covered in dark organic compounds, with dust-sized particles mixed in between them.

THE RINGS OF NEPTUNE

Neptune's five rings are seen in this illustration (*below*), where their color, brightness and thickness have been greatly enhanced for clarity. In real life, the rings are so dark that you would probably not be able to see them with the naked eye, especially with the light of Neptune overwhelming your vision. The rings Adams and Le Verrier are narrow, but still wider than the rings of Uranus, while Galle and Lassell stretch for some 1500 miles (2500 km) and 2500 miles (4000 km) respectively. The outermost edge of the ring system, marked by Adams, lies at a distance of about 2.5 Neptune radii from the center of the planet.

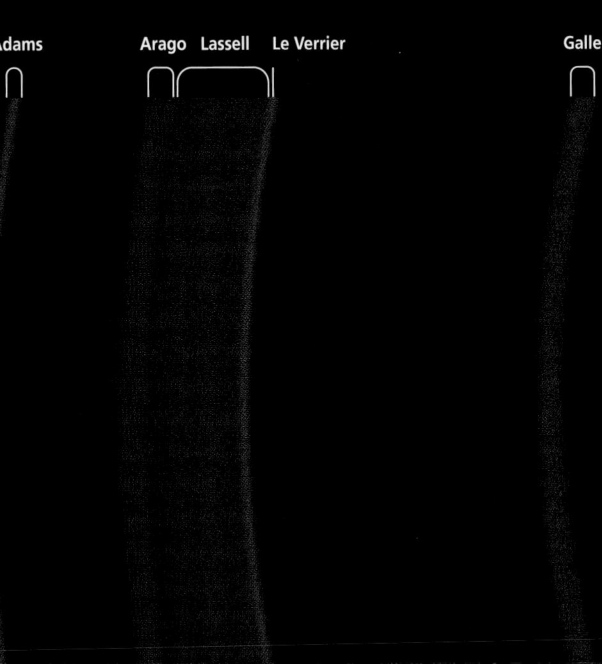

← **The Neptunian ring system** is seen in its entirety in this photo provided by the Voyager 2 probe during its Neptune flyby in 1989. The planet is masked as its comparative brightness would obscure Neptune's dark and delicate rings.

THE RING ARCS OF NEPTUNE

There is something decidedly odd about Neptune's Adams ring. There are three distinct regions along its circumference where the particle density is apparently higher than elsewhere. From a distance and with the correct camera settings, a photograph shows only these enhanced regions of the ring, giving the appearance of a series of incomplete arcs rather than a single solid ring. Astronomers puzzled over this for a while, but they now believe that this strange and unique ring structure is caused by the presence of a nearby tiny moon. The moon is called Galatea, and it orbits Neptune just on the inside edge of the Adams ring. Although Galatea is only 100 miles (160 km) across, with very slight gravity, it is so close to the Adams ring that its gravitational pull is more than enough to herd the ring fragments into piles, preventing them from spreading evenly along the orbit.

→ **The arcs in the Adams ring** are well defined in the top portion of this Voyager 2 photograph. The image has been vastly overexposed to reveal the otherwise invisible rings.

Neptune: satellites

Not many Neptunian moons are known because the planet is so far from the Sun that detections are difficult. There are 13 in all, of which five have yet to be named. Triton is the only large satellite. It is the seventh largest in the Solar System, and measures two-thirds the size of Earth's Moon. Triton is very strange. It orbits Neptune backward compared to other planets' satellites, and at a very high angle to Neptune's equator. Moreover, despite its very cold surface, Triton is active. Geysers dot the surface, spewing frigid substances miles above the ground. Triton was almost certainly not formed in orbit around Neptune, but was likely captured into its irregular orbit. The orbits of some of the smaller satellites are also strange, and may be the result of gravitational disturbances caused by Triton's entry into Neptune's system.

↓ **The partial disk of Neptune's largest moon by far, Triton,** as captured by the Voyager 2 probe in 1989. The dark plumes visible on the surface are thought to be the result of geyser-like activity.

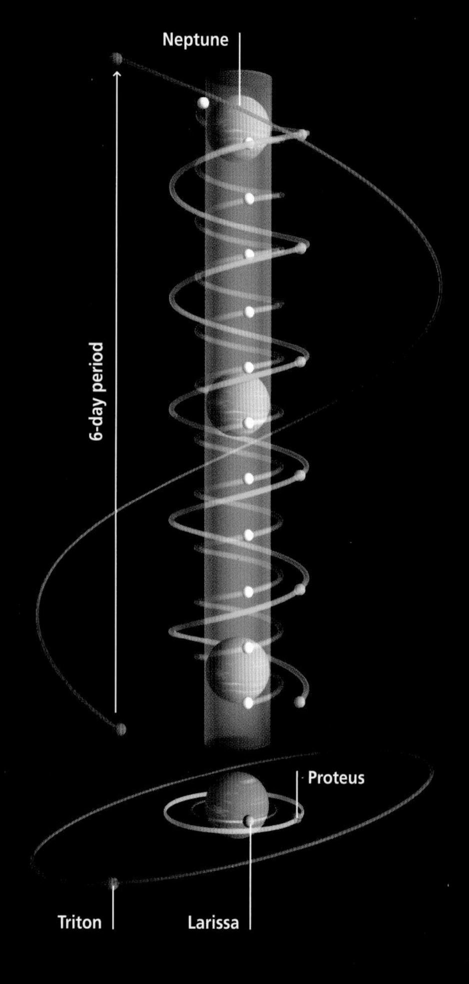

SATELLITES IN ORBIT

Neptune has dozens of satellites, but only 13 are known. This illustration shows three—Triton, Larissa and Proteus. Triton orbits backward and in a plane greatly inclined to the planet's equator. Nereid (not shown) has the most elongated known orbit, which takes it between 800,000 and 6 million miles (1.3 million and 9.5 million km) from Neptune.

SATELLITE DATA

Satellite	Orbit period	Distance from Neptune
Larissa	0.55 day	45,700 miles (73,500 km)
Proteus	1.12 days	73,000 miles (118,000 km)
Triton	-5.88 days	220,000 miles (355,000 km)

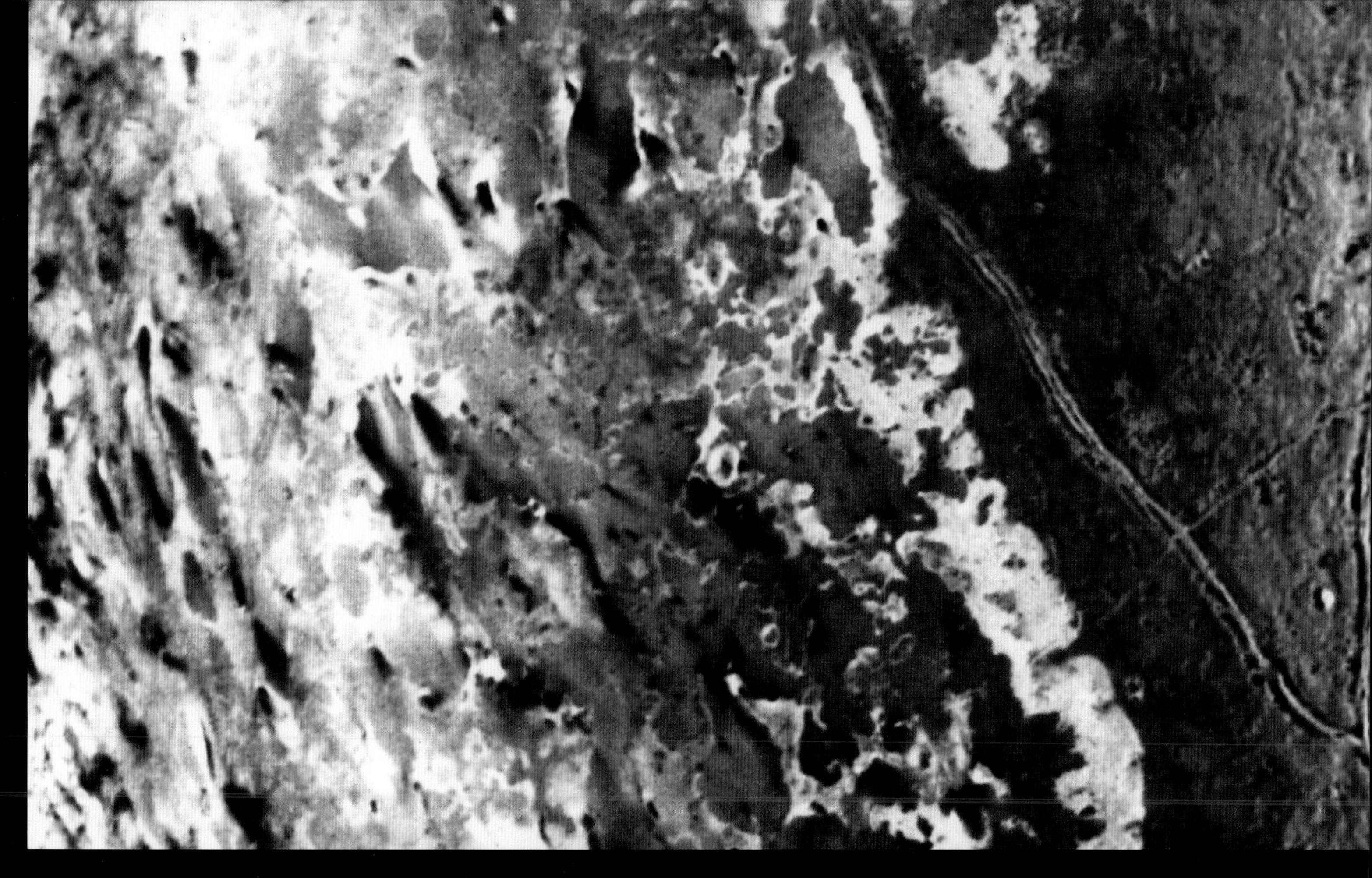

SURFACE ACTIVITY ON TRITON

Triton's surface came as a total surprise to planetary scientists when they first saw it. The last thing they expected from a world with the coldest known surface temperature (-391°F / -235°C) was a form of volcanism. But Triton's volcanoes do not expel lava—they are cryonic. That is, they vent very cold organic compounds, somewhat like frigid geysers. Sunlight filters through the thin ices that coat Triton's surface, and vaporizes pockets of ice beneath. As these vapors expand, they erupt through the surface and are ejected vertically about 5 miles (8 km), where they are caught by high-altitude winds and blown along for 65 miles (100 km). It was another surprise to find that Triton should have an atmosphere. It is extremely sparse, but nevertheless it has a measurable density, and is composed of nitrogen with some methane. There is also a polar cap of solid nitrogen.

↑ **This view of Triton** reveals strange surface features not found on any other world.

← **Neptune's second-largest moon** was originally known as 1989N1 to indicate the year in which the Voyager 2 probe first discovered it. Astronomers have since renamed the satellite Proteus. It is a fair-sized world, but not large enough to be perfectly spherical, with a diameter around 270 miles (430 km).

Pluto

The last planet is tiny Pluto, much smaller than even our Moon. It is mainly rocky, but its surface is coated in a significant blanket of ice—probably water-ice, with scatterings of methane, nitrogen and some carbon monoxide ice. Pluto orbits the Sun in a region known as the Kuiper Belt, home also to icy asteroid-like bodies that are the leftovers from the formation of the Solar System. If Pluto had been discovered more recently, astronomers would call it a large Kuiper-Belt object—and not a planet at all.

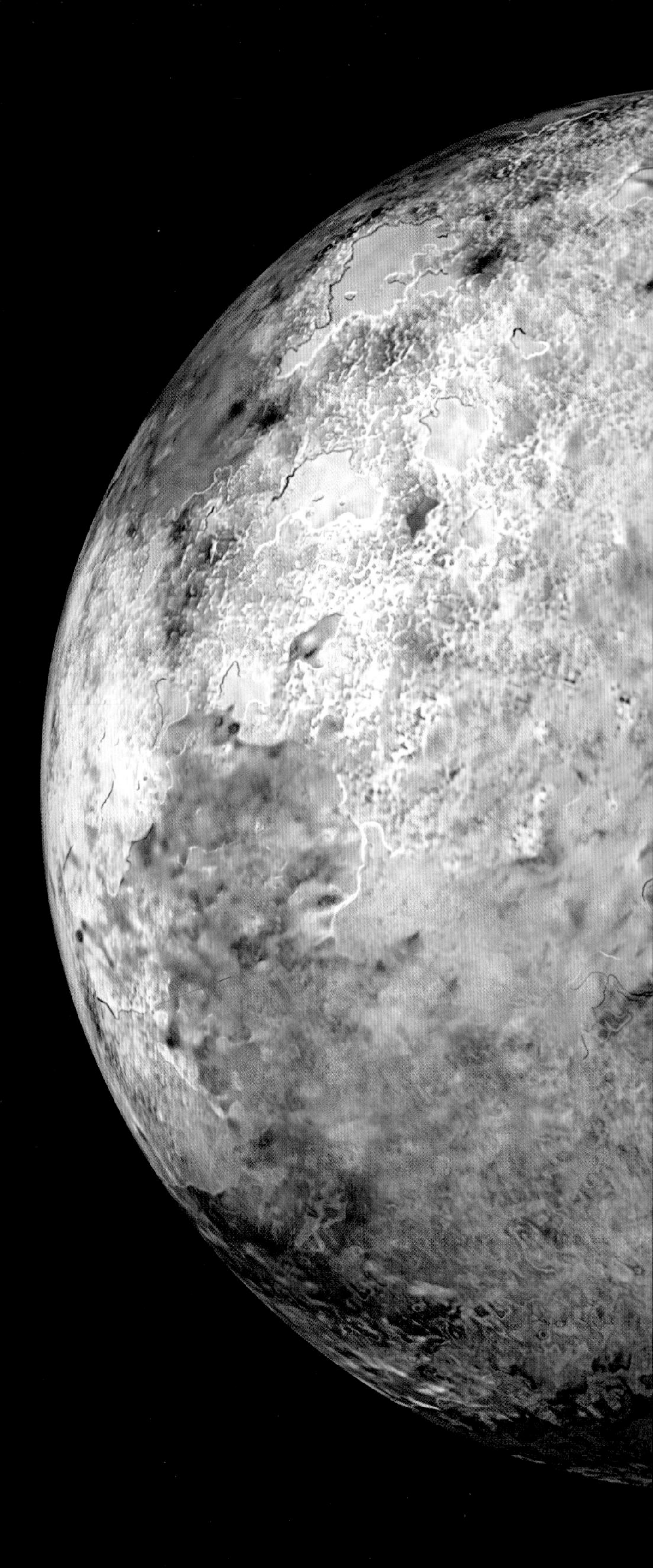

ATMOSPHERE

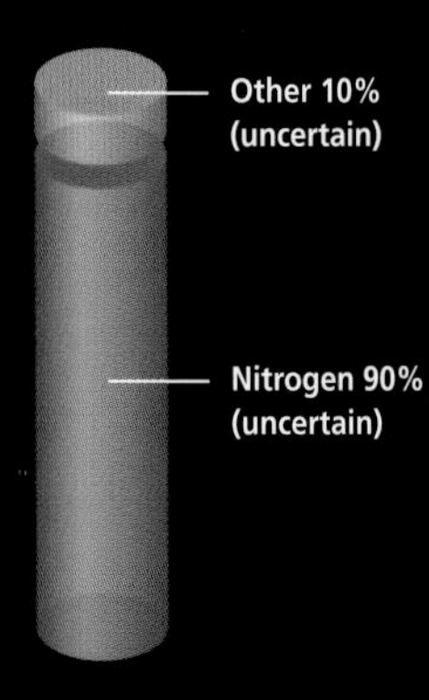

PLANET STATISTICS

Origin of name *Pluto*, Greek god of the underworld
Discovered February 1930, by Clyde Tombaugh
Diameter 1429 miles (2300 km), 18.0% of Earth's
Mass 0.002 x Earth
Volume 0.01 x Earth
Farthest from Sun 4586 million miles (7380 million km)
Closest to Sun 2763 million miles (4447 million km)
Mean surface temperature -380°F (-230°C)
Sunlight strength 0.04–0.1% of Earth's
Apparent magnitude +13.7
Surface gravity 0.06 gee (6% of Earth's)
Magnetic field strength Unknown
Number of satellites 1 (Charon)

↓ **This image of Pluto and its satellite Charon** was taken using the Hubble Space Telescope, and is one of our clearest pictures of the pair.

THE LAYERS OF PLUTO

Pluto's interior is a comparative mystery, not surprisingly, since it remains the only planet not yet visited by a probe. However, its density at least tells us that this world is made of about 70 percent rock and 30 percent ice. This is similar to Neptune's moon Triton, which some astronomers believe to be a good indication what Pluto will look like.

Crust Pluto's crust extends to an unknown depth, and the surface shows bright and dark patches. The bright regions probably consist of various substances frozen solid, such as water, nitrogen, methane, ethane and carbon monoxide. The dark areas could contain organic material originating from the birth of the planets.

Mantle Immediately under the crust there could be a mantle. If it exists, it is likely to be made up of ices, and may contain a large amount of water-ice.

Core Beneath the mantle lies Pluto's core, a huge ball of rock taking up much of the planet's volume. Little is known about the core, except that it must be rocky and massive.

SIZE COMPARISON

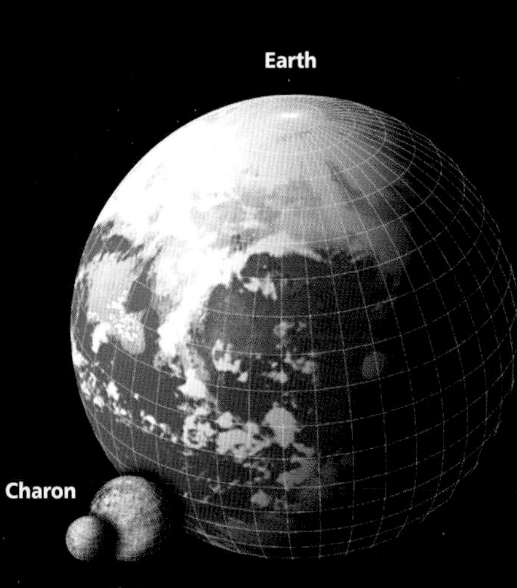

ORBIT STATISTICS

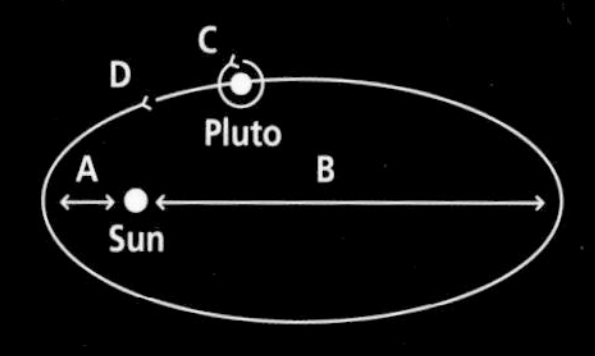

A 2763 mil. miles (4447 mil. km)
B 4586 mil. miles (7380 mil. km)
C rotates in 6.39 Earth days
D orbits in 248.6 Earth years

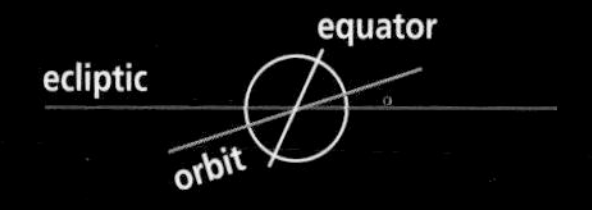

Axial inclination 119.6°
Angle of orbit to ecliptic 17.1°

The Sun

A giant ball of hydrogen and helium, the Sun makes up 99.9 percent of the Solar System's mass. This medium-size, middle-aged star is fueled by nuclear reactions in its core, where hydrogen atoms fuse together to make heavier helium atoms. The fusion process releases an enormous amount of energy that slowly travels through the Sun's layers to the surface. Here, it escapes into space as the heat and light that allow life to exist on Earth.

SUN STATISTICS

Origin of name *Sunne*, from proto-Germanic
Diameter 865,000 miles (1,392,000 km)
Mass 332,946 x Earth
Volume 1.3 million x Earth
Surface temperature 9900°F (5500°C)
Core temperature 27,900,000°F (15,500,000°C)
Time of rotation 25 Earth days at equator, 34 Earth days near poles
Magnetic field strength 2 gauss (645% of Earth's)

The Sun's outer atmosphere is known as the corona. A coronagraph blocks the Sun's disk to reveal activity. Here, two coronal mass ejections on opposite sides of the Sun are sending charged particles into space.

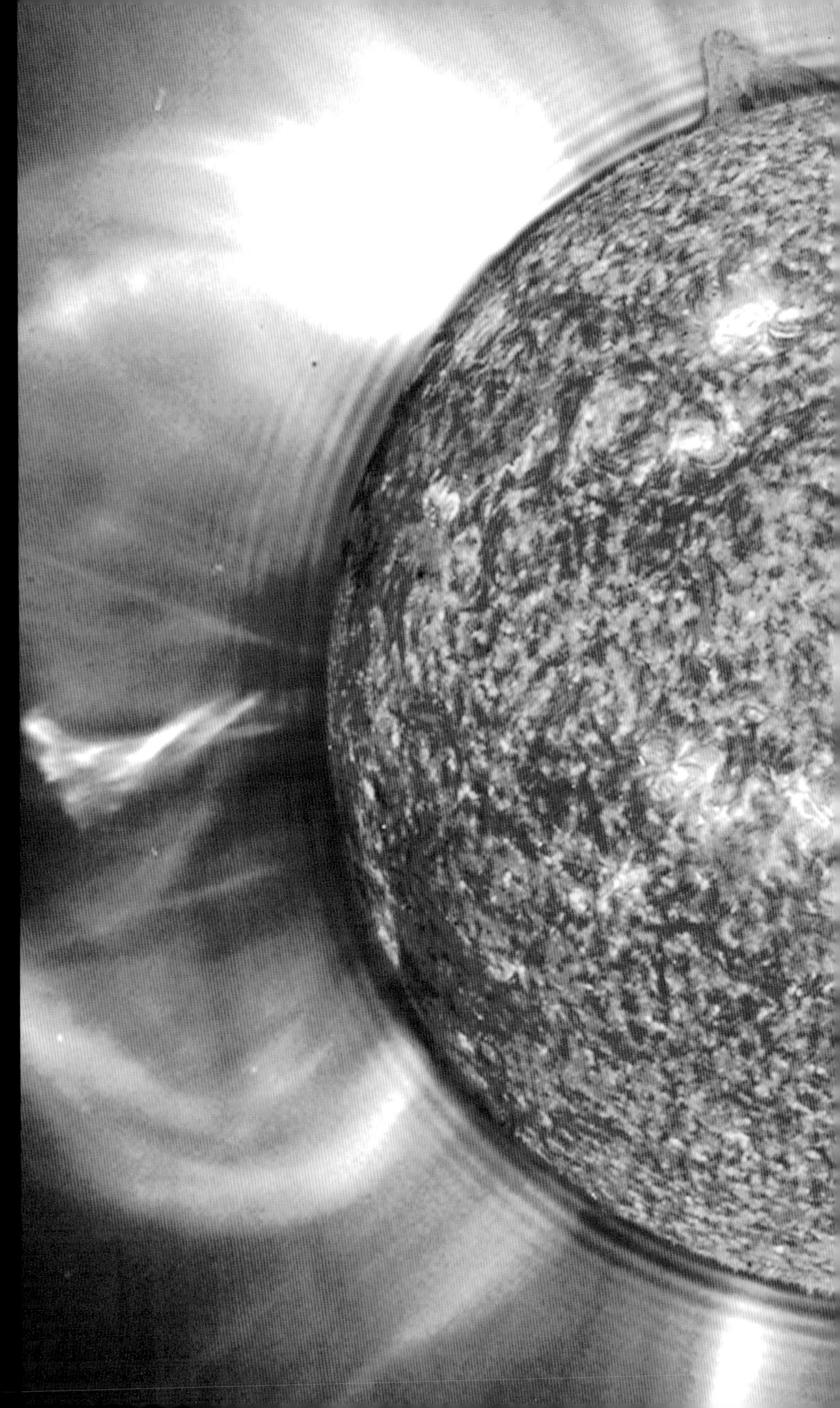

THE LAYERS OF THE SUN

Scientists have used sound waves to study the Sun's structure. This image reveals the speed of sound inside the Sun—sound travels faster through the hot red areas and more slowly through the cooler blue areas.

Photosphere The Sun's visible surface is known as the photosphere. This is a thin shell of gas, only 250 miles (400 km) thick. This is where solar features such as sunspots, faculae (bright areas) and granules (cellular features) form.

Convective zone About three-quarters of the way to the surface, the solar gas becomes cooler and undergoes a boiling, convective motion. Just below the surface, the big upwelling cells of gas break down into smaller ones, creating the photosphere's granular appearance.

Radiative zone The energy that radiates from the core travels through this hot zone in the form of photons (parcels of light). The photons bounce between the layer's particles so many times that each photon takes 200,000 years to reach the convective zone.

Core Containing merely 7 percent of the Sun's volume, but half its mass, the core is the "engine room" where solar energy is created by the process of nuclear fusion. The very center of the core has a temperature of 27,900,000°F (15,500,000°C).

SIZE COMPARISON

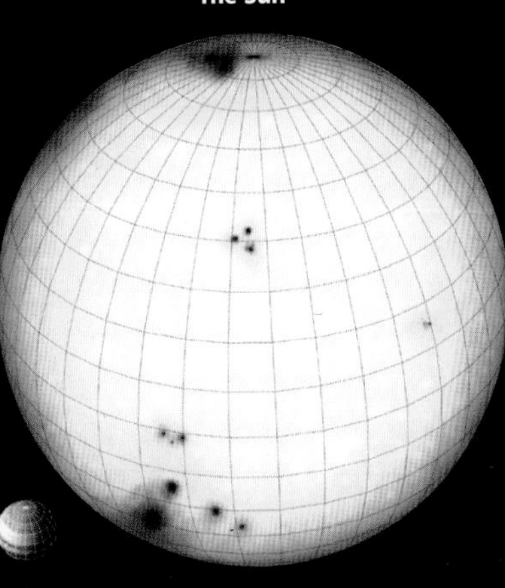

ORBIT STATISTICS

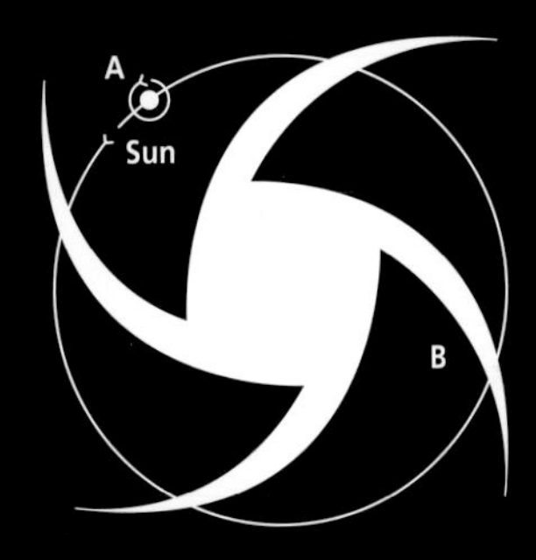

A Sun orbit period around Milky Way galaxy: 225 million years
B Distance from Sun to Milky Way galaxy center: 30,000 light-years

Solar activity

When astronomers first began to photograph the Sun, they realized that its visible surface (the photosphere) is covered in dark regions. These have come to be known as sunspots. As we now know, sunspots, as well as many other solar phenomena, are driven by the Sun's rotation. Like a planet, the Sun spins on its axis. However, the time it takes to do so depends on latitude, with the rotation being fastest at the equator. As a result of this so-called differential rotation, the Sun's magnetic field lines, which flow between the magnetic poles, become twisted and buckled as time progresses. Where the twisting is greatest, the field becomes locally magnified, inhibiting the flow of gases and making the surface appear dark—forming a sunspot. Occasionally, flares erupt near sunspots, throwing vast packets of material high above the photosphere at speeds approaching that of light. This magnetic activity varies on a cycle of 11 years, with the number of sunspots and active regions increasing as the cycle nears its completion.

↑ **The Sun's atmosphere (corona)** is seen throwing off a huge blast of plasma (*bottom left*) in an eruption called a coronal mass ejection.

← **In a coronal hoop,** gas is held high above the surface of the Sun, entrained in a field of magnetic force.

SUNSPOTS

Sunspots (*right*) are dark regions on the photosphere of the Sun associated with enhanced regions of magnetic field strength. Their temperature is about 2500˚F (1400˚C) cooler than the rest of the photosphere, and it is for this reason that they appear comparatively dark. As the 11-year solar cycle progresses, sunspots tend to form at greater and greater latitudes, reaching their greatest latitude, around 40 degrees, at the end of the cycle. After that, the Sun's magnetic field switches polarity—the north magnetic pole becomes the south pole, and vice versa—and the spots appear again near the equator. However, the spots never actually form on the equator itself. Sunspots vary greatly in size, darkness and duration. The smallest, called pores, are just a few hundred miles across. The largest are far more complex, spanning up to 60,000 miles (100,000 km), or several times the diameter of Earth, and lasting several months.

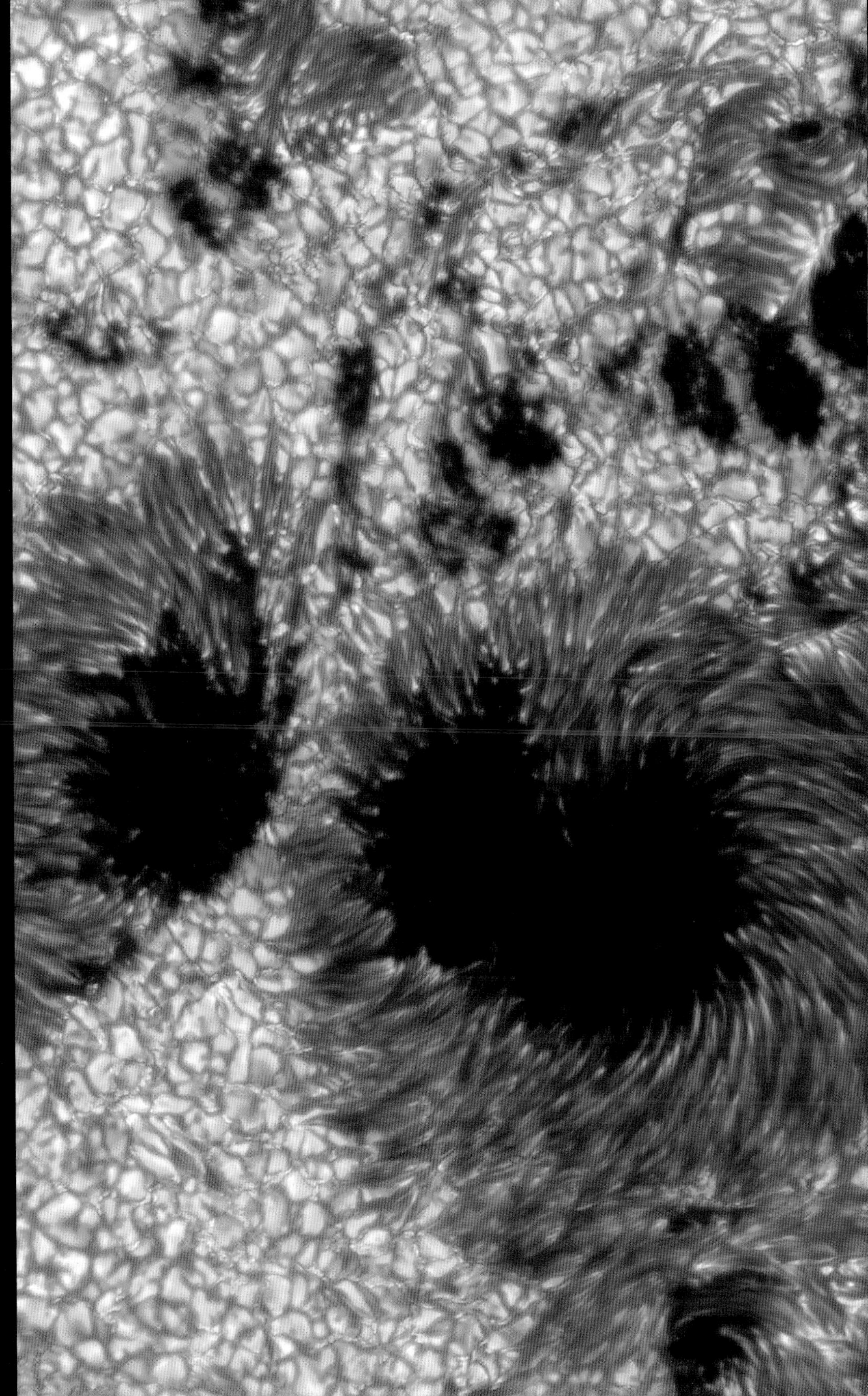

↓ **This image is a so-called chronograph—**the disk of the Sun has been blocked to reveal the detail and activity in its surrounding corona.

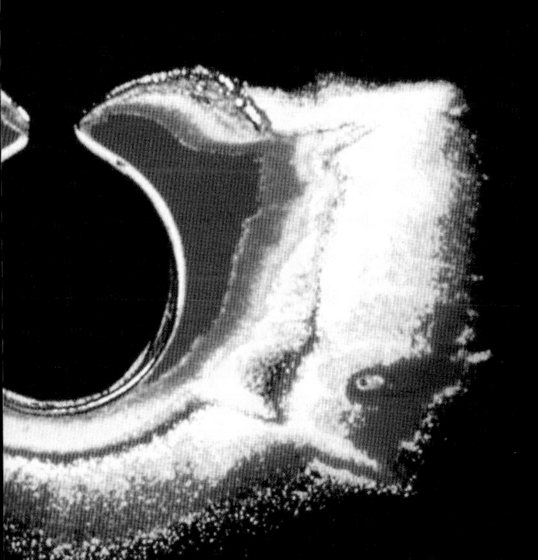

The Sun: lifecycle

The Sun is a typical yellow star 35 to 45 percent of the way through its fuel. The first stage in its creation might well have been the demise of another star, the shockwaves from a supernova causing a nearby cloud of gas and dust to collapse under its gravity somewhat like a cosmic avalanche (1). The collapse leads to the formation of a disk of gas and dust with a dense central knot, known as a protostar—or protosun in this case (2). A million years later, the protosun, shrinking and growing hotter, develops a strong wind of charged particles. This blusters away from the disk's surface to form the two jets that characterize the so-called bipolar outflows (3). The Sun does not become a true star until 30 to 50 million years have passed (4). This is the stage our Sun is still at today, 4600 million years later. Known as the main sequence, it is where a star converts hydrogen into helium (5). About six billion years from now, the hydrogen will run out and the Sun will expand to become a cooler, brighter star called a subgiant (6). One billion years after that, the Sun will balloon enormously into a red giant more than 160 times its present radius (7). Then, the Sun will lose these loose outer layers to space, where they will form a glowing planetary nebula (8). In the end, all that is left is a white dwarf destined to fade forever (9).

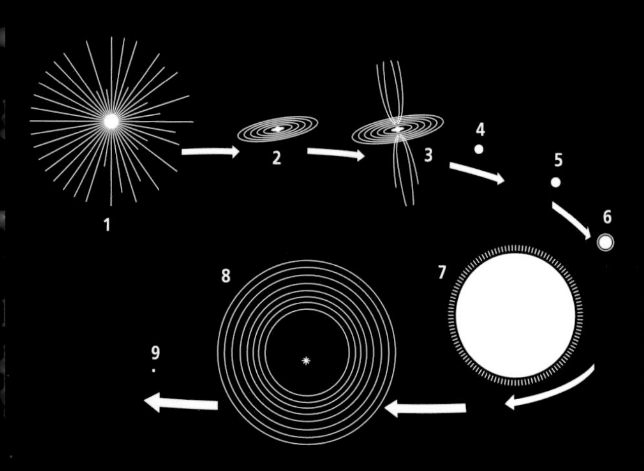

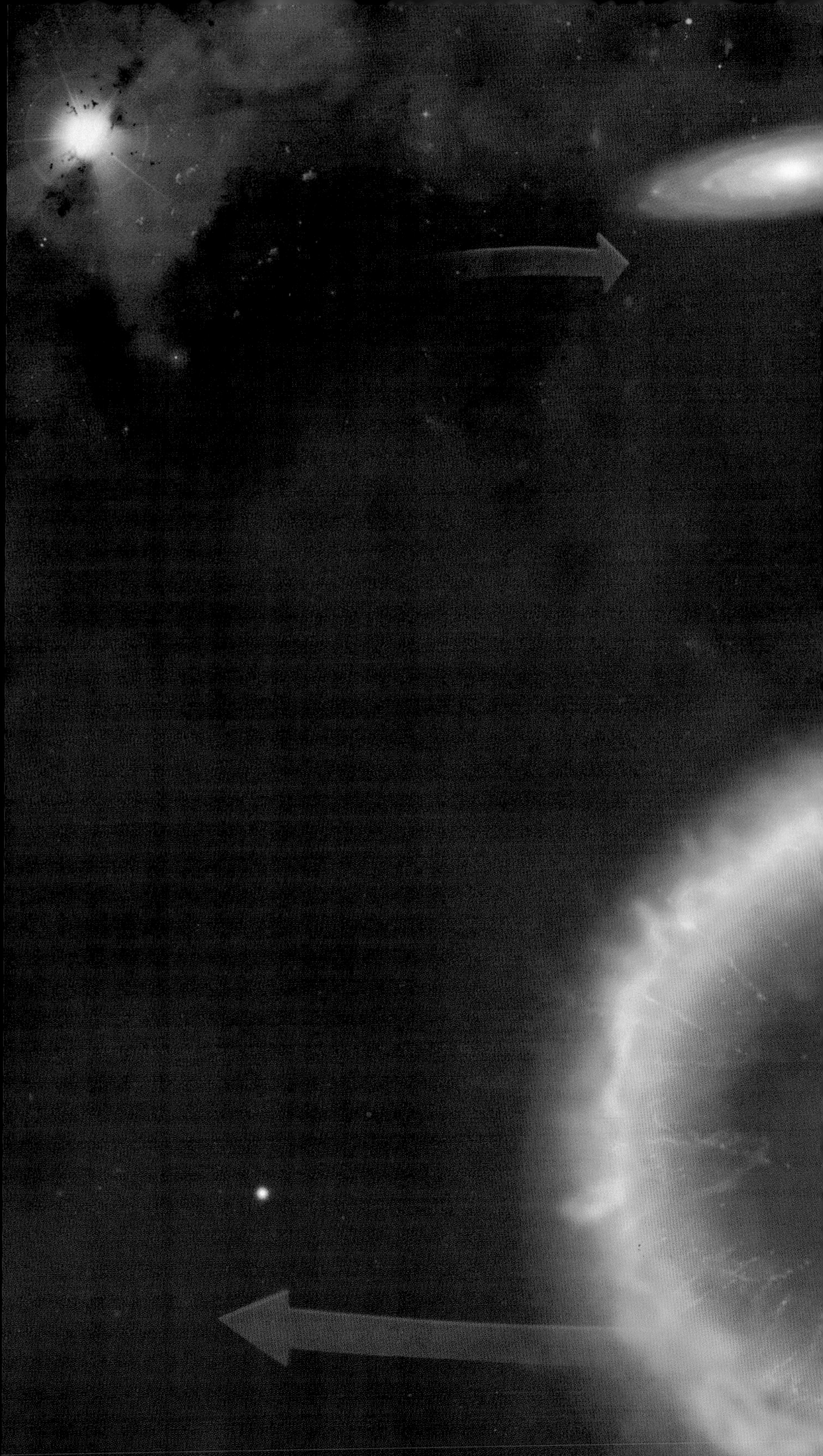

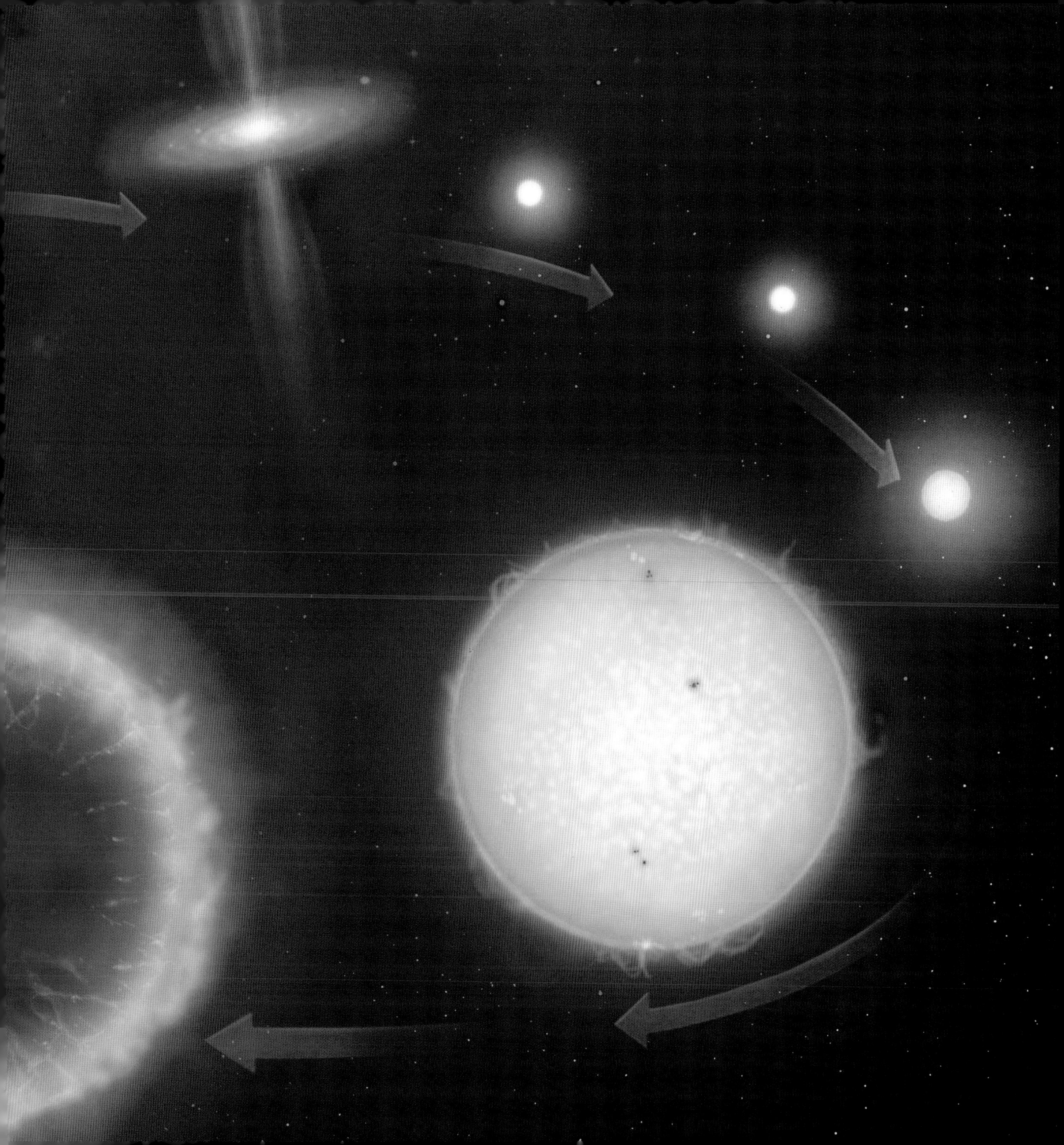

The Moon

The Moon is the fifth-largest satellite in the Solar System, with a sizable diameter compared to its parent planet, Earth. In fact, in some regards Earth and the Moon comprise a binary planet system. In contrast to Earth, the Moon is a lifeless, airless, barren sandpit—an orbiting fossil of a world. Liquid water has never existed there, and so the only forces that have contributed to its geology are the fall of meteorites and episodes of volcanic activity—albeit billions of years ago when the Solar System itself was still young.

ATMOSPHERE

No atmosphere

PLANET STATISTICS

Origin of name *Mænon*, "month" from proto-Germanic
Diameter 2160 miles (3476 km), 27.2% of Earth's
Mass 0.012 x Earth
Volume 0.02 x Earth
Mean distance from Earth 238,900 miles (384,400 km)
Farthest from Earth 252,700 miles (406,700 km)
Closest to Earth 221,500 miles (356,400 km)
Mean daytime temperature 225°F (107°C)
Mean night time temperature -244°F (-153°C)
Maximum surface temperature 253°F (123°C)
Minimum surface temperature -387°F (-233°C)
Apparent magnitude -12.7 (when full)
Surface gravity 0.16 gee (16% of Earth's)
Magnetic field strength <0.003 gauss (<1% of Earth's)

↓ **Craters upon craters typify the lunar surface**, but to an even greater degree on the farside, which is not visible from Earth due to the Moon's orbit and rotation.

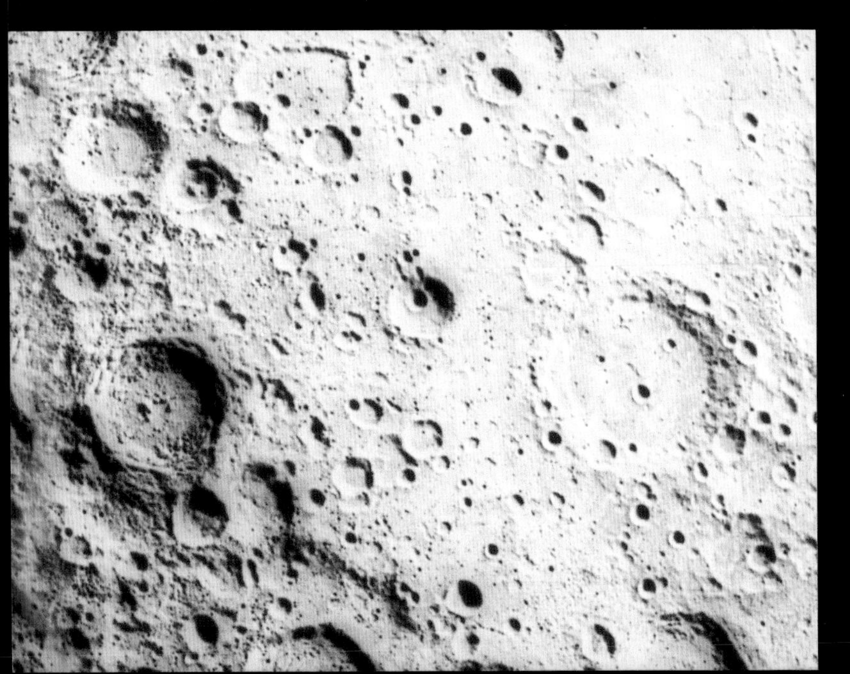

THE LAYERS OF THE MOON

The interior of the Moon is not very well known, but scientists have made their best guesses using seismology reports returned from the Apollo missions. These show that moonquakes occur at a depth of about 500 miles (800 km). However, they are so weak that if you stood on the surface of the Moon during a quake, you would not feel a thing.

Crust The Moon's crust is relatively thick. The craters on the surface are an indicator of great age, and the Maria—seas of lava that once flooded the Moon and then solidified—though younger, are still billions of years old.

Mantle The Moon's mantle is rich in rocky substances but poor in metals such as iron. It is thought to be about 700 miles (1100 km) deep.

Core The Apollo missions found evidence for heat slowly escaping from the Moon's surface and into space—heat that had come from the interior. This indicates the presence of a hot lunar core with a temperature of about 840°F (1500°C). Whether the core is solid or not remains to be seen.

SIZE COMPARISON

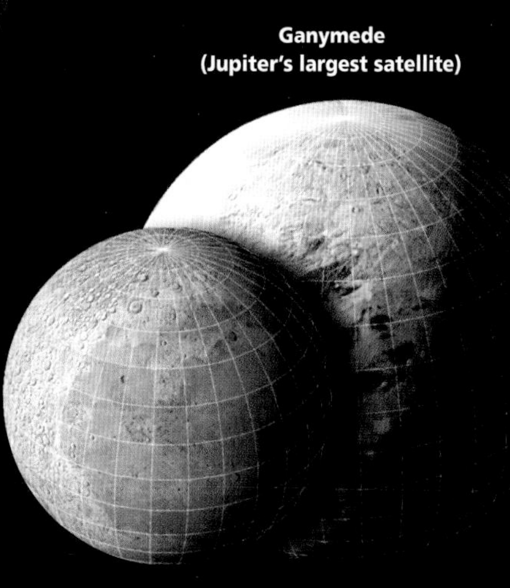

ORBIT STATISTICS

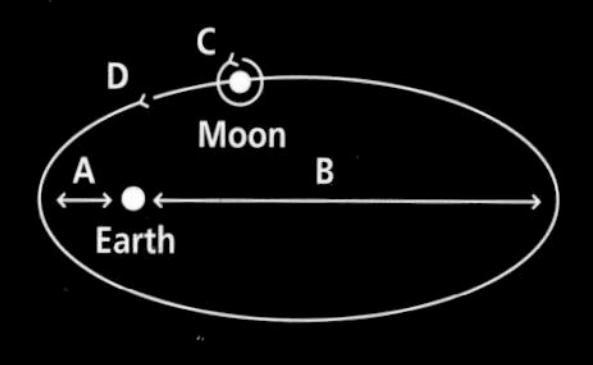

A 221,500 miles (356,400 km)
B 252,700 miles (406,700 km)
C rotates 27.32 Earth days
D orbits in 27.32 Earth days

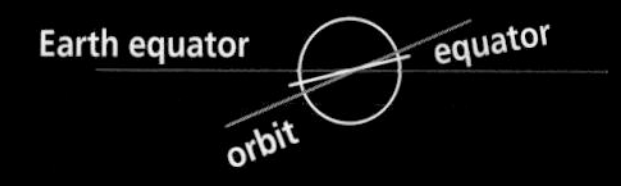

Axial tilt to orbit 6.7°
Orbital tilt to Earth equator 18.3°–28.6°

The Moon: surface

Even a cursory glance at the Moon will show two seemingly different regions—a dark part and a lighter part that, taken together, create a certain pattern that some visualize as a "Man in the Moon." The Man is an illusion, of course, but the pattern of light and dark is not. The bright areas are called the Highlands. They are heavily cratered as they formed billions of years ago during the period of heavy bombardment—a torrential rain of gigantic meteorites that marked the final phase of the construction of the planets. The dark areas are the Maria (singular Mare), which in Latin means "seas." They are not seas as we know them, but are rather vast plains of lava that, eons ago, oozed onto the surface and then solidified. They are smooth because they flooded a great number of craters and have received few meteorite impacts since.

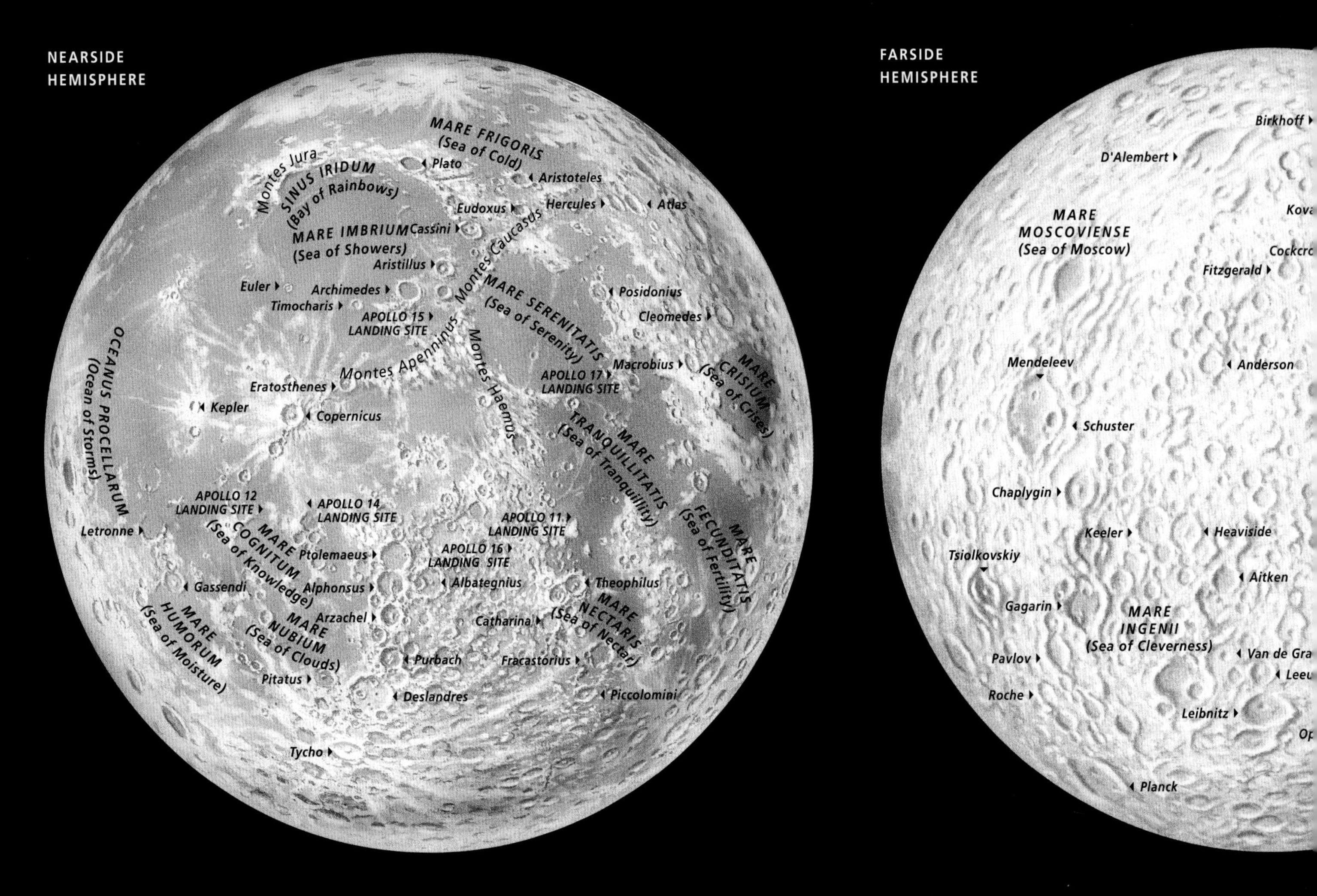

← **This boulder-stewn lunar landscape** is the result of an impact that caused the crater known as Camelot. The impact threw up vast quantities of ejecta that still litter the scene around 70 million years later— a blink of an eye in geological terms. The photo was obtained in 1972 during the Apollo 17 mission to the Moon.

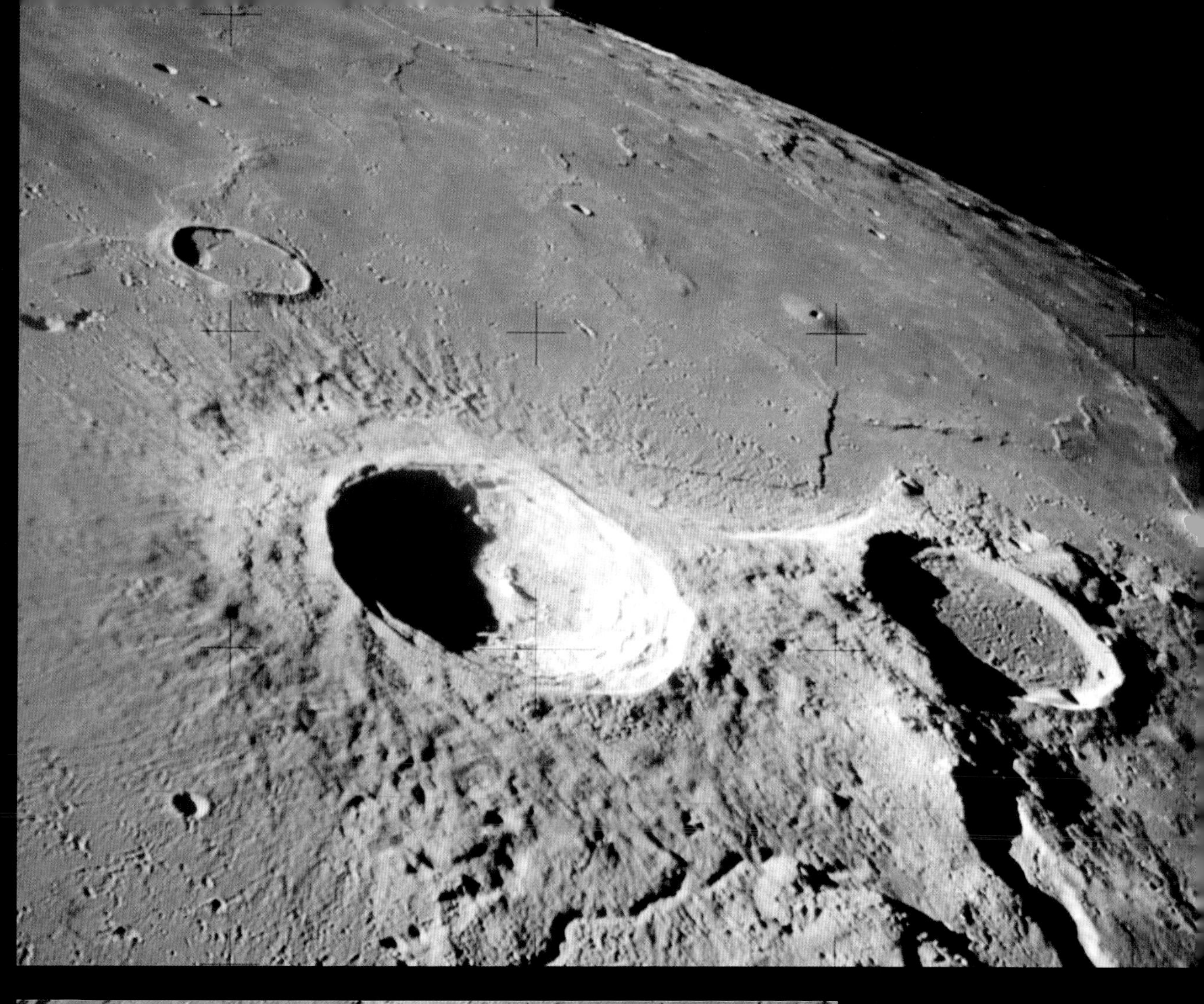

↑ **This view of the lunar surface** shows the Hadley–Apennine region of the Moon as photographed during the Apollo 15 mission in 1971. The large crater in the center is Aristarchus.

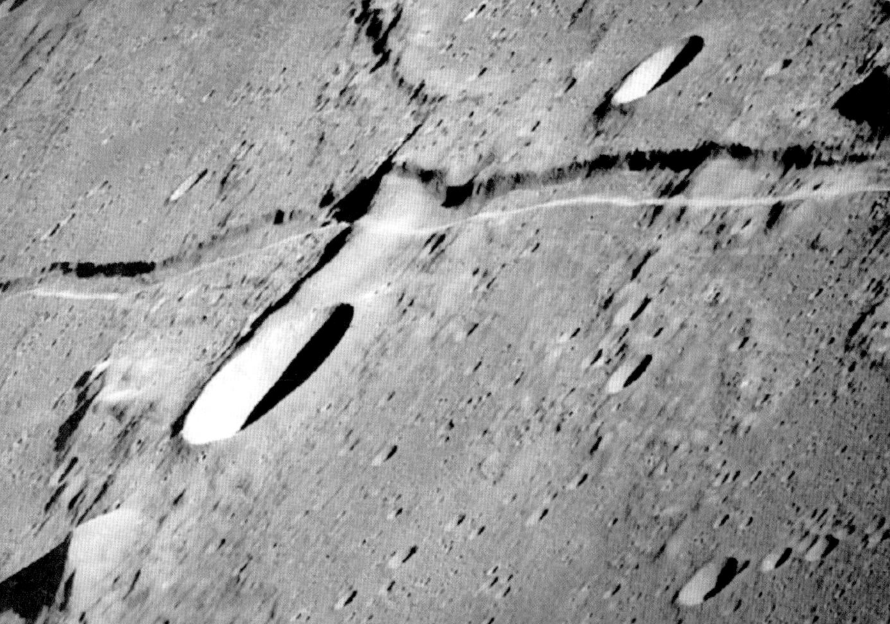

← **Rilles, first discovered** some 200 years ago, are extensive gully-like grooves, which appear across the lunar surface. Rilles (rhymes with "pills") are thought to have formed from ancient lava flows.

Landau

Mach

Hertzsprung

MARE ORIENTALE
(Eastern Sea)

Galois

Phases and tides

Even though the Moon is so far away and its gravity relatively feeble, it is still able to exert a considerable gravitational pull on Earth. Together with the inexorable pull of the Sun—weaker, because it is much farther away—the effect is to raise and lower the surfaces of our oceans to produce the tides. When the Moon is full or new—that is, when Earth, Sun and Moon line up—the gravities of the Moon and Sun reinforce one another to produce strong, spring tides. Water levels are very high when the tide is in and very low when it is out. Neap tides, by contrast, occur when the Moon, Sun and Earth form a triangle in space—when the Moon is seen at quarter phase. At this time, the gravities of the Moon and Sun cancel each other to some degree, and the water levels are then at their least extreme.

THE TIDES OF THE SUN AND MOON

These diagrams (*right*) illustrate how the positions of the Sun and Moon relative to Earth combine to produce strong spring tides and weaker neap tides. The Sun is much bigger and more massive than the Moon but it is 400 times farther away. As a result its ability to raise tides is only one-third as strong as the Moon's. Spring tides result when the three bodies form a line in space (top). When they form a triangle, with the Moon at first or last quarter, they produce neap tides (bottom).

TIDES ON EARTH

Spring tide

High tide

Low tide

High tide

Low tide

Neap tide

High tide

Low tide

Low tide

High tide

↓ **These two photos show how tides** can dramatically affect the appearance of Earth seen from above. The first image shows Whitehaven Beach—on Whitsunday Island off the northeastern coast of Australia—at low tide.

↓ **By contrast, at high tide,** water covers much of the sand and coral to form beautiful turquoise lagoons. Whitsunday Island measures 12 by 8 miles (19 by 13 km), and lies 6 miles (10 km) from the mainland in the Coral Sea.

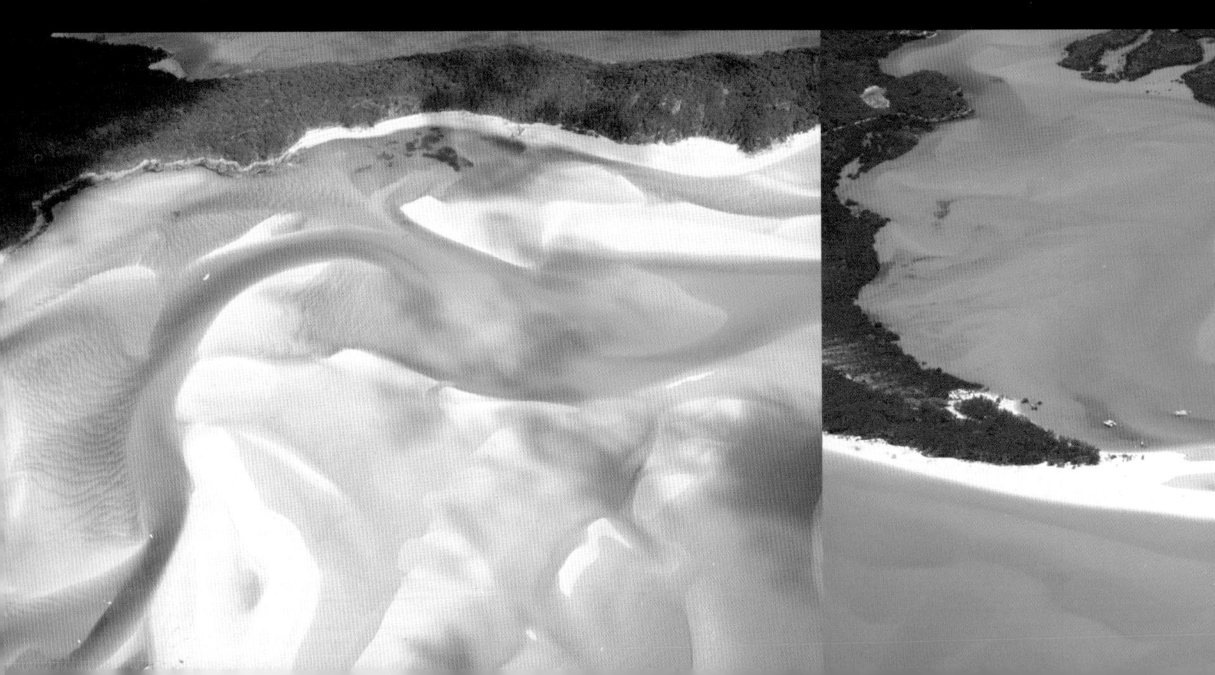

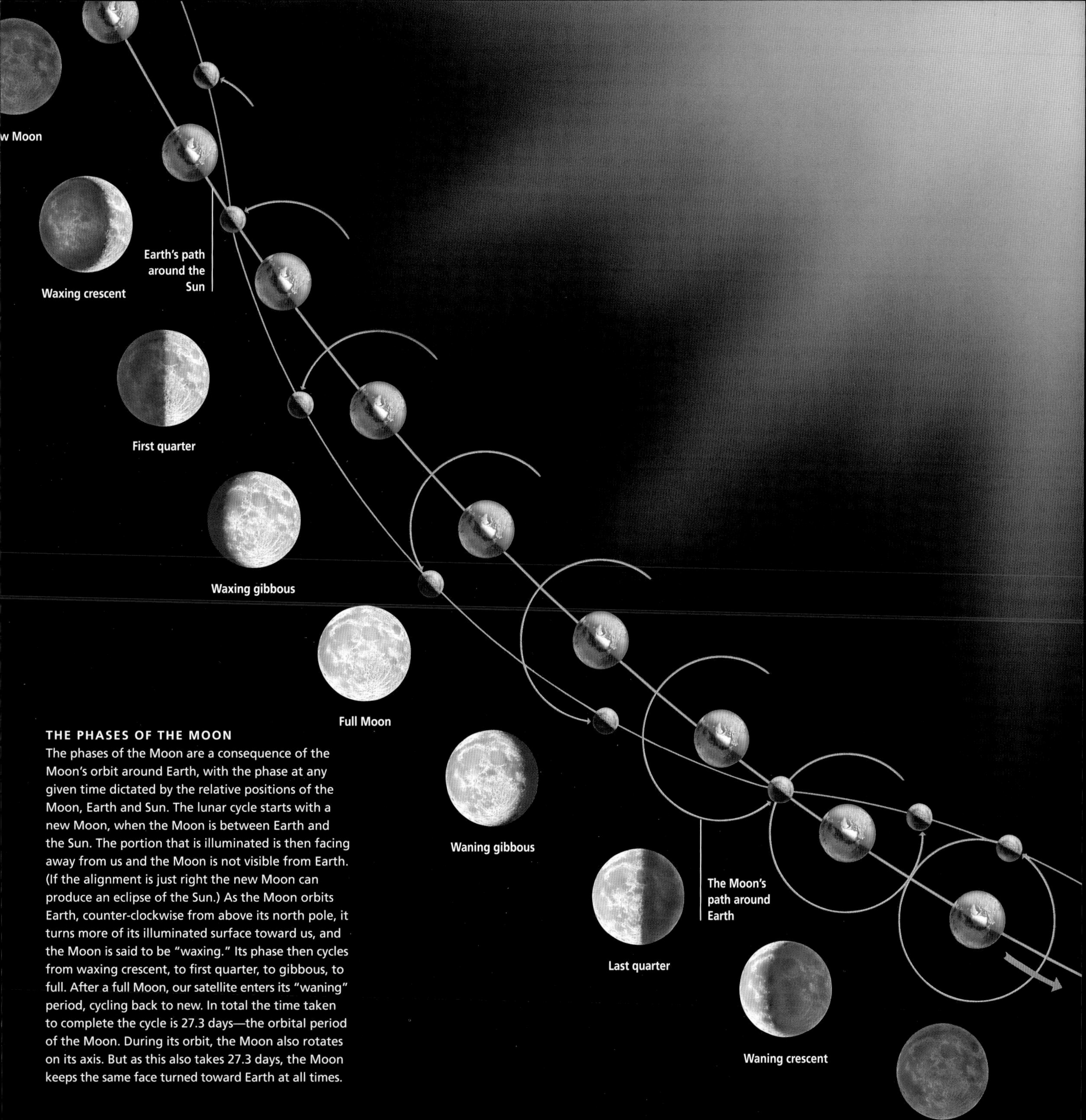

THE PHASES OF THE MOON

The phases of the Moon are a consequence of the Moon's orbit around Earth, with the phase at any given time dictated by the relative positions of the Moon, Earth and Sun. The lunar cycle starts with a new Moon, when the Moon is between Earth and the Sun. The portion that is illuminated is then facing away from us and the Moon is not visible from Earth. (If the alignment is just right the new Moon can produce an eclipse of the Sun.) As the Moon orbits Earth, counter-clockwise from above its north pole, it turns more of its illuminated surface toward us, and the Moon is said to be "waxing." Its phase then cycles from waxing crescent, to first quarter, to gibbous, to full. After a full Moon, our satellite enters its "waning" period, cycling back to new. In total the time taken to complete the cycle is 27.3 days—the orbital period of the Moon. During its orbit, the Moon also rotates on its axis. But as this also takes 27.3 days, the Moon keeps the same face turned toward Earth at all times.

Exploring the Moon

In 1969, following years of frantic competition between the USA and the Soviet Union, the Americans finally achieved their goal of landing a man on the Moon and returning him safely to Earth. Thus did Neil Armstrong and Buzz Aldrin of Apollo 11 become the first humans to walk on the surface of our nearest neighbor in space—but they were not the last. Six more Apollo missions were launched before the Apollo series was scrapped in 1972. Apollo 13 was the only failure, so in total, 12 people have regarded Earth from the Moon's surface. These missions brought a wealth of astronomical data and Moon samples, and have helped enormously in our understanding of the Moon, its history and how it formed.

↑ **The lunar rover** sets off across the Moon's surface with astronaut James Irwin of Apollo 15 in the driver's seat. The battery-powered rover was equipped to communicate directly with Earth.

← **Another lunar rover** standing beside the lunar landing module during the Apollo 14 mission in 1971.

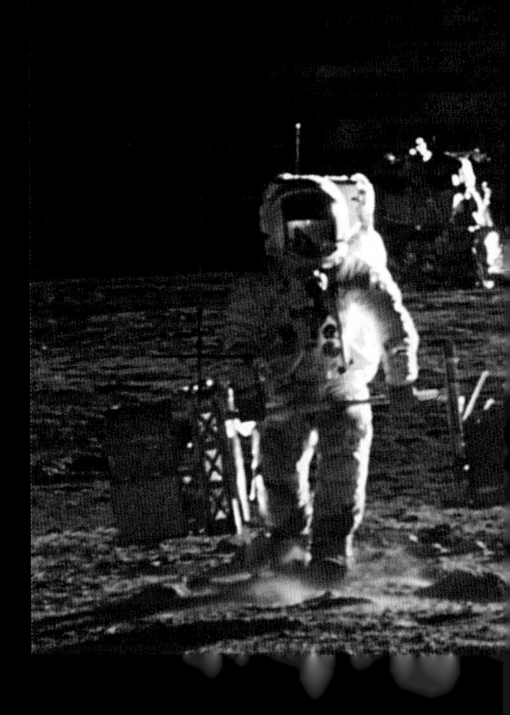

→ **Taken during the Apollo 12** mission in 1969, this photograph shows an astronaut deploying a package of lunar instruments on the Moon. Many experiments were carried out during the Apollo missions, which have greatly expanded our understanding of the Moon. (The red color in this image is due to lens flaring.)

→ **These footprints** were made by the Apollo 15 astronauts. The lunar soil is exceedingly fine and powdery.

→→ **No, not a round** of low-gravity golf! An astronaut from Apollo 17 in 1972 is seen here taking a sample of the lunar surface.

Solar eclipses

When the Sun and Earth are aligned, with the Moon between, the Moon can obscure part or all of the Sun's rays to produce an event known as a solar eclipse. There are three different types: partial, total and annular. The most spectacular is a total eclipse, which occurs when the Moon exactly covers the Sun's photosphere (its "surface"). The Moon casts a shadow on Earth, and within the dark central part of that shadow, the umbra, an observer will witness a total eclipse. In the brighter part of the shadow, the penumbra, the Moon hides only part of the solar surface, and a partial eclipse ensues. An annular eclipse arises when the Moon is farther from Earth—thus the Moon appears slightly smaller than the Sun's disk in the sky and so does not cover it completely.

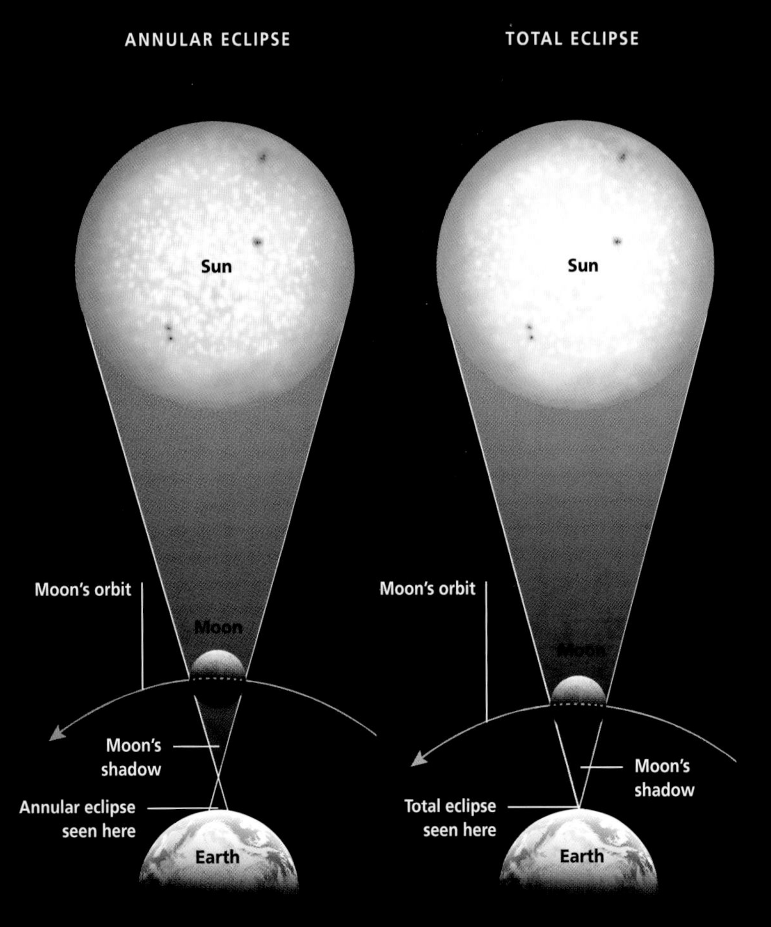

← **A total solar eclipse,** such as this one from August 1999, is the only way on Earth we can view the outer atmosphere of the Sun—the corona—seen here flowing along the Sun's magnetic field lines.

→ **Mauritanians prepare themselves** for an eclipse in 1973 that will last for a satisfying 6 minutes and 20 seconds—close to the maximum possible duration of 7 minutes and 31 seconds.

↓ **The development of a partial eclipse** is revealed in this time-lapse photo—here the Moon passes just in front of the Sun before moving away again.

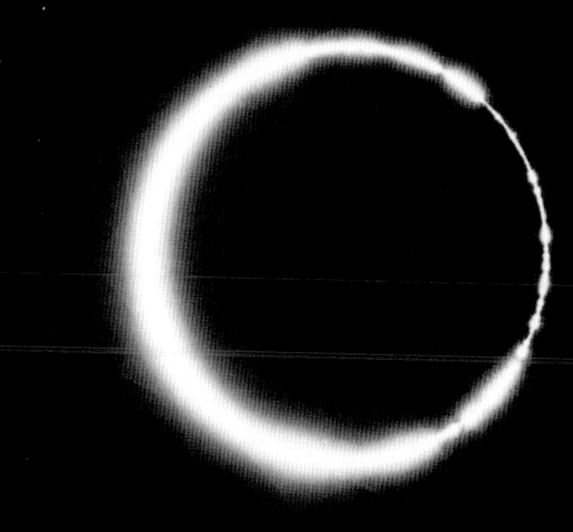

OBSERVING SOLAR ECLIPSES

A total solar eclipse is the only occasion during which you can safely look at the Sun unaided. Before totality (when the eclipse is partial) or during an annular eclipse, you should look through a piece of Mylar or a special pair of eclipse glasses with Mylar filters for "lenses." Even during an annular eclipse the glare is too bright to look at without filters, despite the fact that the vast majority of the solar disk is obscured. When totality occurs, you can safely observe with the naked eye, or even with binoculars. But do take extreme care! Make very sure that you do not observe too long, or the Sun will blind you when it reappears from behind the Moon.

↑ **The Sun shining behind** the Moon during an annular eclipse forms an effect known as a "diamond necklace." Annular eclipses occur when the Moon is farther away from Earth than during a total eclipse.

Lunar eclipses

Just as the Moon can pass between Earth and the Sun to produce a solar eclipse, so Earth can get between the Moon and Sun to create a lunar eclipse. Seen from space, the Moon, as it orbits, passes into the shadow of Earth cast through space by the Sun. As with a solar eclipse, Earth's shadow has two components as a result of the fact that the Sun is not sufficiently distant to appear as a point. The dark central component is the umbra and the brighter part surrounding it is the penumbra. During a total or "umbral" eclipse of the Moon, the Moon is wholly within the umbra. It is totally blocked from direct solar illumination and can appear very dark. Exactly how dark depends on the clarity, at the time, of Earth's atmosphere—which will bend and transmit a little sunlight. If the Moon enters only the penumbra, a partial or "penumbral" eclipse ensues.

→ **This photo shows the progression** of a partial lunar eclipse, due to the passage of the Moon through the shadow of Earth cast through space.

TOTAL ECLIPSE

Sun

Earth

Moon's orbit

Eclipsed Moon

Earth's shadow

← **This time-lapse photo** reveals the stages in a total lunar eclipse, when the Moon passes fully into the darkest part of Earth's shadow, the umbra.

→ **During a partial lunar eclipse** such as this one, the Moon appears a coppery red. This is because it is receiving sunlight that has been filtered by Earth's atmosphere.

OBSERVING LUNAR ECLIPSES

Lunar eclipses are fairly common events. Unlike solar eclipses, which are limited to local geographic regions, an eclipse of the Moon is visible from everywhere—so long as it's clear! The diagram above shows how the eclipse arises, with the Moon passing through the shadow of Earth. The maximum theoretical duration for a lunar eclipse is 1 hour and 47 minutes. Through binoculars a lunar eclipse can be quite a sight. Sunlight filtered by Earth's atmosphere stains the Moon a coppery red during a partial eclipse, while during a total eclipse—if the Moon is still visible, not blotted out completely by the shadow—it can appear a steely gray color.

Meteors and meteorites

There is often confusion between the terms "meteor" and "meteorite" and the lesser-known term, "meteoroid." In fact, each is often a manifestation of the same object—the different names reflect how the object is seen. Meteoroids are fragments of asteroids or comets, often no larger than a grain of sand, which pervade the Solar System. When these enter our atmosphere, they burn up and leave bright streaks. They are then called meteors or "shooting stars." Lastly, if the meteor is not completely destroyed by the atmosphere, it will hit the ground, in which case we call it a meteorite. If the impact speed is slow, the meteorite will survive. But if the object is very massive, it will impact so quickly that it will form a crater, evaporating both the ground and itself.

MAJOR METEORITE IMPACTS

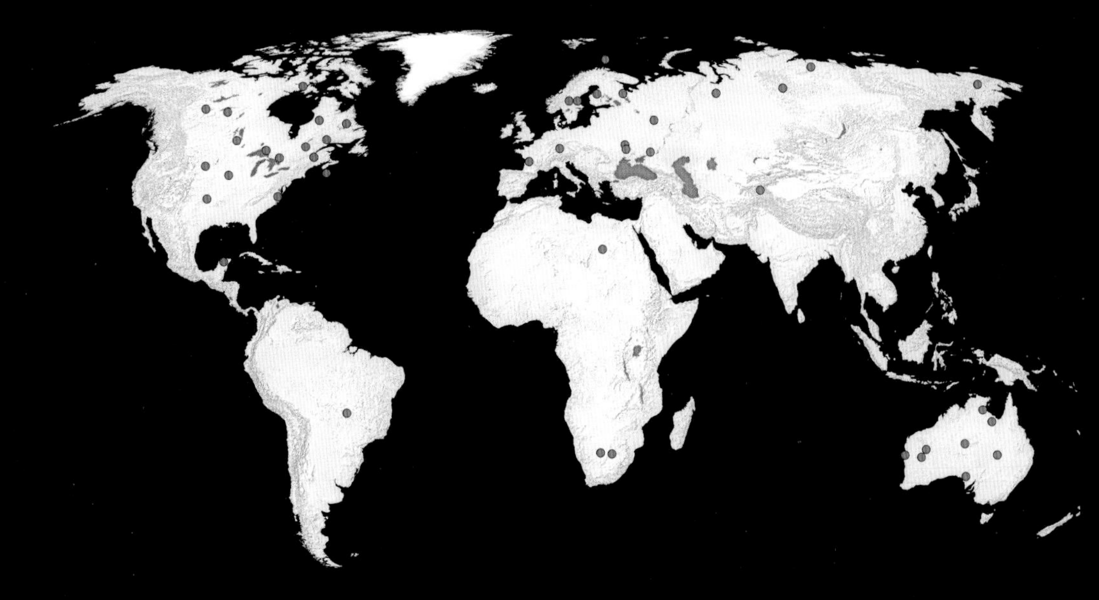

↑ **A "fireball" (a very bright meteor),** captured in this time exposure photo, is seen against a curtain of light—the aurora borealis.

↖ **Craters on Earth are rare** as the impact scars are eradicated almost as soon as they form, in geological terms, by plate tectonics, volcanism, and wind- or water erosion. But there are around 100 large-scale craters and many more smaller ones. This map shows the locations of some of the better known sites.

← **One of best-preserved craters** on Earth is the Barringer Crater near Flagstaff, Arizona, in the United States. The crater is ¾ mile (1.2 km) in diameter and 570 feet (175 m) deep, and was formed about 50,000 years ago. The perpetrator was a nickel-iron meteorite perhaps measuring only 160 feet (50 m) across.

← **The Leonid meteor shower,** as seen in this time-lapse photograph, occurs each year around the middle of November and lasts for several days. It is the result of Earth's orbit taking it regularly into the trail of debris left around the Sun by the comet Tempel–Tuttle. As Earth enters the debris trail, the particles enter our atmosphere and we see them as meteors, or "shooting stars."

Asteroids

Between the orbits of Mars and Jupiter lies a cosmic trash pile. This is the realm of the asteroids—shards, pebbles, boulders and country-size chunks of rock and metal, battered and ancient, all but the very largest irregular in shape. Astronomers used to refer to the asteroids as the vermin of the skies, but today they are less judgmental. They have realized that these haphazard fragments are the leftovers from the accretion process that built the planets more than 4600 million years ago. The largest asteroid is Ceres, which was also the first to be found, in 1801. It is around 650 miles (1000 km) across. Most, however, measure less than 30 feet (10 m). The asteroids were unable to gather to form a single large planet because of the presence of nearby Jupiter, whose gravity tugs at them and keeps them from sticking together.

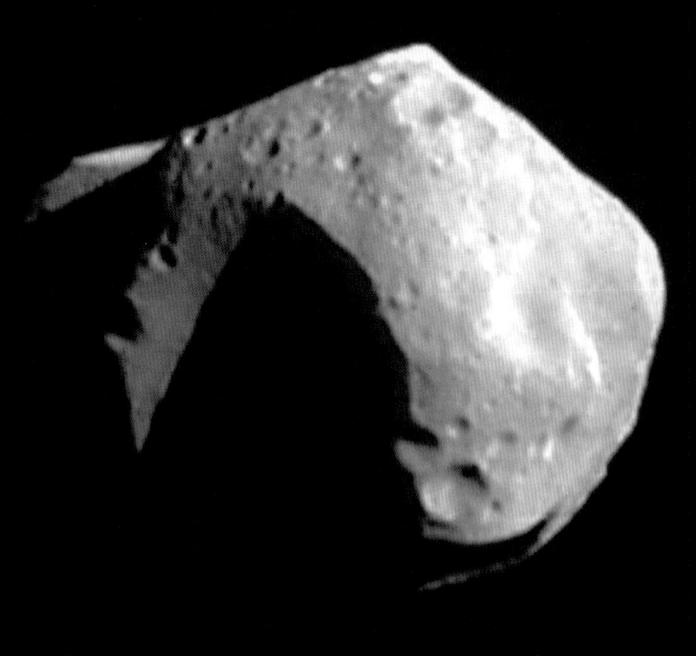

↓ **The asteroid Ida, seen here,** as photographed by the Galileo space probe. Ida was the first asteroid known to have its own moon—a much smaller body called Dactyl (not shown) measuring only about 1 mile (1.6 km) in diameter.

↑ **Mathilde is a rather large asteroid** measuring some 37 by 29 miles (59 by 47 km). It is very dark, reflecting less than 4 percent of the sunlight that falls on its battered, ancient surface, and spins once in 17 days.

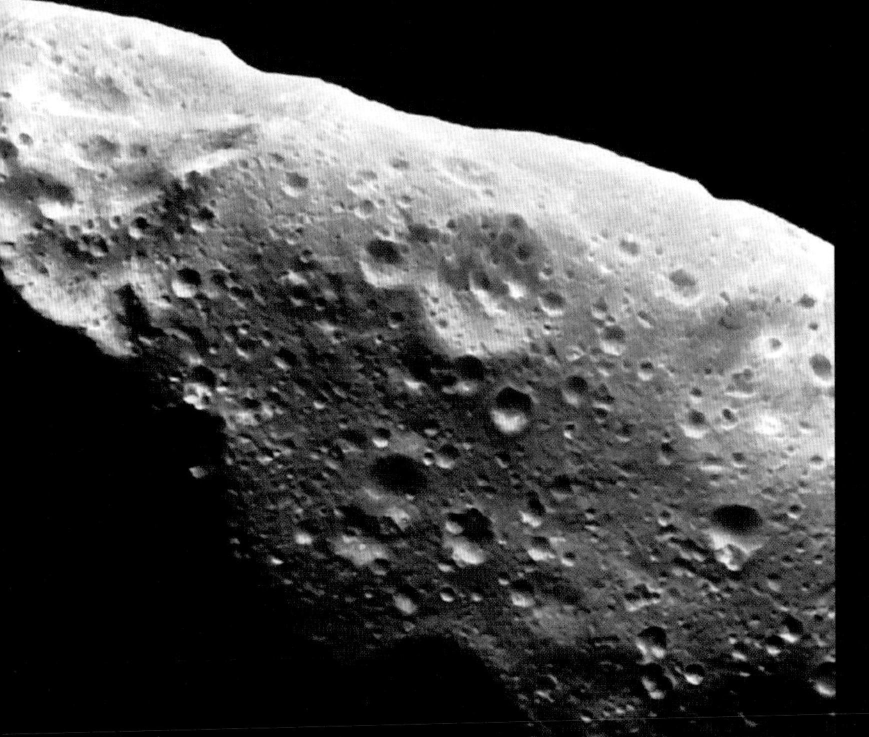

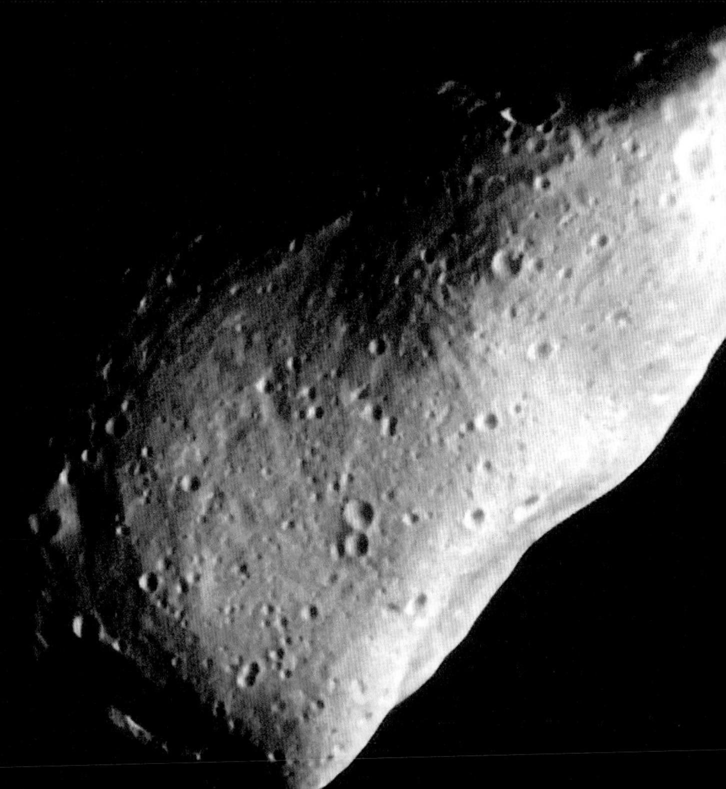

← **Gaspra, photographed here by the Galileo spacecraft** on October 29, 1991, is a sizable asteroid measuring approximately 12 x 7½ x 7 miles (19 x 12 x 11 km) with a rotation period of about 7 hours. Gaspra, like other asteroids, is covered in craters—in this image alone there are more than 600 craters in the range from 330 to 1650 feet (100–500 m) in diameter. This small world also has a few linear depressions about 330 to 1000 feet (100–300 m) wide, thought to be fractures created during energetic impacts with other asteroids. Similar grooves are also seen on the Martian moon Phobos, and are thought there to have a similar origin.

THE ASTEROID BELT

Most asteroids orbit the Sun between Jupiter and Mars. This is a region known as the asteroid belt, and it stretches from 2.15 to 3.3 astronomical units (AU) from the Sun, where 1 AU is the distance between Earth and the Sun (93 million miles / 150 million km). At the inner edge of the belt, the asteroids take just over three Earth years to orbit the Sun, while at the outer edge the period is roughly double that. There are also several gaps, know as the Kirkwood gaps, which are regions where the gravity of Jupiter maintains a low asteroid density. There are no reliable estimates of the total number of these bodies, but their combined mass is thought to be about one-twentieth of the mass of our Moon. The asteroid belt is often depicted in movies as packed with countless, spinning boulders, jostling-up alongside each other and even colliding. Indeed, there is no doubt that asteroids do collide. However, the illustration above has been exaggerated for clarity. In reality, the asteroid belt is so sparse that, if you were stationed on one, you would be unlikely to see another with the unaided eye for your entire lifetime. The reason is simple: while there are many asteroids, there is a lot of space out there. They also have slow rotation periods, often taking hours or even days to rotate once, much slower than is often depicted.

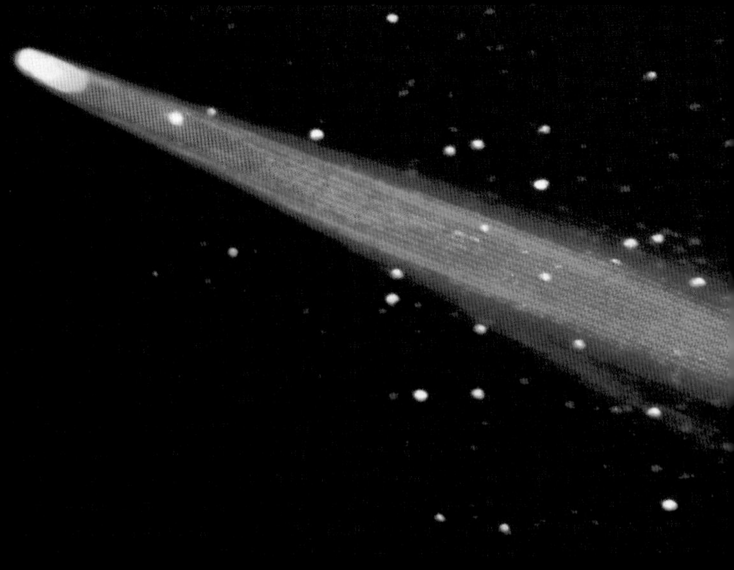

Comets

Comets are among the most primitive bodies in the Solar System, the icy leftovers from the formation of the outermost regions 4600 million years ago. They are loose aggregations of water-ice and rock, a few miles across, and are often referred to as "dirty snowballs." Although most famous for their tails, which can stretch for more than 100 million miles (160 million km), most comets do not actually have one. They are usually so far from the Sun that they are frozen and inert. It is only when they enter the inner Solar System, courtesy of their elongated orbits, that comets come alive. There, the Sun's heat begins to melt the comet's surface ice. As these gases slip away, they form the coma—the hazy head of the comet—and the tail that makes them so spectacular.

↓ **This diagram shows the paths** of some famous comets in the inner Solar System: Encke, Hyakutake, Hale–Bopp and Halley. All have very elongated elliptical orbits that are highly inclined to the ecliptic—the plane close to which most planets orbit.

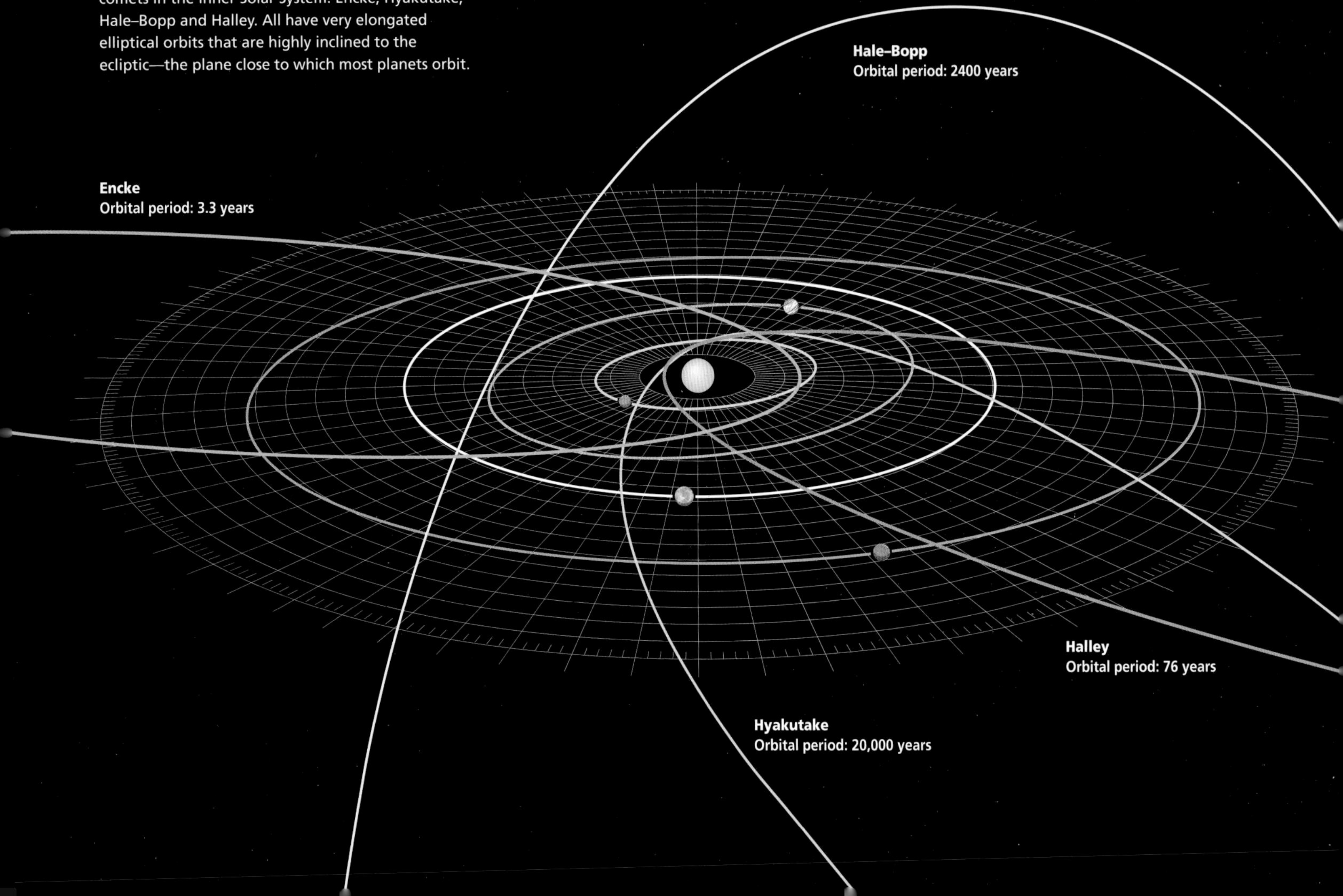

← **Probably the most famous comet is Halley.** It is a regular visitor to the inner Solar System, gracing Earthly skies every 76 years, most recently in 1985/86.

→ **Comets are most likely very porous.** They are made of chunks of ice held together with rocky gravel, and possibly a solid core of rock.

↓ **The most recent bright comet was Hale–Bopp,** which in 1997 reached magnitude -0.5. Its tail stretched some 20 degrees across the sky, a length equivalent to the distance from Earth to the Sun.

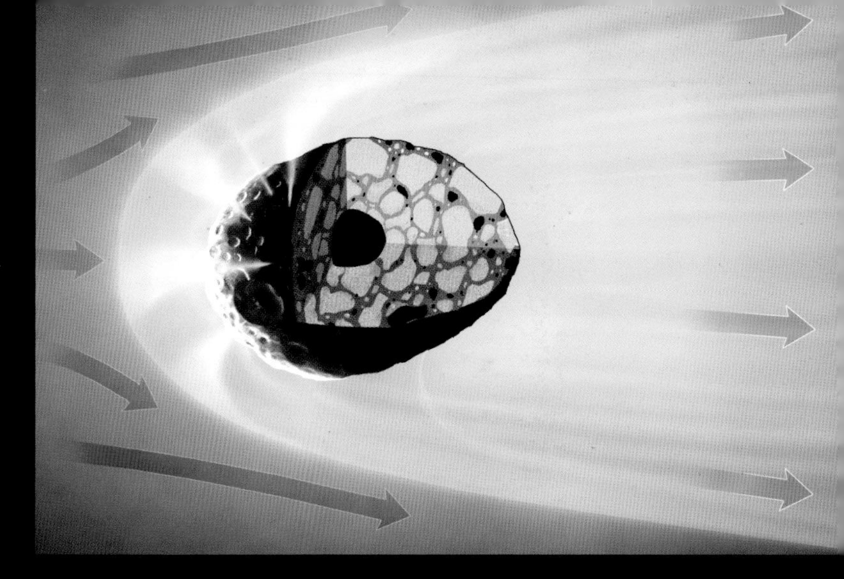

Beyond Pluto

Pluto is the farthest planet from the Sun, but it does not mark the boundary of our Solar System. Far from it. Beyond Pluto we encounter the most extensive region in our planetary system, the realm of the comets. We see comets most brightly when they come close to the Sun and their tails form, as their surface ices evaporate. But trillions of inert comets are thought to surround the Sun in two vast reservoirs.

Comets come in two varieties: long-period and short-period, with the long-period comets taking more than 200 years to orbit the Sun. A great number of long-period comets have orbits that are highly inclined to the plane in which the planets orbit, the ecliptic. Since these comets come from so far from the Sun and they do so at all manner of angles, astronomers believe that they must reside in a vast, spherical shell that surrounds the Sun at huge distances. This is known as the Oort Cloud. Closer to home but still beyond Pluto, a smaller, donut-shaped cometary deep-freeze called the Kuiper Belt can be found—the source of the short-period comets.

↑ **Jan Hendrik Oort** (1900–1992), an astronomer based in the Netherlands, is mo famous today for the comet reservoir that bears his name, the Oort Cloud. Though it has never been detected, astronomers believe it must exist du to the nature of come

← **In 1951, Gerard P. Kuiper** (1905–1973) predicted the presence of a donut-shaped zon of frozen comets beyor Pluto. Two years earlie another scientist, Kenneth Edgeworth, had foreseen the same Today, this zone is known as the Kuiper o Edgeworth-Kuiper Bel

CROSS-SECTION
OF THE OORT CLOUD

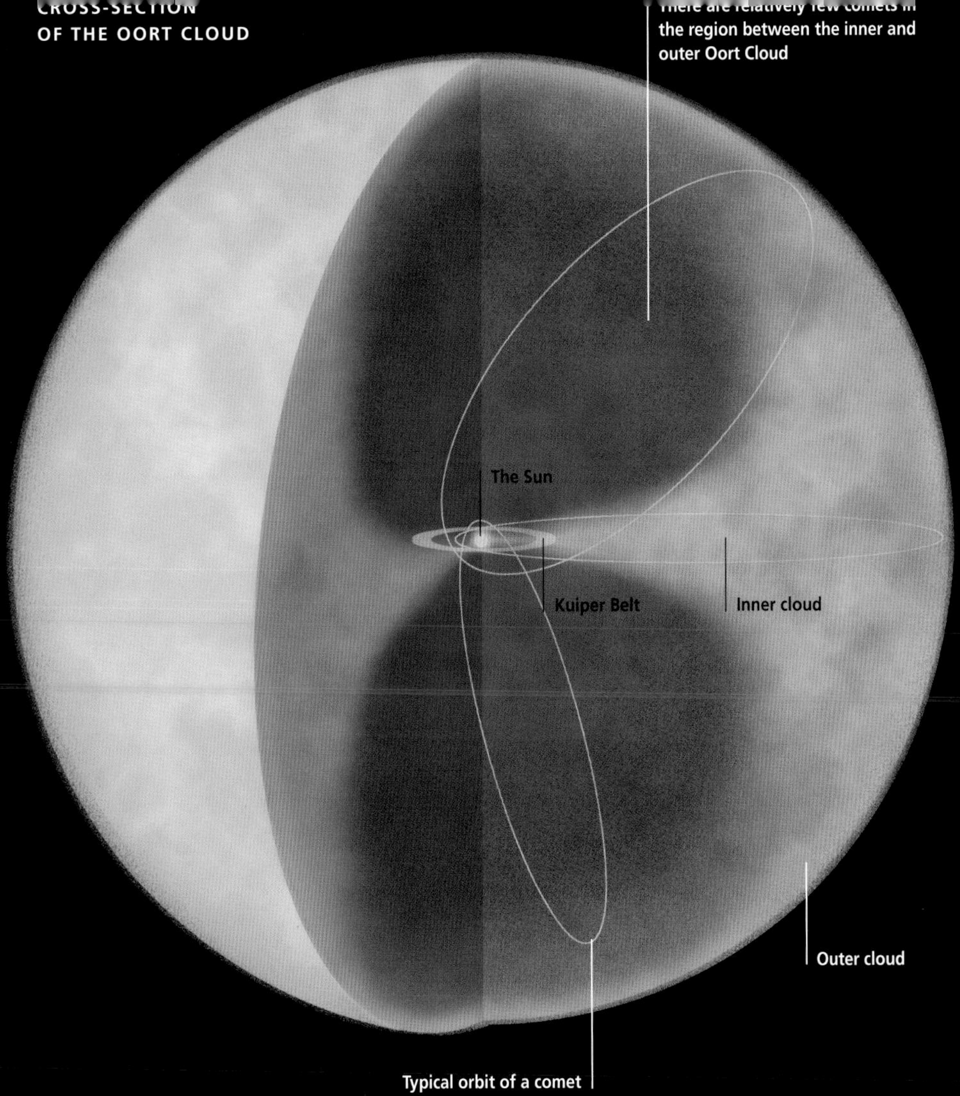

There are relatively few comets in the region between the inner and outer Oort Cloud

The Sun

Kuiper Belt

Inner cloud

Outer cloud

Typical orbit of a comet

Since comets evaporate a little each time they approach the Sun, they ought not last more than a few tens of thousands of years. Yet billions of years since the Solar System was born, it is still replete with comets. How can this be? The answer, according to Jan Hendrik Oort, was that a vast reservoir of frozen comets must surround the Sun, its diameter possibly greater than 2 light-years. As the Sun orbits the Milky Way, once in about 220 million years, it frequently drifts within a few light-years of nearby stars. When this happens, these stars exert a tidal force on the Solar System (as the Moon does on Earth). During the Sun's close encounters with other stars, some of those comets will be wrenched free of the Oort Cloud and sent into the inner Solar System. From our point of view, these long-period comets, their journeys taking thousands to millions of years, seem to come from all manner of directions in space. Thus the Oort Cloud theory neatly explains how comets have persisted for so long, and how they can have such highly inclined orbits.

↓ **The largest Kuiper-Belt object** so far found, in 2002, is called Quaoar (pronounced *kwa-whar*), seen here in this time-lapse photograph. Quaoar, the most distant object in the Solar System discovered by a telescope, measures about half the size of Pluto.

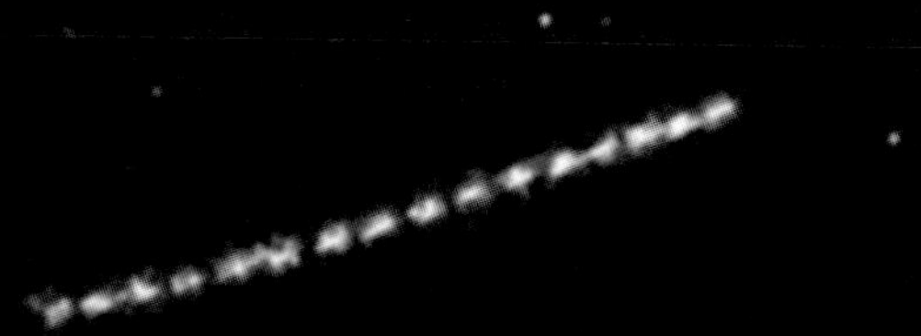

THE KUIPER BELT

Some comets, such as Halley and Encke, have short periods. It is possible that some had longer periods but were captured into tighter orbits. But most short-period comets have orbits constrained to the ecliptic, in contrast to the highly tilted orbits of long-period comets. So the short- and long-period comets must have different origins. Today, astronomers know that a smaller comet pool shaped like a donut, the Kuiper Belt, must also surround the Sun beyond Pluto. Supporting this theory, many hundreds of giant icy worlds have been found beyond Neptune, some larger than the largest asteroids. These are called Kuiper-Belt objects (KBOs) or Trans-Neptunian objects (TNOs). The Kuiper Belt's radius is uncertain, but it is likely that it is only a few times the size of Pluto's orbit. The Oort Cloud, by contrast, is a thousand times larger

Stars, galaxies and celestial light

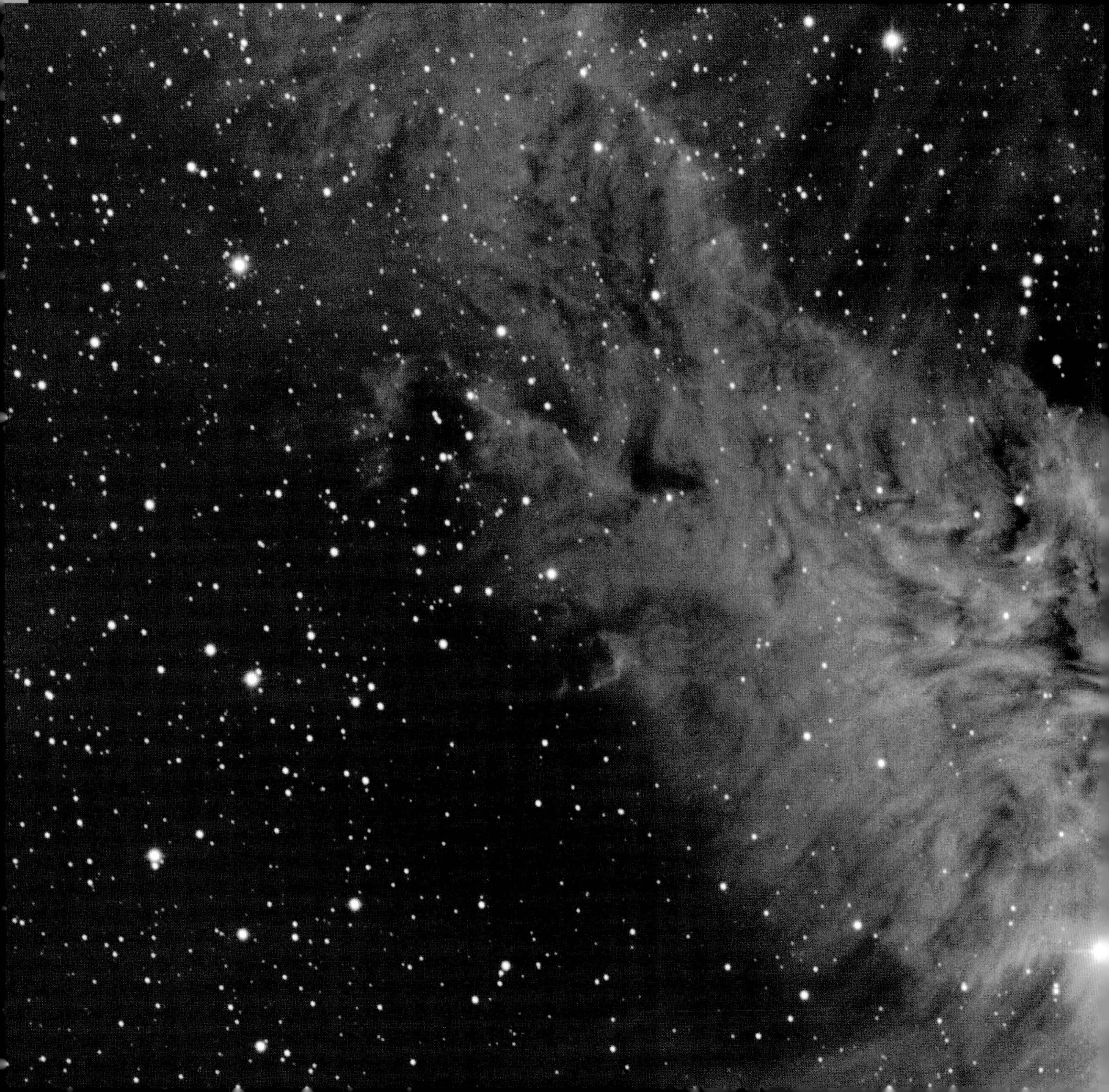

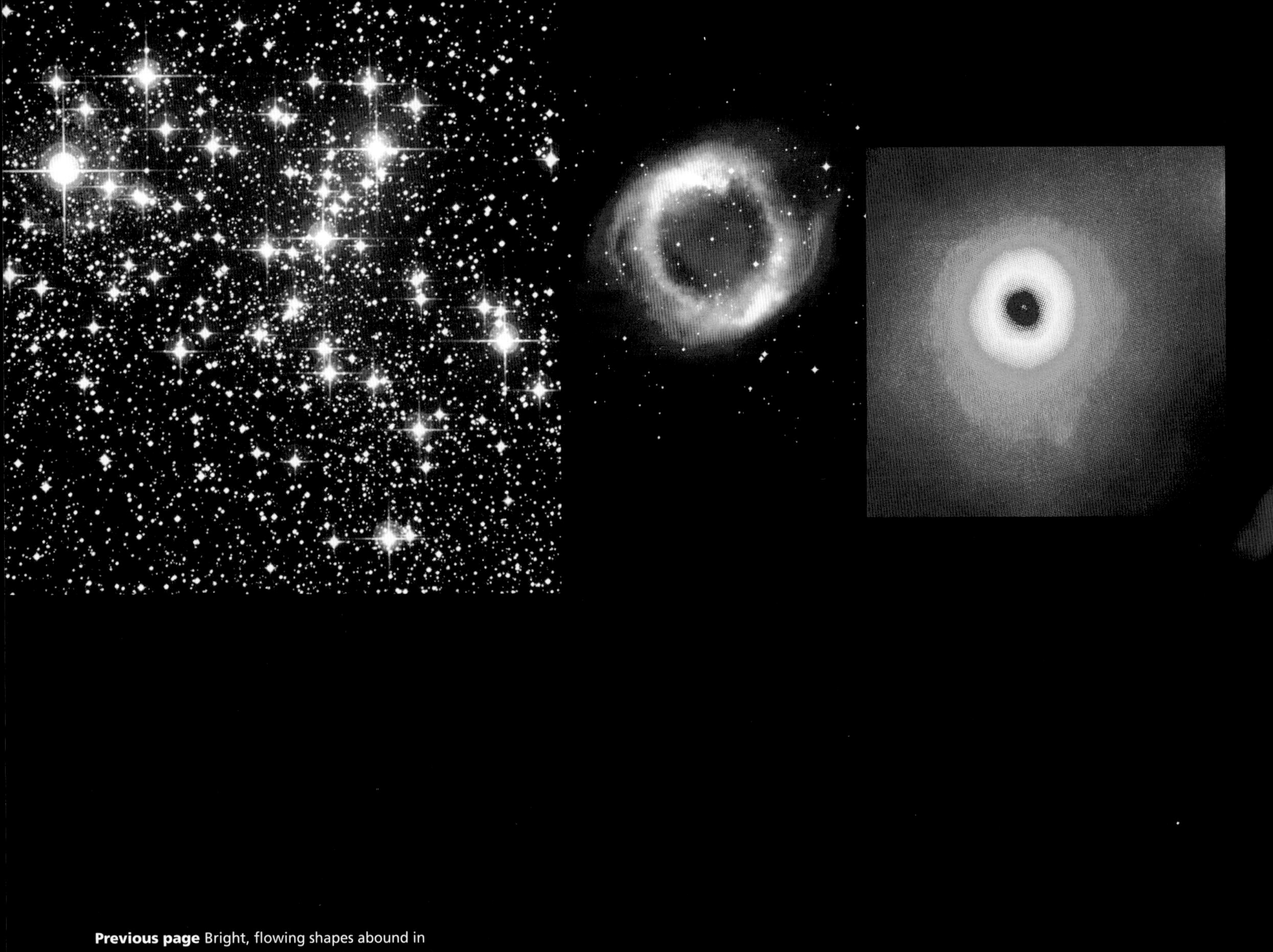

Previous page Bright, flowing shapes abound in stellar nurseries, such as the Cone Nebula, pictured.

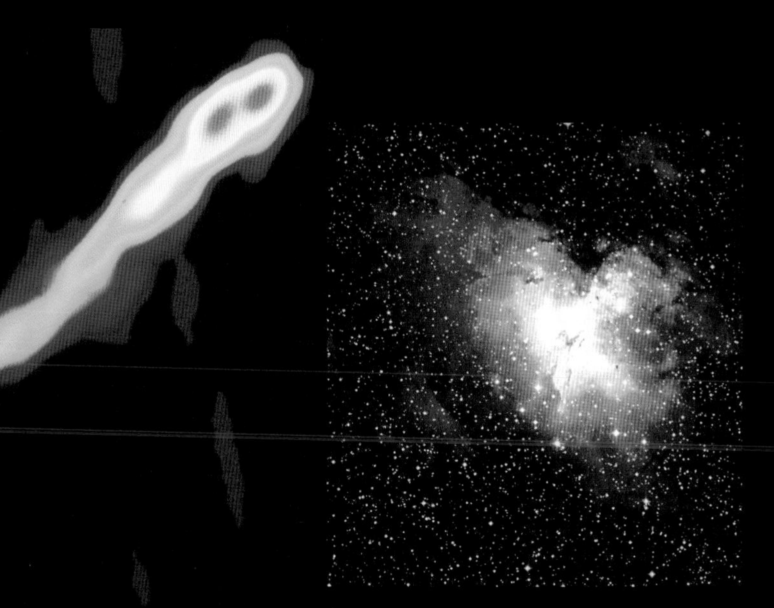

Stars, galaxies and celestial light

Beyond the Solar System, things take on an entirely new scale. Stars are strewn across the Milky Way Galaxy like grains of sand many miles apart. They share their home with a whole host of celestial lights called nebulae. And then, beyond our own galaxy, there are hundreds of billions of others.

Celestial clouds

The Milky Way is filled not only with stars and planets, but also with raw materials for making more of the same—the nebulae. These celestial clouds come in many different sizes, densities and colors. The smallest are dense, inky patches called globules, and span around 1 light-year. But the largest, giant molecular clouds, are 100 times larger and can contain enough material to make tens of millions of stars. Both globules and molecular clouds are made chiefly of molecules—in particular, molecular hydrogen. This is a form of hydrogen in which the atoms are glued together in twos to make simple H_2 molecules. The rest is a mixture of helium and "dust"—solid grains made from silicon and carbon. Where a nebula contains a new-born star, the star's light ionizes the gases and produces a kind of bright nebula called an HII region. Lastly, those nebulae that merely reflect the light of nearby stars are called reflection nebulae.

↓ **Molecular clouds** emit no visible light. They appear as dark clouds set against a bright background of stars or luminous nebulae. These dark nebulae are common. Small molecular clouds called globules, typically a light-year across, are thought to be cocoons, inside of which new stars are forming.

← **Some interstellar clouds** emit no light of their own. Instead, they merely reflect or scatter light from nearby stars or nebulae. These are known as reflection nebulae and are often found close by emission nebulae. The light produced is typically blue, in the same way that sunlight scatters in Earth's atmosphere and colors our sky.

↓ **When new stars** are born, the ultraviolet radiation they emit floods into the remains of the star's birth cloud and produces a dazzling display of colors that we see as an emission nebula. The light is produced in a process called recombination or de-excitation.

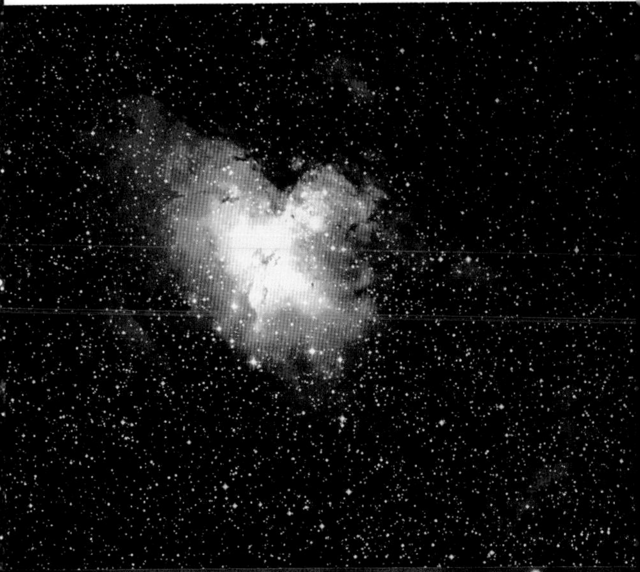

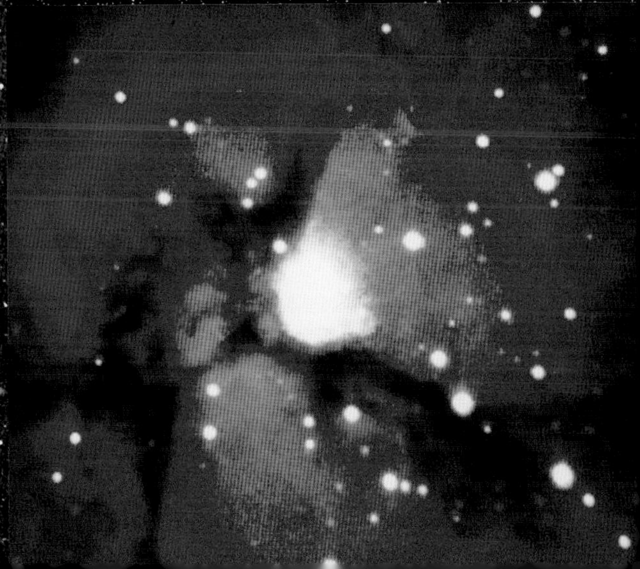

What is a star?

A star is a vast sphere of glowing, ionized gas, the majority of which is made up of the two most abundant elements in the Universe—hydrogen and helium. So massive are these gas balls that they are constantly in danger of shrinking under their own weight—a process called gravitational contraction. However, they remain balanced against this inward force because of the presence of nuclear reactions in their centers, the core. There, fierce pressures and temperatures fuse hydrogen nuclei together to make helium—exactly as occurs inside a thermonuclear bomb. These reactions generate a great outward pressure that exactly balances gravity and keeps the star shining for up to tens of billions of years. A star's color depends on how hot it is, and their sizes vary enormously. Our Sun, for example, (seen below in ultraviolet light) is a typical yellow star of moderate size and mass.

Stars are born in molecular clouds—giant patches of interstellar fog like the Eagle Nebula (*right*), comprised also of hydrogen and helium. Some parts of these gas clouds are denser than others, and in time these regions pull in gas from neighboring territories by virtue of their gravity, until they form a dense blob called a protostar. Later, as the protostar sucks in more material, it contracts, spins and heats up until the density and pressure in its core become so great that nuclear reactions ignite for the first time, and a new star lights up the heavens.

ARGE STARS

Iow does the Sun
neasure up against the
argest stars? Sirius A,
he brightest in the
night sky, is comparable
o the Sun in scale,
but it is hotter and
herefore blue. Rigel,
n Orion, is hotter still,
nuch larger and vastly
nore luminous—as
bright as 150,000 Suns!
3ut the largest type of
tar, a red supergiant,
dwarfs even this.
Antares in Scorpio is
ust one example.

SMALL STARS

While there are stars
bigger than the Sun,
o there are a great
number that are
smaller. In fact the
nost common stars, red
dwarfs, are quite a bit
smaller, cooler and less
nassive than the Sun.
Proxima Centauri, the
nearest star to Earth, is
an example. Smaller still
are white dwarfs, barely
arger than Earth, such
as Sirius B. But the
smallest are neutron
stars—too tiny to be
seen on the scale of
his diagram.

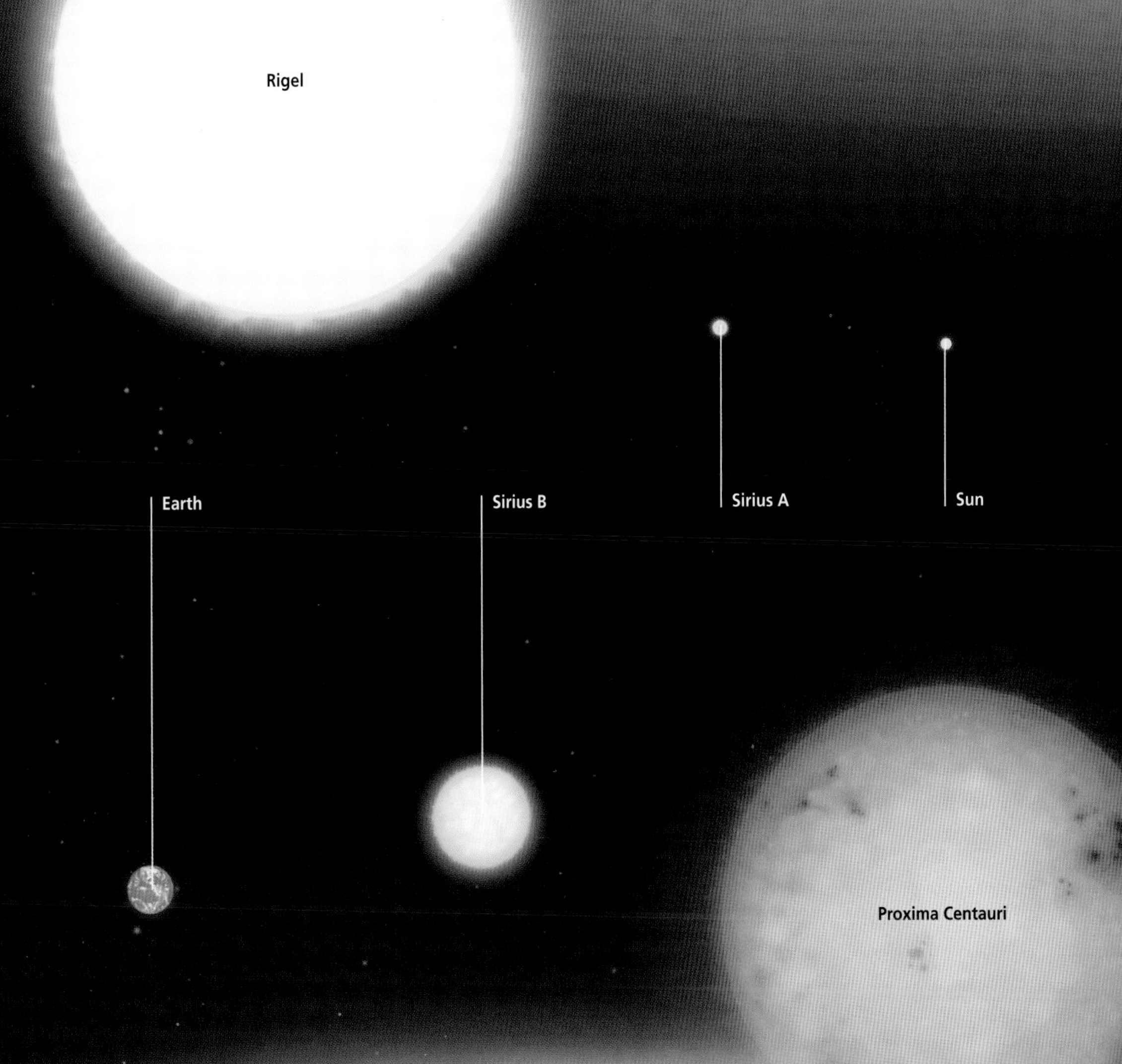

Lifecycle of a star

Stars are not alive but they have well-defined patterns of existence. All stars go through a long period on the main sequence. Main-sequence stars—also called dwarfs—all do exactly the same thing: they convert hydrogen in their cores into helium, to generate the outward pressure needed to keep them balanced against gravity. But this process cannot last indefinitely. As a star ages, it grows lower on reactionable fuel. Soon, its core is no longer balanced and it contracts slightly. This heats the core so reactions can return to normal. The energy pushes the star's outer layers outward and it becomes a subgiant—growing a few times larger. Later, when the star again runs low on fuel, the same process occurs, but this time the outermost layers expand by a much larger factor to form a red giant or supergiant. What happens then, though, depends on the mass of the star. Either a white dwarf, a neutron star, or a black hole will result.

↑ **The Butterfly star cluster** (M6), in the constellation Scorpius, is replete with main-sequence stars.

HERTZSPRUNG-RUSSELL DIAGRAM

The Hertzsprung-Russell diagram is named after the two astrophysicists who first used one, albeit independently, in the first decade of the twentieth century: Ejnar Hertzsprung and Henry Norris Russell. It is a plot of star color against luminosity. When they drew up the first such diagrams, Hertzsprung and Russell saw that the majority of stars formed a characteristic S-shaped line on the graph, running from top left (hot and bright) to bottom right (cool and dim). This is a graphical representation of the main sequence. All dwarf stars lie on this line, their exact position governed by their temperature and mass, which determine the luminosity. Giants, supergiants and white dwarfs all have their places on the diagram also. Often, spectral class rather than temperature is plotted along the x-axis. The spectral class determines a star's temperature and color. The main classes are, from hottest to coolest: O, B, A, F, G, K and M—which you can remember with the now woefully dated but still useful mnemonic, "Oh, Be A Fine Girl, Kiss Me."

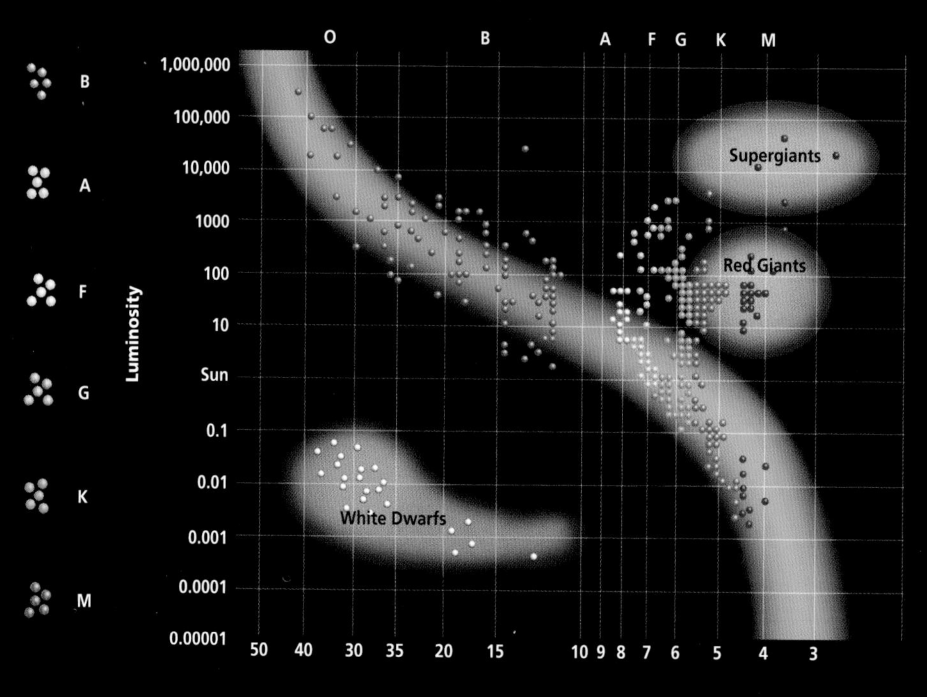

1

2

3

4a

4b

5

6

7

LIFECYCLE
This diagram charts the lifecycles of two separate types of star: a massive blue star (*left track*) and a yellow, Sunlike one (*right track*). Despite its larger mass, the blue star uses up its fuel very rapidly. This is because it has such an enormous central pressure under all that weight, that it must burn its nuclear fuel extremely vigorously in order to generate a sufficient internal pressure to avoid gravitational collapse. In perhaps a million years, this star runs out of hudrogen (1) and then expands into a huge and luminous red supergiant (2, not to scale in this diagram), before exploding as a supernova (3). After that, depending on the mass of what's left over, a black hole (4a) or a neutron star (4b) will result. The Sunlike star lives thousands of times longer—for billions of years. It expands to form a red giant rather than a supergiant (5, not to scale), puffs off its outer layers to become a brief planetary nebula (6), and leaves behind its former core—a shining ember known as a white dwarf (7).

Binaries, multiples and variables

Around half of the Milky Way's stars are binaries: two stars orbiting a common center of mass, invisibly tied together by a gravitational tether. Binary stars are distinct from double stars, in which the two components merely appear close together on the sky, but are physically separated. True binaries are common because of the way in which stars form: from an interstellar gas cloud contracting under gravity. These clouds contain so much mass that they frequently form twins—and even multiple systems. As the various components in a given binary or multiple system move around, one star can temporarily pass in front of the other as seen from Earth, varying the total light output we receive. Still other stars are intrinsically variable—they physically change the amount of light they emit over time.

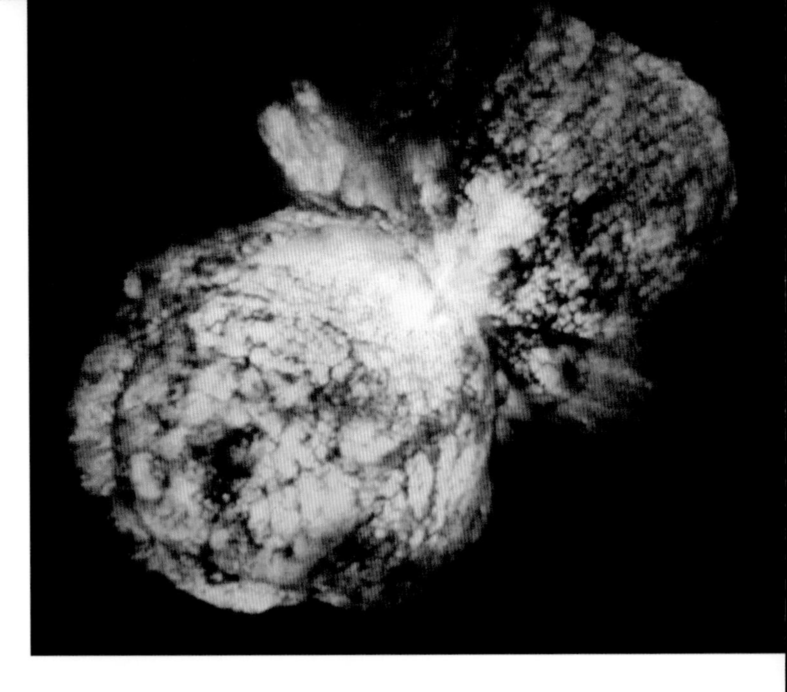

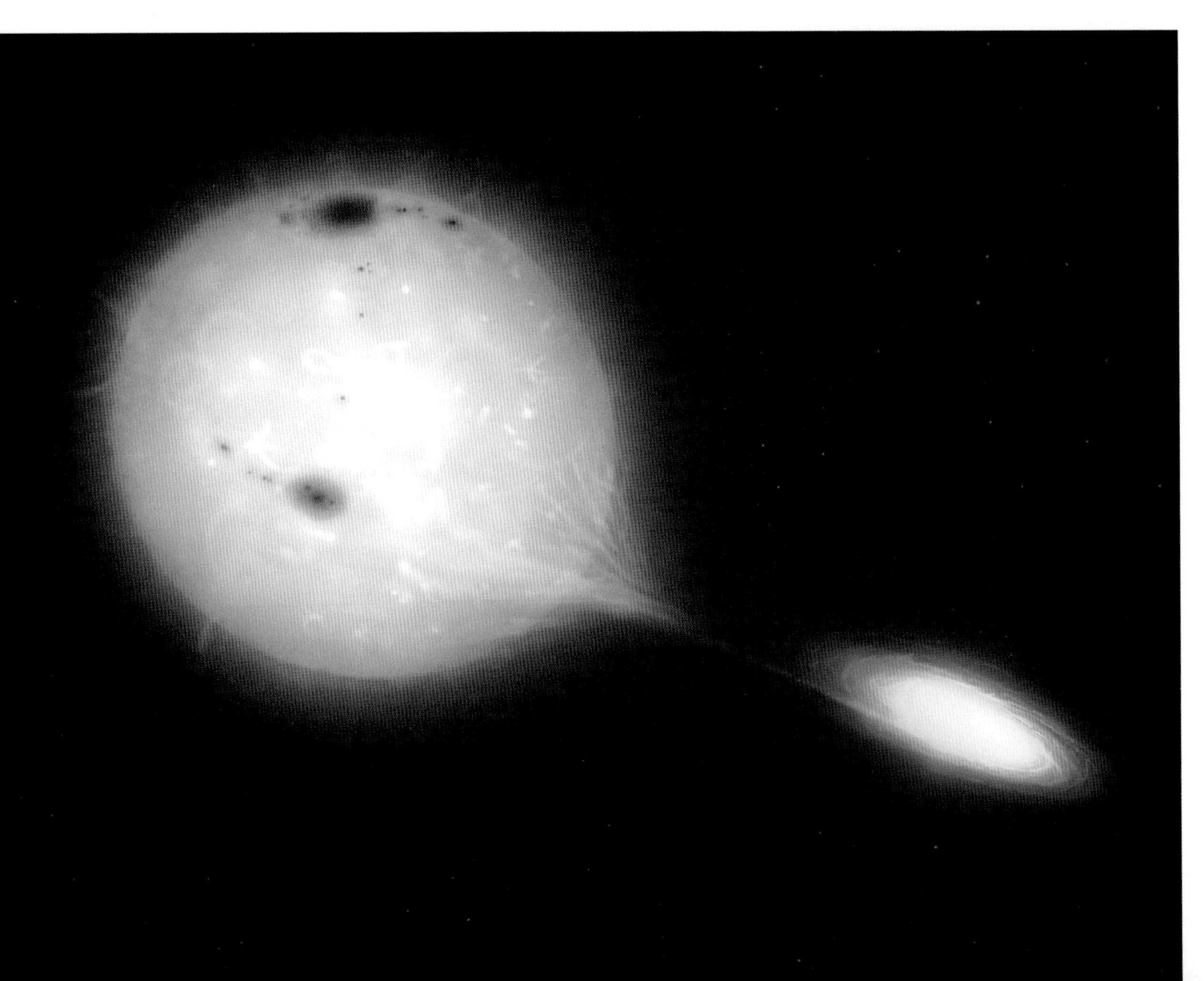

↑ **Eta Carinae is a hypergiant variable** star which, at a mass of 100 Suns, is the most massive known. In 1843 it flared up to become the second-brightest star in the sky, throwing off the shells of gas seen in this dramatic image.

← **This is an impression** of an interacting binary system in which a red dwarf star (*left*) is distorted by the gravity of its close companion, a white dwarf, typically a solar diameter away. The white dwarf is surrounded by a disk of gas stolen from its unlucky companion. In time, this material builds up, becomes unstable, and undergoes a thermonuclear explosion called a nova. It is because of this abrupt and dramatic change in brightness that these binaries are called cataclysmic variables.

BINARY AND MULTIPLE STARS

Binary stars come in many forms. The individual stars may be similar or wildly different, close together or far apart. Some binary components are so close together that one actually pulls gas from the other and slowly devours it. This exchange of mass is often explosive and can lead to dramatic stellar eruptions called novae and, sometimes, supernovae. In other close binaries, each star reaches out to touch the other, forming a so-called contact binary.

Astronomers estimate that as many as half of the known binary stars are actually multiple systems containing three or more members. Triple systems are the most common, and even some six-member systems are known.

VARIABLES

There are two general ways in which stars can vary their brightness: by mechanical means such as rotation, or intrinsically via some other mechanism. Eclipsing variables, for example, use mechanical means. As seen from Earth, the stars alternately cross in front of each other, and this modulates the total light output we receive, just as happens when the Moon passes in front of the Sun during an eclipse.

There are several different kinds of intrinsic variable stars, most of which vary by either pulsating in and out or erupting unexpectedly. Pulsating stars are old ones whose diameters—and therefore total light output—change with time as a consequence of stellar evolution. Because brightness depends on surface area—simply because a larger surface area can emit more light—pulsating stars vary their output as their outer layers throb radially in and out. In eruptive variables, by contrast, the variability is unpredictable, and arises because of flaring and other stellar activity.

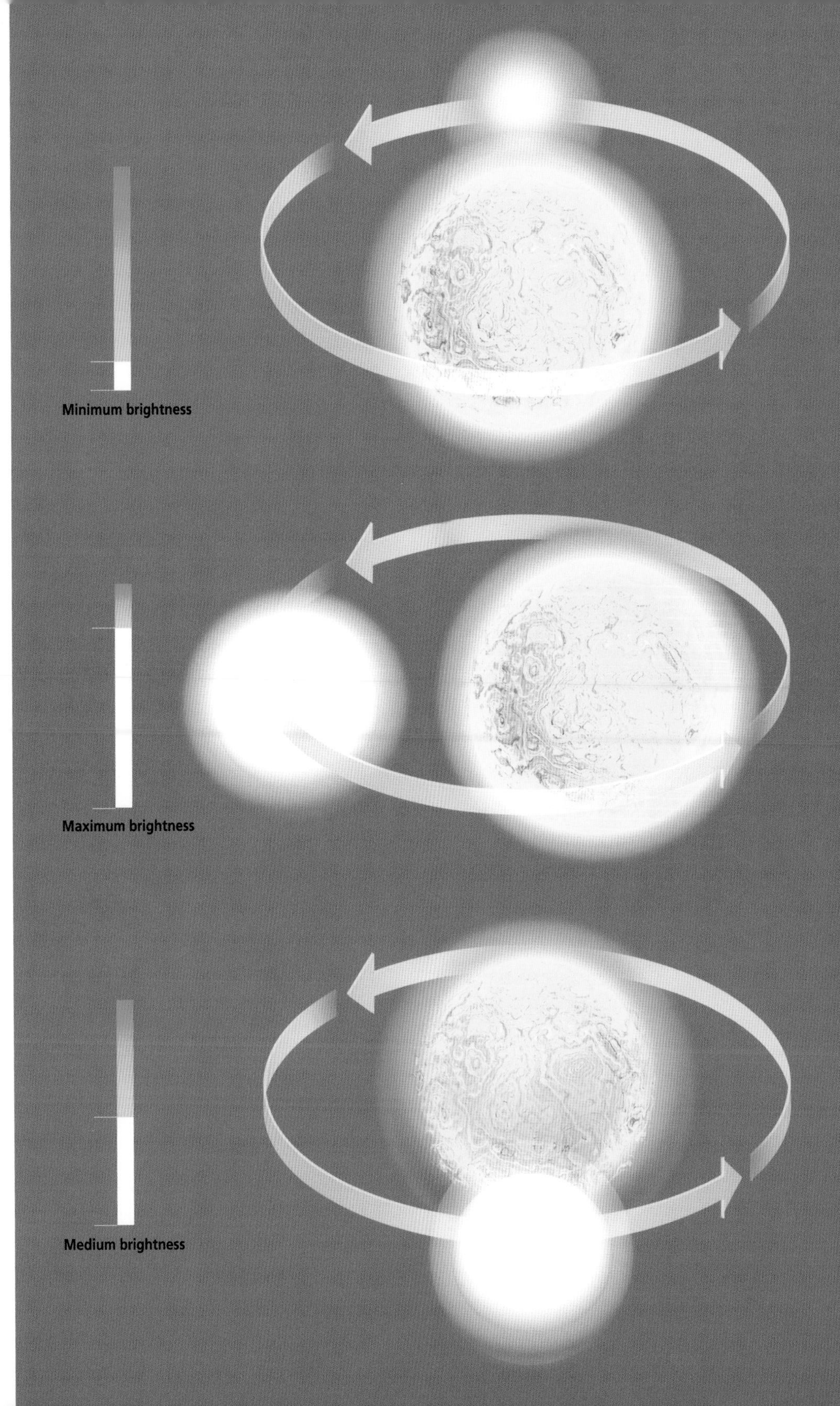

→ **In some binary stars,** such as the famous Algol system pictured here, one member eclipses the other every orbit, causing periodic dips in the overall system light intensity. In the top frame, the larger but cooler and dimmer star eclipses the brighter one, and the light drops down low. In the middle frame, both stars are visible and the light output is at a maximum. And in the last frame, the smaller star eclipses the cooler one, reducing the light just a little.

Star clusters

The Universe, it seems, likes multiplicity. While more than half of the stars are binary or multiple systems, there are larger groupings of stars that nature has to offer. These are known as globular and open clusters.

GLOBULAR CLUSTERS

The largest star clusters are called globular clusters. As their name suggests, these objects are roughly spherical. They span over 100 light-years and are home to 10,000 to greater than one million stars—comparable to the smallest, dwarf galaxies. In contrast to the young blue stars that colonize the much smaller open clusters (*see facing page*) globular clusters mainly contain old, and therefore red, stars. These ancient objects can only have formed alongside, or very nearby, their host galaxies themselves, billions of years ago. However, so densely packed are these clusters that stellar collisions are comparatively frequent. Such events lead to the occasional formation of new, blue stars, which astronomers have dubbed blue stragglers. Globulars do not inhabit spiral arms—and indeed they are often found in elliptical galaxies, which have no spiral structure. Instead, they surround their galaxies in vast swarms known as haloes, orbiting in highly elliptical paths. Our Milky Way has around 140 globular clusters in its halo. The most luminous of these is Omega Centauri (NGC 5139, *right*), and 47 Tucanae (NGC 104, *below*). Slightly farther afield, at 25,000 light-years, is M13 in Hercules (*bottom right*).

OPEN CLUSTERS
Sometimes, a whole group of a few dozen to a few hundred stars will form from a common gas cloud, all at once. These become open clusters or galactic clusters—loose groupings of stars strewn randomly over a region of space spanning not more than 50 light-years and bound together through gravity (M35, *left*, is a typical example). Open clusters are a beautiful sight, but they do not last long. Eventually, as they orbit the galaxy, their members become more and more dispersed owing to the gravitational tugs they receive from other stars they pass on their journey, and because of the rotation itself. It's rather like what happens when you pour coffee whitener into a cup of swirling coffee. Dispersion usually happens quickly, within a few hundred million years. Older open star clusters are therefore rare, and they all contain relatively young stars.

Star death: planetary nebulae

Even the shortest-lived stars have life spans far longer than we humans, with our fleeting existences, can ever fully appreciate. But stars are not eternal. As with living things, the end must eventually come—and they must face death. For stars, this happens when they exhaust the hydrogen in their cores and can no longer maintain thermonuclear reactions. Stars make their exit in one of two ways, depending simply on their mass. The really massive stars blow themselves apart in star-shattering explosions called supernovae. But the vast majority of stars are comparatively lightweight—including the Sun—and can look forward to more dignified endings. When their time comes, they transform themselves into beautiful sculptures of light and ionized gas called planetary nebulae—but not for long. Eventually, these disperse into space, leaving behind tiny stellar leftovers called white dwarfs, destined only to fade forever.

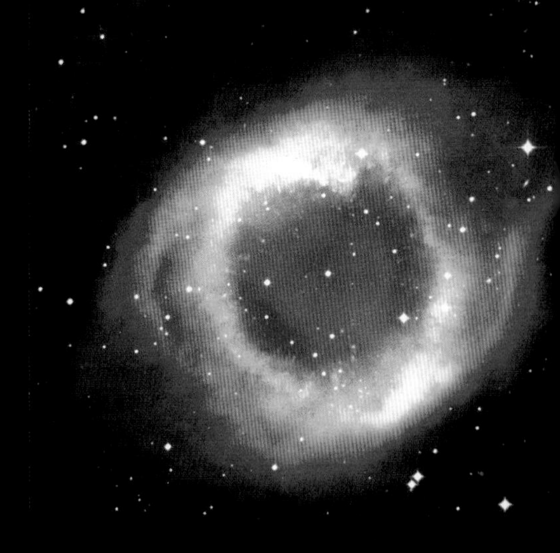

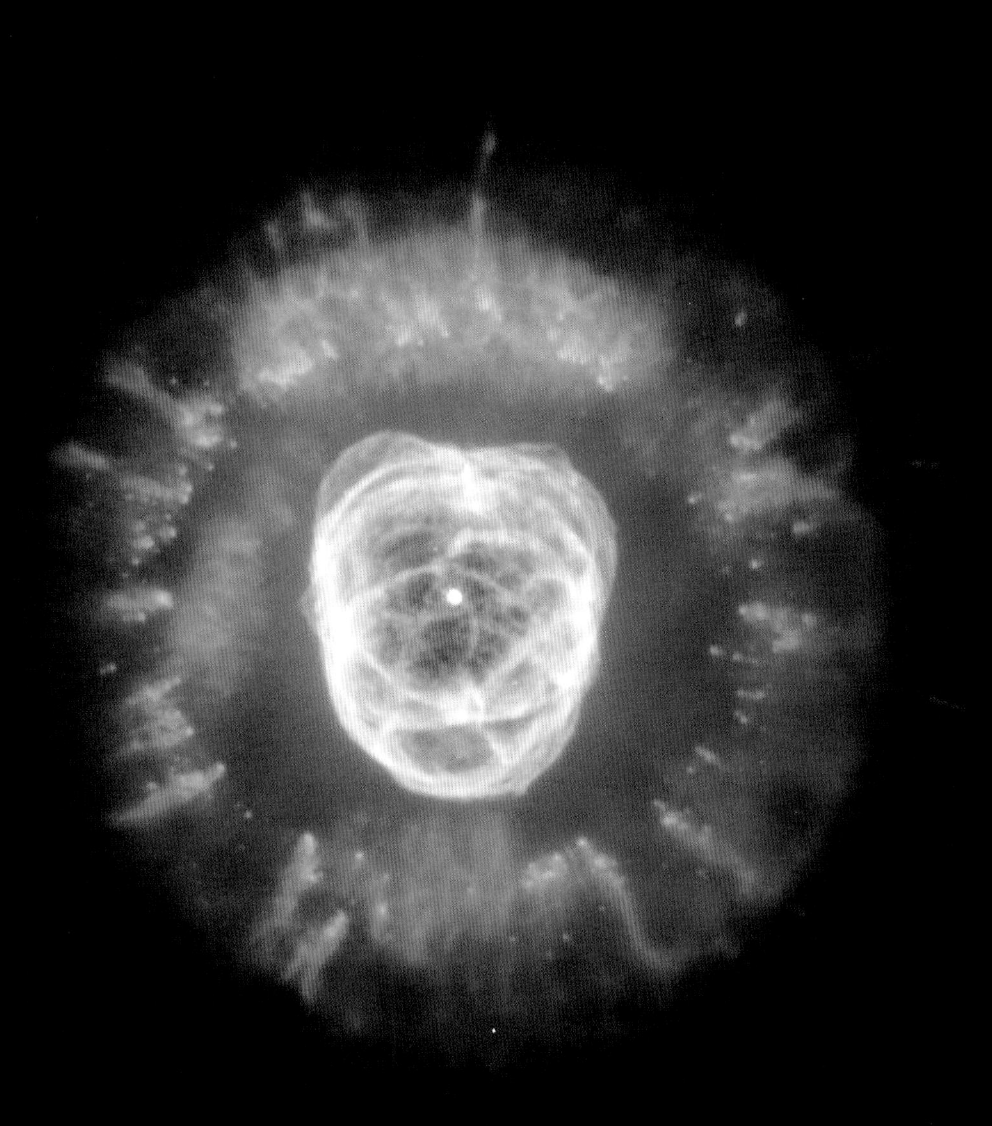

PLANETARY NEBULAE

For most of its life, a star holds itself in check against the crush of gravity by fusing hydrogen into helium in its core to generate a compensating outward pressure. Eventually, though, the hydrogen will run low and reactions begin to shut down. When a star reaches this point, its core has no choice but to contract. In doing so, the core grows hotter and denser, and its vibrant energy plows through the rest of the star—rapidly puffing it up like a giant balloon. If the star—now a red giant—is of low mass (less than three times as massive as the Sun) it will form a planetary nebula. This happens when the expanding star becomes so large that its weak gravity can no longer hold onto its outermost layers, and they slip away into space to form a surrounding shell of gas and made to glow by the now exposed core.

Planetary nebulae come in a striking variety of forms, with some so complex that astronomers do not understand how they form. Sadly, these beautiful objects do not last long. At maximum extent, the largest planetaries may span more than a light-year. But after around 10,000 years of expansion they mingle with the interstellar medium and gently fade.

↑ **The Helix Nebula**, NGC 7293, is the nearest planetary nebula to Earth, lying around 450 light-years away in the constellation Aquarius.

← **The Eskimo Nebula,** NGC 2392, is a complex planetary nebula in Gemini. This image reveals the material that gives the nebula the appearance of an Inuit's hood. The white blob at the center, over-exposed, is the central white dwarf.

→ **Intense colors** of the planetary nebula IC 4406 are well revealed in this Hubble image. Similar to other planetaries, IC 4406 exhibits a symmetry of shape.

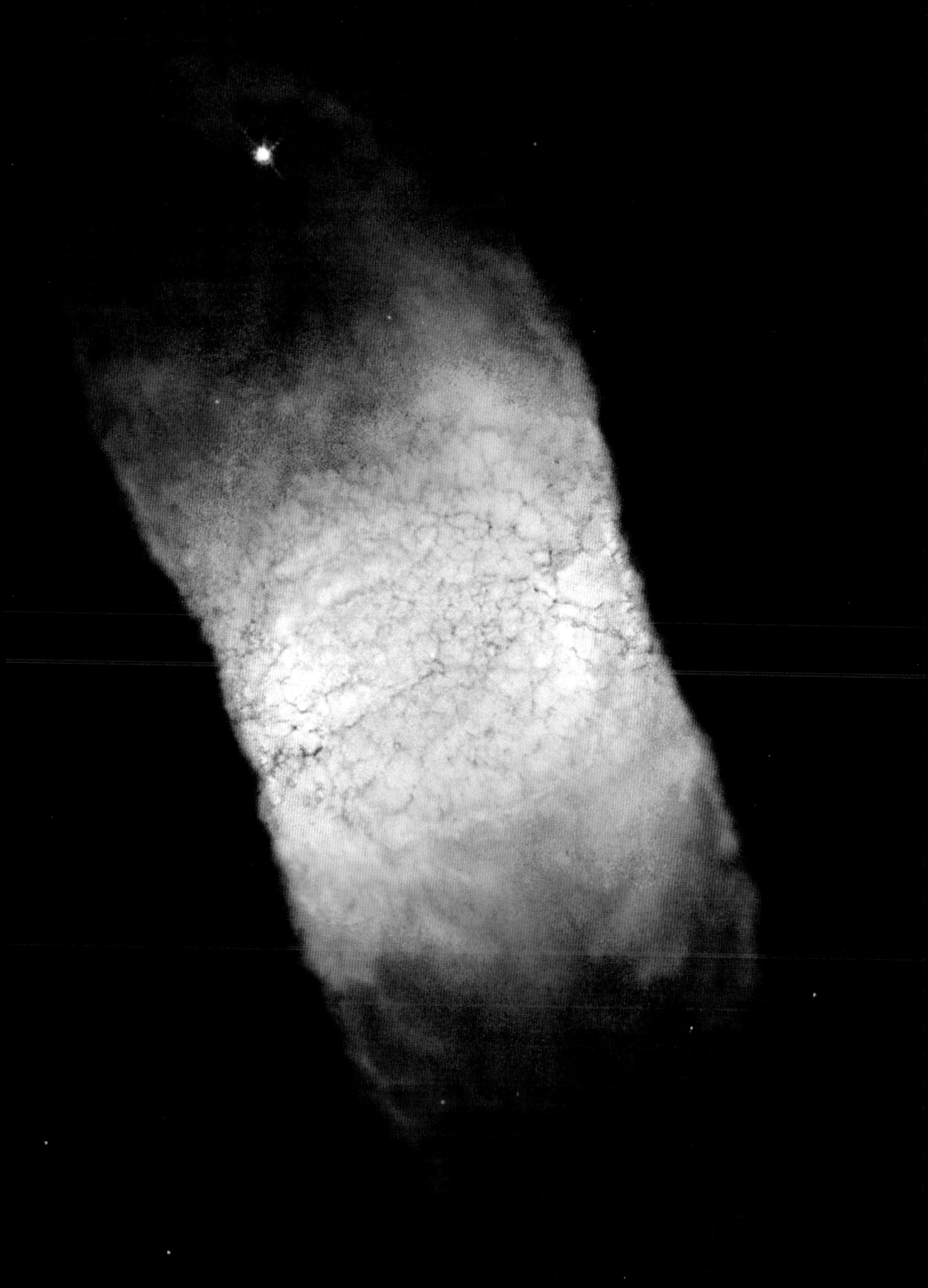

WHITE DWARFS

A planetary nebula does not mark the end of a star's life. Even after the nebula has dispersed, the exposed core of the original star remains. Astronomers call these exotic objects white dwarfs. These stars are exceedingly dense, slowly compacted under gravity over hundreds of millions of years while the former star was consuming its last dregs of fuel. A fully compressed white dwarf, with half the mass of the Sun, is only 50 percent larger than Earth—the original star is squeezed into a volume one-millionth of its former size.

White dwarfs do not shine through nuclear reactions, since they are inert. Instead, they shine by releasing the energy stored inside them while they were still active. The immense density ensures that this energy release takes a tremendously long time, during which the white dwarf gradually cools and fades. It can take tens of billions of years—far longer than the current age of the Universe—for a white dwarf to completely cool down. This is the ultimate destiny of any low-mass star: a dark stellar corpse the size of an average planet, known to astronomers as a black dwarf.

Star death: supernovae

There is no stellar death more spectacular than a supernova. This is the gigantic explosion that signals the end of a very massive star. Like planetary nebulae, supernovae also leave behind a stellar corpse—neutron stars, or sometimes black holes. But while white dwarfs are very dense, neutron stars take compression to unprecedented levels. Astronomers have identified two broad classes of supernova: type I and II, each with further subclassifications. Type I events always occur in binary systems in which a white dwarf draws matter from a nearby companion star. The stolen material accumulates on the surface of the white dwarf, until the additional weight becomes so great that the luckless star suddenly buckles inwardly and collapses. The collapse continues until the star cannot become any denser, and abruptly rebounds. This recoil sends out a gigantic series of shockwaves, blowing the star to pieces, and is what we witness as the actual supernova. Type I events are so powerful that they often involve the total annihilation of the original white dwarf star—nothing remains. Sometimes, though, a tiny neutron star is left behind. Unlike type I events, type II explosions do not involve white dwarfs, but occur when a star with initially more than about eight times the mass of the Sun runs out of nuclear fuel. Moreover, type II supernovae always leave behind a neutron star—or a black hole.

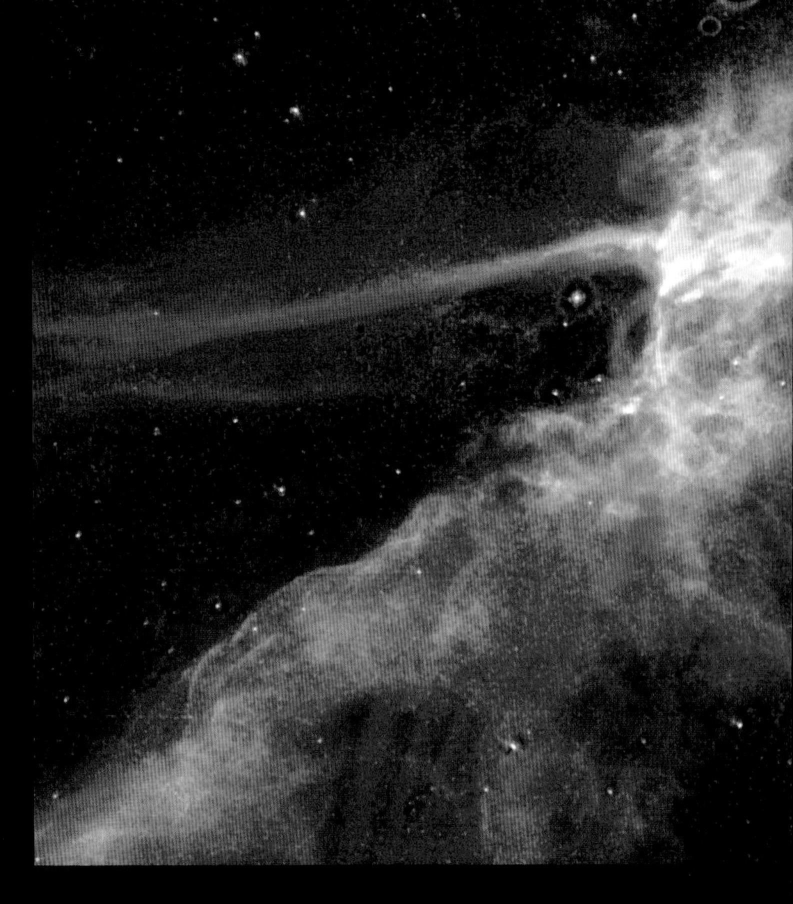

NEUTRON STARS

Neutron stars are the densest large-scale objects in the known Universe. Though they may measure a mere 20 miles (32 km) across, they pack in as much material as the entire Solar System. Put another way, it's like sticking the entire human race of six billion people into a sugar cube. Neutron stars are so called because they are made almost completely of subatomic and electrically neutral particles called neutrons. Normally, atoms contain neutrons (and protons) in their nuclei, with electrons orbiting the nucleus. But when neutron stars form, the compression is so great that matter literally breaks down. Electrons and protons combine to form more neutrons, and the star becomes a gigantic ball of fused subatomic particles. The compression has a side effect too—it massively magnifies any magnetic field the star might have had. Normal neutron stars have magnetic field strengths up to 50 billion times more powerful than a domestic fridge magnet. Occasionally, even more powerful fields form that are up to 1000 times stronger still. These super-magnetic neutron stars are known as magnetars. If you placed a magnetar where the Moon is, it would render all credit cards on Earth useless.

NEUTRON STAR AND PULSAR

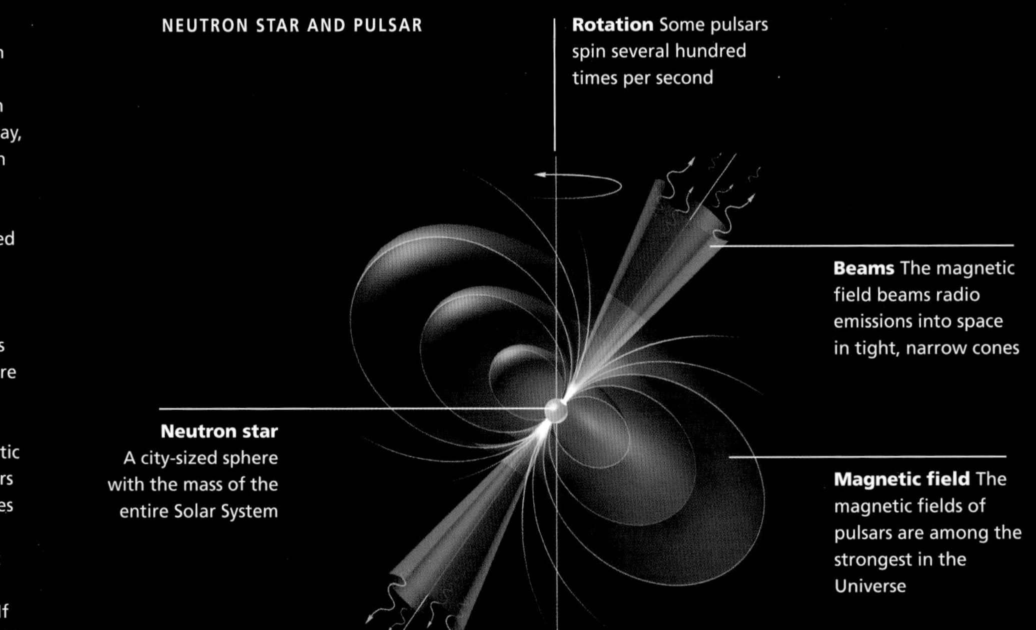

Rotation Some pulsars spin several hundred times per second

Beams The magnetic field beams radio emissions into space in tight, narrow cones

Neutron star A city-sized sphere with the mass of the entire Solar System

Magnetic field The magnetic fields of pulsars are among the strongest in the Universe

On February 24, 1987, an exploding star in the Large Magellanic Cloud gave modern astronomers their first-ever close-up view of a supernova. Called Supernova 1987A, it reached magnitude 2.8, making it the first naked-eye supernova since 1604.

The Cygnus Loop supernova remnant marks the edge of a blast wave from a colossal stellar explosion occurring some 15,000 years ago.

The Crab Nebula is a famous remnant of a supernova explosion that was observed and documented by Chinese astronomers in 1054.

PULSARS

Despite their small size, certain neutron stars are conspicuous even across thousands of light-years. This is partly because of their extreme magnetism. The magnetism accelerates charged particles in the "atmosphere" of the neutron star—a thin layer of particles ½ inch (1 cm) or so high—and as a result they emit radio waves. This in itself is not enough to render the object observable, because the emission is emitted only in a very narrow cone, which may not be directed at Earth. However, in addition, neutron stars rotate very quickly—some as fast as several hundred times every second. As they spin, they flash radio waves across the Universe like cosmic lighthouses. If Earth lies within the sweep of the beam, astronomers can detect the emission as a periodic series of radio bursts, modulated at the spin period of the neutron star. These objects, discovered in 1967 and confirming the existence of neutron stars, are called pulsars. They are the most stable clocks in the known Universe. In time, though, pulsars do slow down. The radio-emission mechanism ceases to function—since it requires rapid rotation—and the pulsar becomes inert. Our galaxy contains approximately 100,000 of these exotic objects.

Star death: black holes

Throw a ball up and it will always fall back down. Hurl something fast enough, though, and it will attain "escape velocity" and leave the planet altogether. You need to provide even more thrust to get the object into deep space. The higher the gravity, the faster the speed you need to overcome it. But what happens if the gravity is so strong that nothing can ever escape, no matter how fast you launch it? What happens when even light, at a speed nothing can ever exceed, is pulled back on itself? This is the basic concept of probably the most exotic known astronomical phenomenon—a black hole.

BLACK HOLE THEORY

When a very massive star dies, the neutron star left over is so massive that it cannot remain balanced against its own gravity. The neutron star shatters. It shrinks to a single point. Its gravity becomes so magnified that it sucks in everything—even light. The object becomes a black hole. Black holes are strange objects indeed. They have no material form—only their gravity remains. Everything else has been compressed to a point of undefined mathematical size at the center of the hole called the singularity. There, the laws of physics fail to operate. Gravity is essentially infinite. As you move away, the gravity falls off until, at a certain radius, the escape velocity drops below that of light. Within this radius, no signal emitted during any event can ever leave. For this reason, this critical boundary is known as the "event horizon." Up close against a bright background, a black hole would appear as an utterly featureless black disk the size of its event horizon, surrounded by massively distorted images of the background.

↑ **The core of M87,** a giant elliptical galaxy, appears in this image taken by the Hubble Space Telescope. The extended object is a jet of charged particles emitted from a disk surrounding a suspected black hole of 2.6 billion solar masses. The white dot marks the very core, where the density of stars is 300 times higher than normal, owing to the attraction of the black hole.

← **Black holes may emit no light,** but there are ways astronomers can spot them. About 8000 light-years away in the constellation Cygnus lies a bright source of X-rays called Cygnus X-1. A telescope reveals only a single star—a blue giant. However, the star is moving in a strange way, as if something invisible, weighing 10 to 15 solar masses, is orbiting it. To be so massive and yet go undetected, this companion object can only be a black hole. The X-rays are emitted not by the hole itself, but by an accretion disk of gas that surrounds it, stolen from the blue giant. Cygnus X-1 is just one example of many suspected binary black hole systems.

→ **In a comparatively weak gravitational field,** but where the gravity is still strong enough to substantially bend light (1), all of the light rays escape from the object but follow a curved path. This is the case with, for example, a neutron star. In stronger gravitational fields (2), only light rays within the region known as the "exit cone" are able to escape. The remaining light is caught in the object's gravitational field. But when the object is a black hole (3), its gravitational force is so strong that the exit cone has shrunk to zero and no light is able to escape.

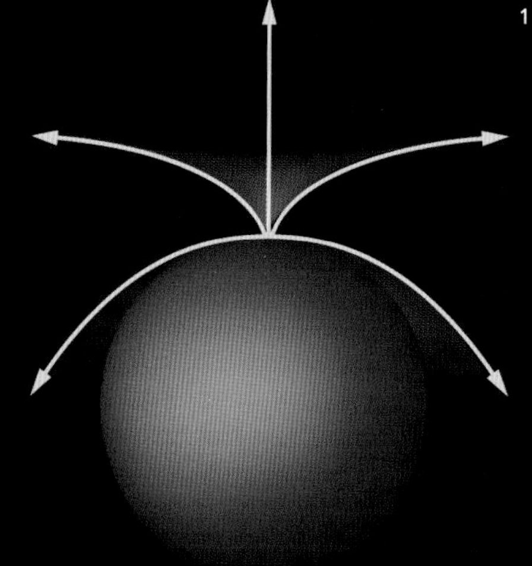

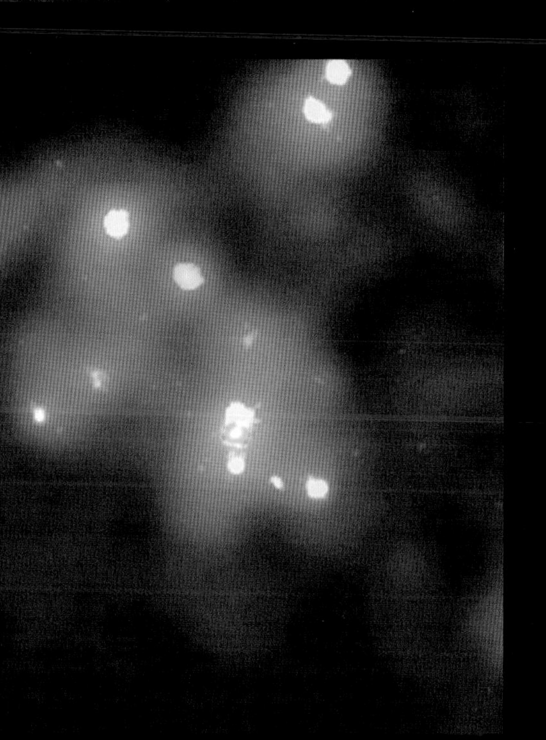

SUPERMASSIVE BLACK HOLES

Some black holes are very massive indeed. There is good evidence to suggest that the centers of many large galaxies, perhaps all, are the hunting grounds of supergiant black holes that feed on entire stars. These objects, billions of times more massive than stars, are called supermassive black holes. Active galaxies, such as quasars, emit so much radiation from a region only the size of the Solar System, that they can only be powered by black holes. The idea is that, as with stellar black holes, bright accretion disks surround the supermassive ones too. These disks are the remains of star systems torn apart by gravity. An example is the nearest large galaxy, Andromeda, whose core is shown in close-up in the X-ray image at left. The blue object is a comparatively small supermassive black hole with an estimated mass of "only" 30 million Suns. The yellow blobs, meanwhile, are X-binary stars like Cygnus X-1. Like Andromeda, our Milky Way almost certainly has a supermassive black hole in its heart, because astronomers have measured that the stars there move as if they are attracted to something very massive which remains unseen.

Galaxy classification

Galaxies are truly vast: incredible islands of stars that even light can take hundreds of thousands of years to cross. But not all galaxies are so large, and there are a great number of other differences between them aside from scale. Astronomers have identified two broad types of galaxy: elliptical (class E) and disk (class D). The flat, disk galaxies are often subdivided further into spirals (class Sa to Sc), barred spirals (class SBa to SBc) and lenticular galaxies (class S0). Meanwhile, galaxies that fit none of these classes are dubbed "irregular" (type I and type II), and very small galaxies in each class are called dwarf ellipticals (dE), dwarf irregulars (dIrr), and so on. Of all galaxies, the most massive are the ellipticals. Spirals and lenticulars, by contrast, although having diameters comparable to those of ellipticals—some larger than 100,000 light-years across—are not so massive because they are flat, shaped like disks, while elliptical galaxies have much greater volumes. Elliptical galaxies also tend to be the most common, representing about 60 percent of all galaxies that astronomers have counted. Disk galaxies comprise 30 percent, and the irregulars make up the final tenth. Clockwise from top-left on the opposite page are examples of spiral, elliptical, irregular and lenticular galaxies.

EDWIN HUBBLE'S TUNING FORK

The American astronomer Edwin Hubble (1889–1953) originally thought that galaxies followed an evolutionary path from one type to another. To illustrate his concept, he devised the so-called Hubble Diagram (*below*). Hubble imagined that galaxies started off spherical and gradually became more elliptical (E), a trend he illustrated on the single arm of the diagram. The very elliptical galaxies gradually flattened and became lenticular (S0). Then the path split, with some galaxies forming spirals (S) and others becoming barred spirals (SB), the spiral arms in both cases opening with age. In fact, we now know that galaxies do not follow this pattern of evolution, but rather evolve through interactions with their neighbors.

HUBBLE DIAGRAM

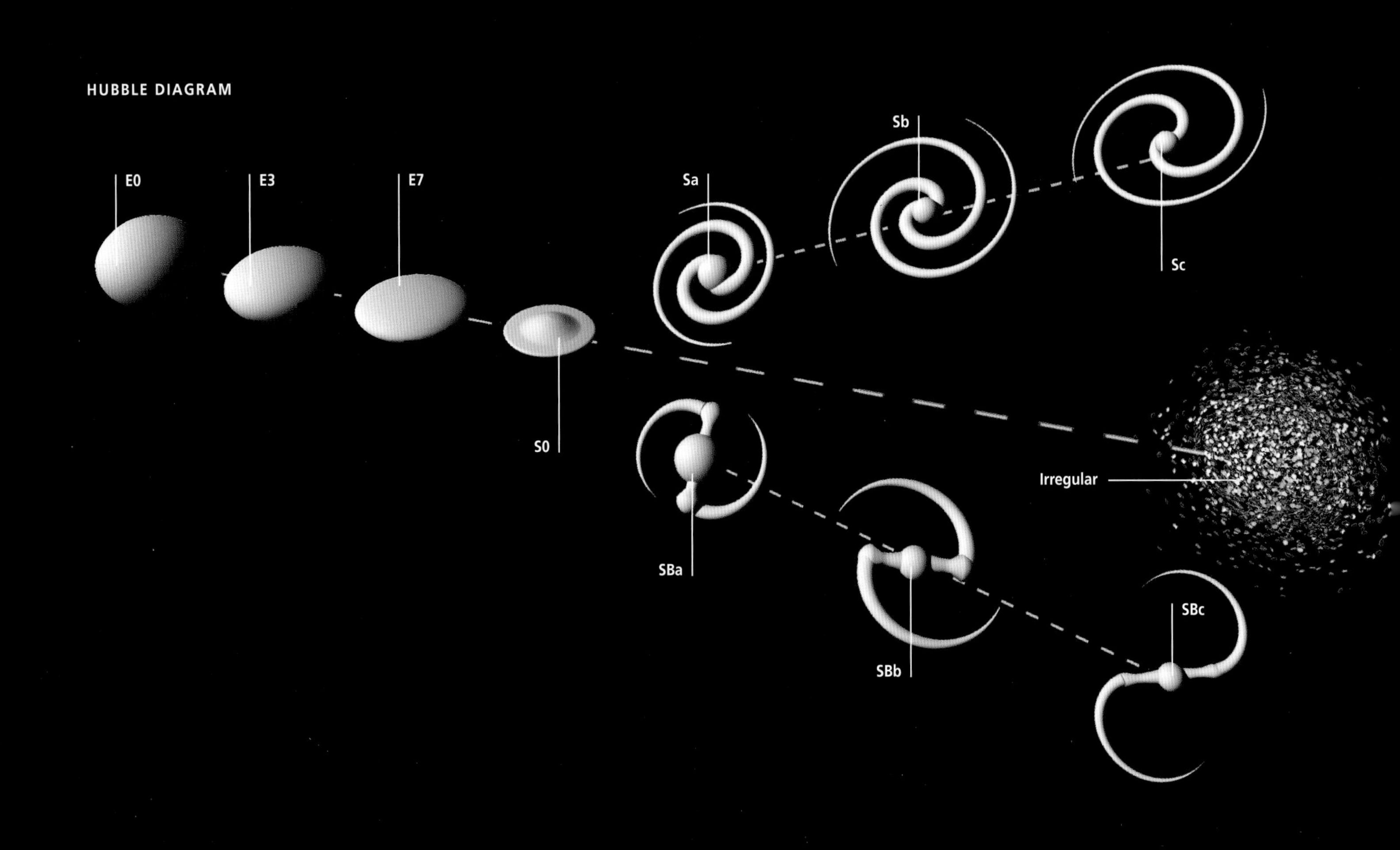

Spiral galaxies

Spirals are the quintessential galaxies, even if they are not the most common. They fall into two general types: normal spirals and barred spirals. In barred spirals, a straight, thick sheet of stars is seen going right through the nucleus, with spiral arms emerging from the ends. In normal spirals, the arms start just on the outside of the nucleus. Other than that, though, both classes are similar. Spirals usually have two arms, but some can have three or four. All are very flat, with a prominent nucleus or central bulge made of old, red stars. The disk, meanwhile, roughly 10 times the diameter of the nucleus, is home mainly to blue stars, and also vast drifts of gas and dust. Because spiral galaxies have so much interstellar material, they exhibit a great deal of star formation, which is why many young (blue) stars populate the spiral arms in the disk. Lastly, both classes of spirals are surrounded by a so-called halo—a roughly spherical swarm of globular clusters, each housing up to a million old, red stars.

In spiral galaxies, shockwaves related to the rotation of the galaxy are constantly passing though the disk. As they spread through the disk, these shocks—known as "density waves"—compress interstellar gas and dust, triggering star formation there and lighting up a spiral pattern made of the new stars left in their wake.

→ **The barred spiral galaxy** NGC 1300 lies 75 million light-years away in Eridanus. The arms are blue with the light of new, young stars, while the bulge and bar are home to older, redder stars.

↓ **M74 is a large spiral galaxy** with wide-open spiral arms, in the constellation Pisces. At about 95 million light-years across, M74 is about the same size as our own Milky Way Galaxy. This image shows the galaxy in false-color.

↓ **The Whirlpool Galaxy, M51,** is a classic spiral seen entirely face-on, and one of the first-ever spiral galaxies identified. This photo shows a close-up of the central regions, as seen by the Hubble Space Telescope.

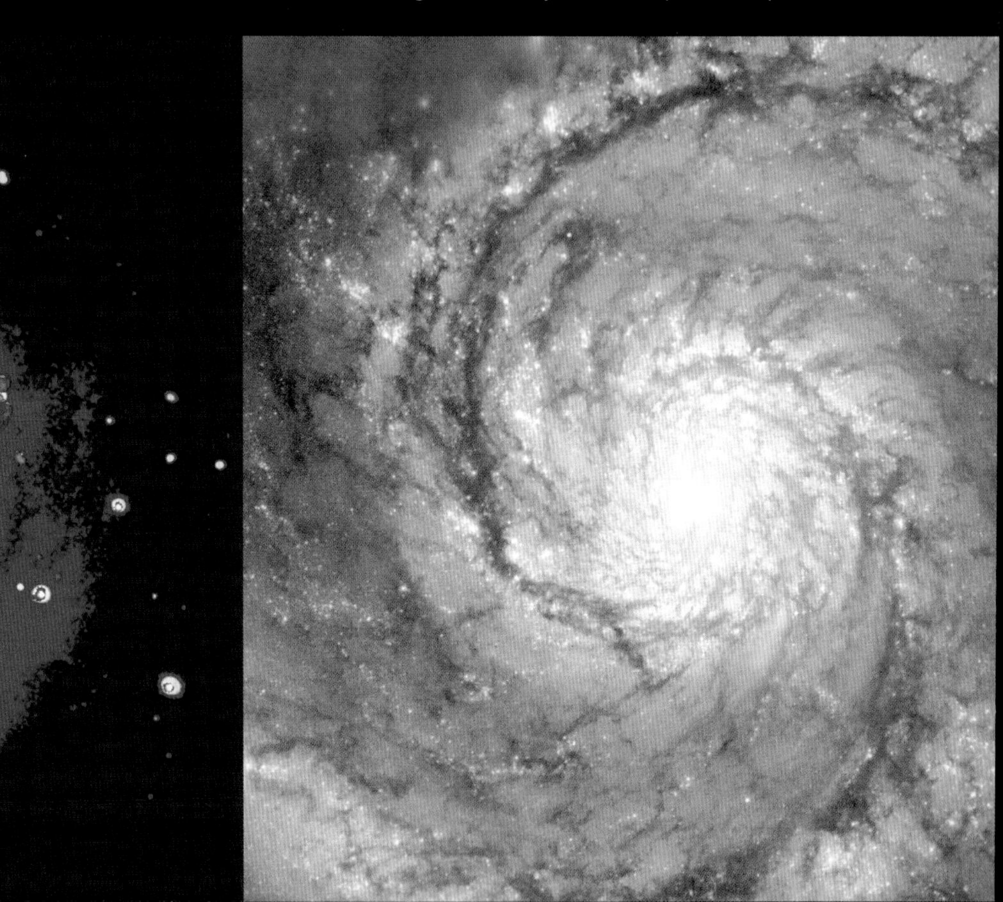

Elliptical galaxies

Ellipticals account for well over half of all galaxies. On the sky, they appear as slightly fuzzy oval blobs, lacking the majesty of their spiral cousins. In reality though, they are more complex. Try to imagine how a rugby ball would look if you squashed it or stretched it. Elliptical galaxies are similar. They are not shaped like proper rugby balls, for while rugby balls have only two unique axes—the length and the diameter—elliptical galaxies have a length, a breadth and a width. Even apparently spherical ones may in reality be long and narrow, seen end-on. However, since a two-dimensional view is all that astronomers have to go on, they classify these galaxies according to how elongated they appear on the sky. Round ellipticals are classed E0, with larger numbers assigned depending on the disparity from circularity, up to E7 for very elliptical.

→ **The giant elliptical galaxy M87** as it appears at optical wavelengths, taken by the Anglo–Australian Observatory. The surrounding large blobs are neighboring galaxies, while the much smaller, starlike points are M87's globular clusters.

↓ **This is an X-ray photo,** taken by the Rosat X-ray telescope, of a pair of galaxies in the Virgo galaxy cluster: M86 (top right) and the much larger elliptical M87 (center). M87 is a typical giant elliptical galaxy with an incredible mass of at least a trillion solar masses, and possibly more.

↓ **The Hubble Space Telescope** took this photograph of M32, one of the satellite galaxies of the famous Andromeda Galaxy and, along with the Milky Way, a member of the Local Group cluster. M32 is a small elliptical galaxy in which there is strong evidence of a central black hole.

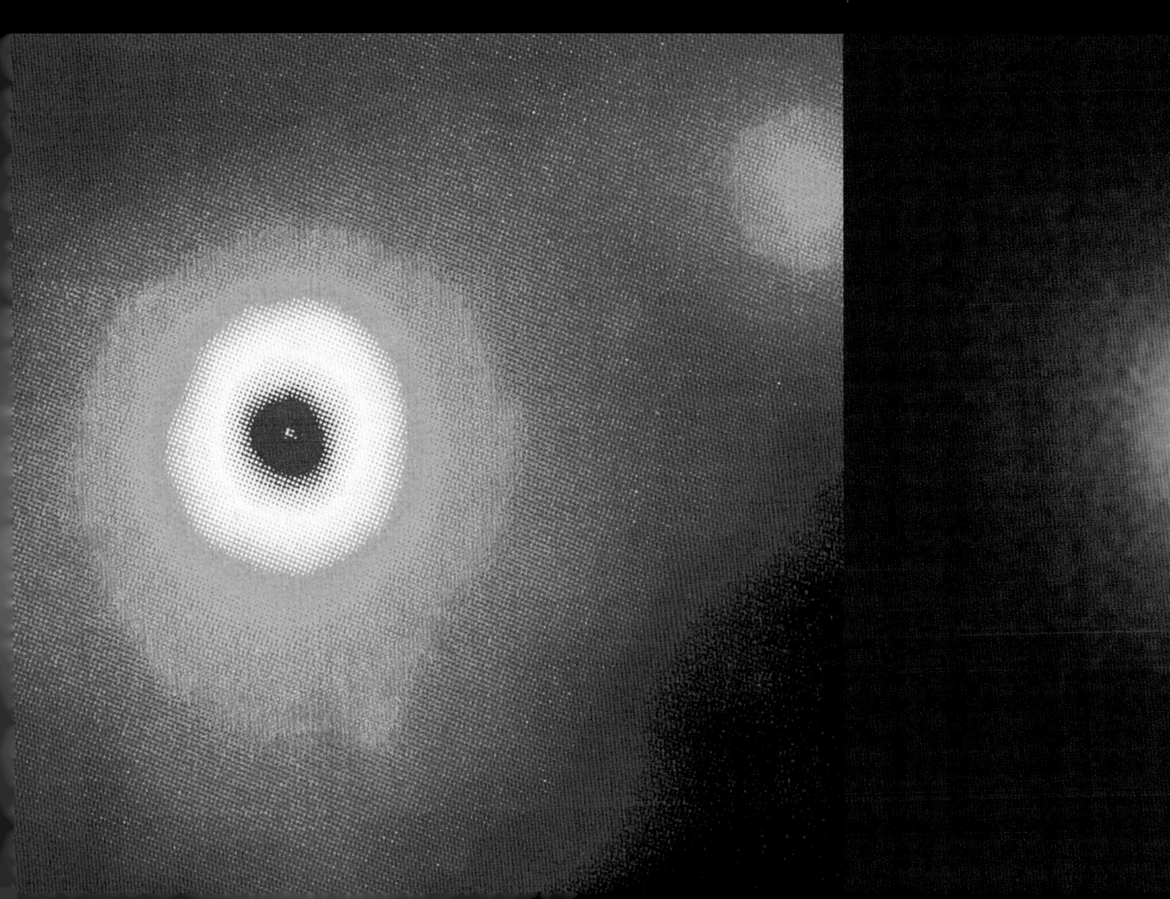

Lenticular galaxies

Lenticular galaxies, classed S0, get their name from their shape, somewhat like a convex lens seen edge-on. These galaxies are hybrids, sharing features of both spirals and ellipticals. Like the splendid spirals, lenticulars have a disk with a prominent central bulge, while some contain clouds of interstellar gas and dust. But they differ from spirals in several ways. Notably, they lack any signs of a spiral structure. The shockwaves found in spiral galaxies, which are responsible for their structures, are absent in lenticulars. Thus there is no spiral structure and no young blue stars—only old, cool, red stars. This is a trait that the lenticulars share with the ellipticals. Additionally, the central bulges in lenticular galaxies are much larger than in true spiral galaxies. While in spirals the nuclear region is typically one-tenth of the disk diameter, in lenticular galaxies the disk and bulge components are of a comparable size. This, and the fact that like ellipticals they contain only old stars, makes some lenticulars look just like elliptical galaxies—indeed astronomers often misclassify them as ellipticals. However, they are not elliptical galaxies either, because these have no disks and contain no interstellar material.

→ **The Hubble Space Telescope** took this image, which shows the lenticular galaxy NGC 2787 in the constellation Ursa Major. The brown rings are lanes of dust, and the bright patches on the outskirts of the image are the galaxy's globular clusters.

↓ **NGC 5866, sometimes called the Spindle Galaxy,** is a lenticular galaxy located 40 million light-years away in Draco. The galaxy is seen almost edge on, with the dark lane of dust that marks its disk neatly bisecting it.

↓ **M86 is an example of a galaxy** that astronomers are not sure how to classify. It looks like a lenticular galaxy, but it may also be elliptical, class E3. M86 is a member of the giant Virgo galaxy cluster.

Irregular galaxies

Irregular galaxies are oddballs. These comprise the 10 percent of galaxies that fit into none of the other major classes. Despite the name, though, not all irregular galaxies are without some sort of structure. Astronomers identify two types of irregular galaxy, called type I and type II. Type I irregulars actually share several things in common with spiral galaxies. Spirals, to recap, are flat with a high gas content and therefore a great deal of star formation taking place. Type I irregulars have these features too. But at the same time, they have no discernible spiral structure; they have even more interstellar material than spirals; and they are not symmetrical. By contrast, type II irregulars are much more misshapen, more deserving of their name. Again they contain large drifts of interstellar material, but there is virtually no sign of any sort of ordered structure—they range greatly in size and shape.

→ **NGC 55** is a nearby irregular galaxy in the Local Group galaxy cluster to which our own Milky Way Galaxy belongs. Located 3 million light-years away in Sculptor, NGC 55 is very similar to a spiral, but its central bulge is located off-center.

↓ **M82, also known as the Cigar Galaxy,** lies some 12 million light-years away in Ursa Major. M82 is a prototypical irregular galaxy with a disk. It has been distorted because of the presence of a neighboring galaxy, M81.

↓ **Here again is M82,** as seen through the X-ray eyes of the Chandra satellite. The bright spots are star-forming regions, triggered by the interaction with nearby M81. The diffuse red cloud is high-temperature gas escaping the core.

Active and radio galaxies

So far the galaxies we have seen can all be considered "normal." The light—more properly known as electromagnetic radiation—that they emit is just what one would expect from a vast assemblage of stars. Taken as a whole, their emissions look stellar in origin. But some of these galaxies produce a much wider range or spectrum of electromagnetic radiation—not just optical, but radio, infrared, ultraviolet light and X-rays. These are known as active galaxies, because the vast majority of their emissions come from violent activity that is definitely non-stellar in origin. This vibrant activity is almost always localized to, or at least generated within, a few light-years of the galaxy's center, the core. For many years, astronomers puzzled over what could be blasting out so much energy, and the debate continues. But many now think that the culprit is a supermassive black hole. They cannot see the holes, but no other object is massive enough—up to as much as a billion stars—and small enough to create such incredible luminosities. The energy comes not from the hole itself, but from the shredded remains of stars that are heated up just before being swallowed.

↓ **A superfast jet** of charged subatomic particles spews from the center of a vast disk of gas, in this artist's impression. Such disks are thought to surround the supermassive black holes that reside at the hearts of active galaxies.

ACTIVE GALAXIES

Active galaxies are not something totally separate from regular galaxies. They are simply those galaxies, such as spirals or ellipticals, which are exceptionally luminous because they contain what astronomers call an active galactic nucleus (AGN). Active galaxies come in several diverse varieties. Radio galaxies form the largest group, while others include Seyfert galaxies, quasars and BL Lacertae galaxies. Some types of activity are almost always associated with one particular galaxy class. For example, with Seyfert activity, the underlying galaxy is almost always a spiral. But Seyfert galaxies are atypical spirals with very bright nuclei that from afar resemble stars.

RADIO GALAXIES

Radio galaxies come in two types called compact and extended sources. In compact sources, the radio emission is concentrated in the galaxy core. In extended sources, additional radio waves come from vast areas of extragalactic space that surround and dwarf the actual galaxy. Extended emission arises when radio galaxies spew jets of charged particles from their cores—the process is not well understood but it is driven by the black hole. As the jets pour into extragalactic space, they collide with the very tenuous extragalactic gas that pervades the space between all galaxies, known as the intracluster medium. Like air inflating a balloon, the jets pump up the extragalactic gas to form vast lobelike structures, rich in radio emission, often spanning millions of light-years. Sometimes, pairs of radio lobes are seen forming a U-shape, with the source galaxy situated at the bottom of the "U." Astronomers attribute this to the motion of the galaxy through the intracluster gas, sweeping back the lobes as the galaxy moves through it.

→ **This false color radio image** shows a typical pair of giant radio lobes. The central galaxy, Cygnus A, is not seen. But twin jets of charged particles emitted by the galaxy have collided with the extragalactic gas in which the galaxy is embedded, to create the two radio hotspots, vastly bigger than the galaxy itself.

← **The giant elliptical galaxy M82,** also known as Virgo A, as it appears in the radio part of the spectrum. The extended object from the center diagonally downward to the right is a jet of charged particles spewing from a suspected black hole at the center of the galaxy.

↓ **Centaurus A is a powerful** radio and X-ray source about 15 million light-years away. A typical radio galaxy, Centaurus A has a pair of radio jets, each about 10,000 light-years long and feeding a vast radio lobe. The largest of the two lobes spans an incredible 1.5 million light-years. Just as Seyfert activity is usually associated with spirals, radio galaxies are typically giant ellipticals, as is Centaurus A. However, there is a conspicuous lane of dust across the middle of this galaxy—very unusual for ellipticals—so astronomers think that Centaurus A is actually the result of a two-galaxy collision.

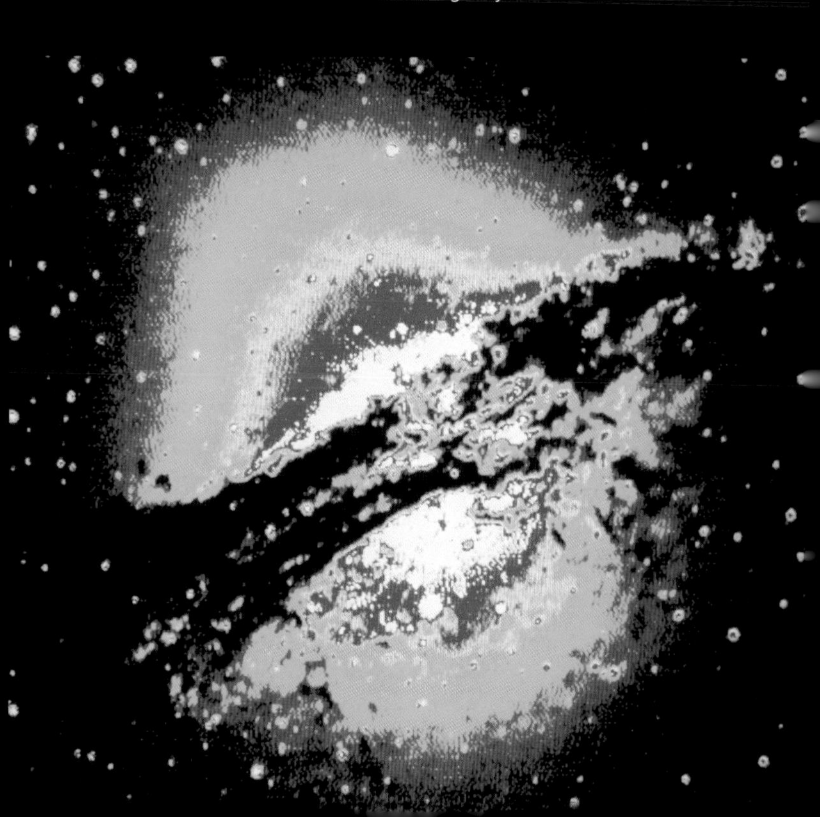

Quasars

The most celebrated of all active galaxies are the mysterious quasars. The name is a contraction of "quasi-stellar," which was applied to indicate their starlike appearances. Quasars appear small and starlike for two reasons: they are exceptionally distant, at billions of light-years, and their activity comes from a very tiny central region of the underlying host galaxy. The vast distance coupled with the observed brightness implies that some quasars are 100 to 1000 times brighter than a typical normal galaxy. Since this energy is all concentrated in a tiny region at the center of the core, the active galactic nucleus of a quasar is almost certainly driven by, as with other active galaxies, a supermassive black hole, feeding on typically one solar mass of material every year.

DISCOVERING QUASARS

Astronomers uncovered their first quasar in 1963. At that time, they knew of many radio objects not associated with optical sources. One such radio source was an object called 3C 273 (*below*). Astronomers were able to watch as the Moon passed in front of this radio source. They timed the moment when the Moon cut off the radio emission, thus enabling them to pinpoint the source's location. Four decades since its discovery, 3C 273 still remains the brightest optical quasar. Its power is almost unimaginable. It pumps out as much power as 40 large normal galaxies, and yet all this comes from a region of space more than 30 million times smaller than the Milky Way. Put another way, 3C 273 emits more energy in a single second than the Sun does in three million years. Like many quasars and most radio galaxies, 3C 273 beams a jet of material deep into space. The radio image below shows the quasar's core in red with emanating jets of radiation.

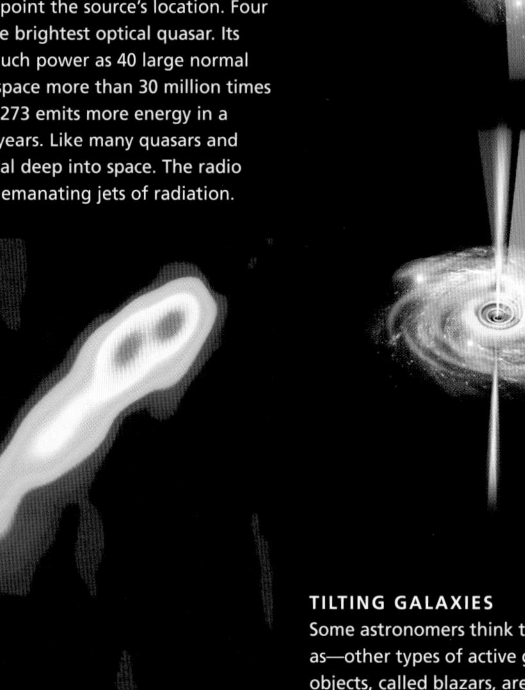

1

2

3

TILTING GALAXIES

Some astronomers think that quasars are very similar to—possibly even the same as—other types of active galaxies, but seen from a different perspective. (1) Some objects, called blazars, are thought to be the jets from active galaxies seen end on. (2) If the underlying host galaxy is seen from an oblique angle, astronomers detect a quasar. (3) Meanwhile, an active galaxy seen from the side appears in the guise of a radio galaxy.

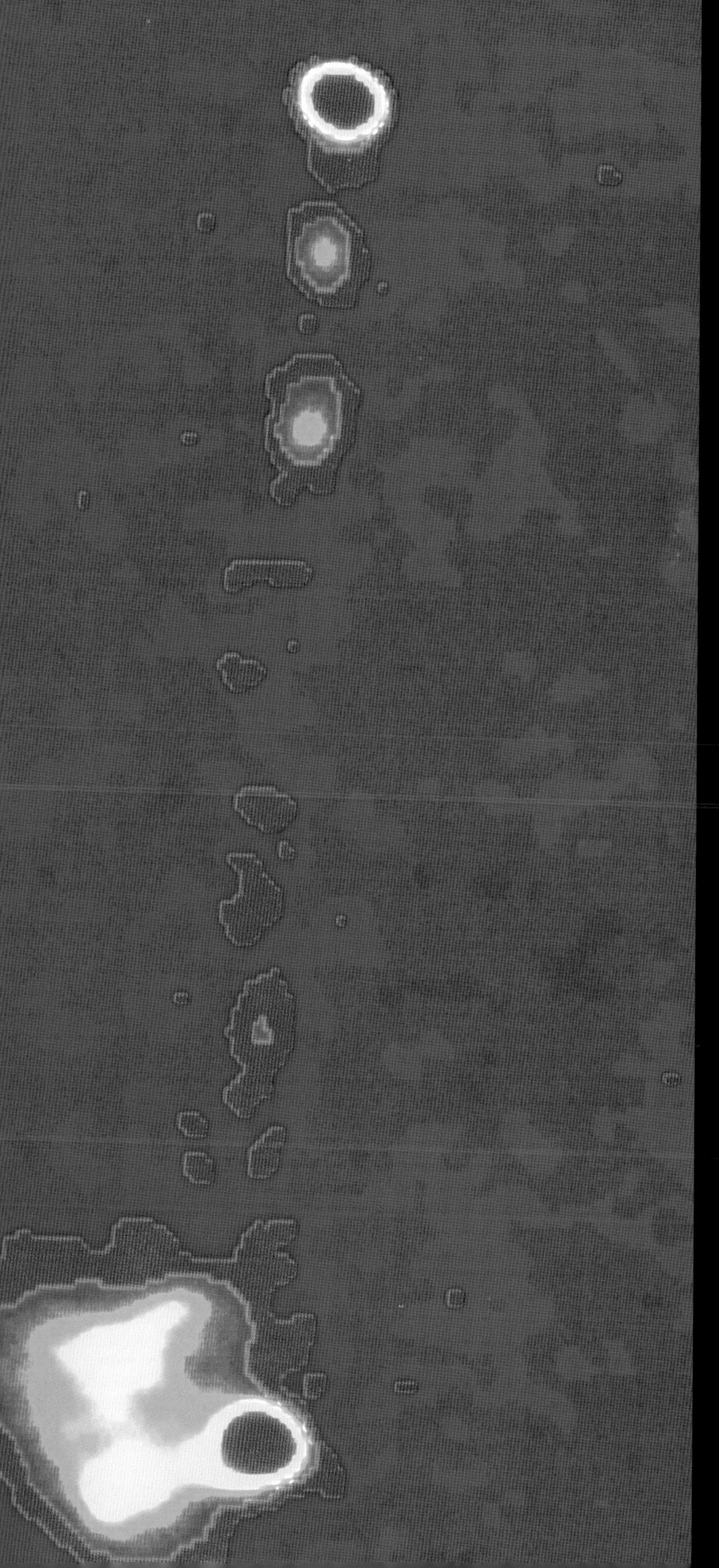

← **This is a false-color photo** of the quasar 1007+417, taken with the Very Large Array radio telescope in New Mexico. The quasar itself is the dense knot at the top, while the elongated extension is a jet of charged particles.

↓ **This abstract photo** of the quasar 0957+061 is taken in polarized light, which tells astronomers a little about the magnetic fields that surround it.

↓↓ **This radio image** of the quasar 0932+004 shows what can happen when gravity bends the light of a distant object. The blue blobs in the middle and at the bottom are actually two images of the same object, the quasar itself.

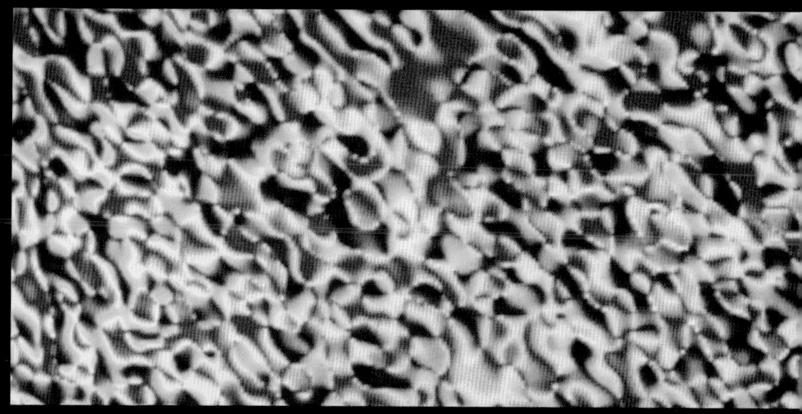

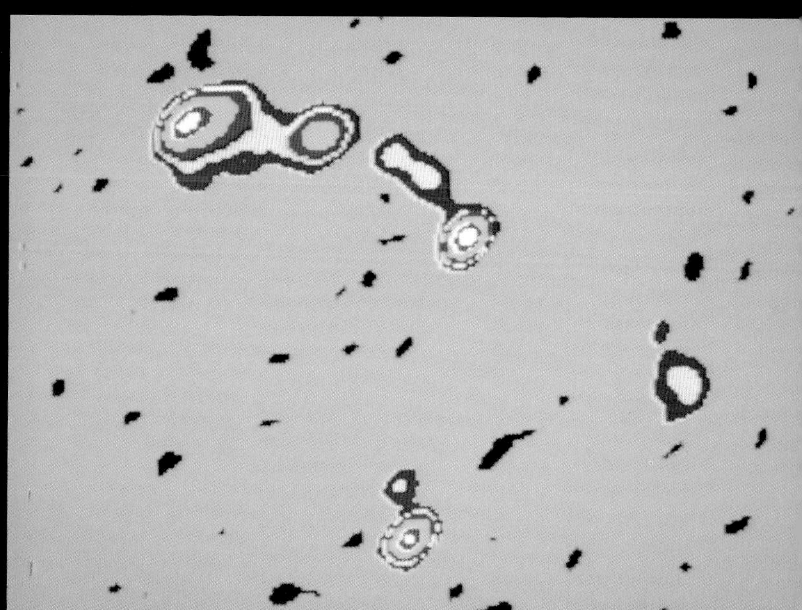

Galaxy clusters

Just as stars gather into vast assemblages called galaxies, so galaxies in turn gather into clusters. The smallest galaxy clusters contain only a handful of members spread across a couple of million light-years or so. The largest, meanwhile, are called Abell clusters. They appear in a catalog of almost 3000 clusters drawn up in 1958 by the American astronomer George Abell (1927–83). Abell clusters are rich—that is, they have a high concentration of at least 50 galaxies, and sometimes as many as several thousand. These, the true monster clusters, are held together by gravity, just as individual galaxies are. But smaller, non-Abell groupings are loosely bound at best and may not be gravitationally tied together at all.

→ **Abell 2218** is a cluster of galaxies some 3 billion light-years away. It is so massive that the gravity of its galaxies is bending the light of galaxies behind it into a series of arcs—an effect known as gravitational lensing.

THE LOCAL GROUP

The Milky Way has its own cluster of course, known as the Local Group. This is a small gathering of about 31 members that extends across 3 million light-years of space. The brightest members are all large spiral galaxies: the Milky Way, the Andromeda Galaxy and the spiral galaxy in Triangulum, M33. The other members are all small. There are some dwarf irregulars, but most are dwarf spheroidal (dSph) galaxies no larger than 10,000 light-years across or so. These are like dwarf ellipticals, but they are not very centrally concentrated. The closest Local Group member to the Milky Way is the recently discovered (1994) Sagittarius dwarf galaxy.

KEY

Galaxy name	Distance	Galaxy type
1 LMC	160,000 ly	Irregular
2 SMC	190,000 ly	Irregular
3 Draco Dwarf	230,000 ly	Dwarf spheroidal
4 Ursa Minor Dwarf	260,000 ly	Dwarf spheroidal
5 Sculptor Dwarf	320,000 ly	Dwarf spheroidal
6 Sextans Dwarf	320,000 ly	Dwarf spheroidal
7 Ursa Major Dwarf	320,000 ly	Dwarf spheroidal
8 Fornax Dwarf	540,000 ly	Dwarf spheroidal
9 Pegasus Dwarf	320,000 ly	Dwarf spheroidal
10 Leo II	870,000 ly	Dwarf spheroidal
11 Leo I	1.1 Mly	Dwarf spheroidal
12 NGC 6822	1.7 Mly	Irregular
13 NGC 185	2.5 Mly	Dwarf spheroidal
14 NGC 147	2.7 Mly	Dwarf spheroidal
15 Andromeda	2.9 Mly	Barred spiral
16 M32	2.9 Mly	Dwarf elliptical
17 M110	2.9 Mly	Dwarf elliptical
18 M33	3.2 Mly	Spiral

SUPERCLUSTERS

Just as galaxies team up into clusters, so do entire clusters of galaxies bunch together to form vast assemblages—clusters of clusters, called superclusters. Superclusters can comprise between 10 and 50 individual clusters, on scales possibly even larger than 300 million light-years. That's 100 times the distance between the Milky Way and its nearest large neighbor, Andromeda. The supercluster to which the Local Group belongs is known as the Local or Virgo Supercluster, since it is centered on the gigantic Virgo cluster of galaxies. In all, the Local Supercluster is approximately 100 million light-years across and accommodates some 10,000 individual galaxies.

← **The coma cluster**, about 350 million light-years away and stretching 16 million light-years across, is a large cluster dominated by two central galaxies: the elliptical galaxy NGC 4889 (*left*) and the lenticular galaxy NGC 4874 (*far right*).

Galaxy birth

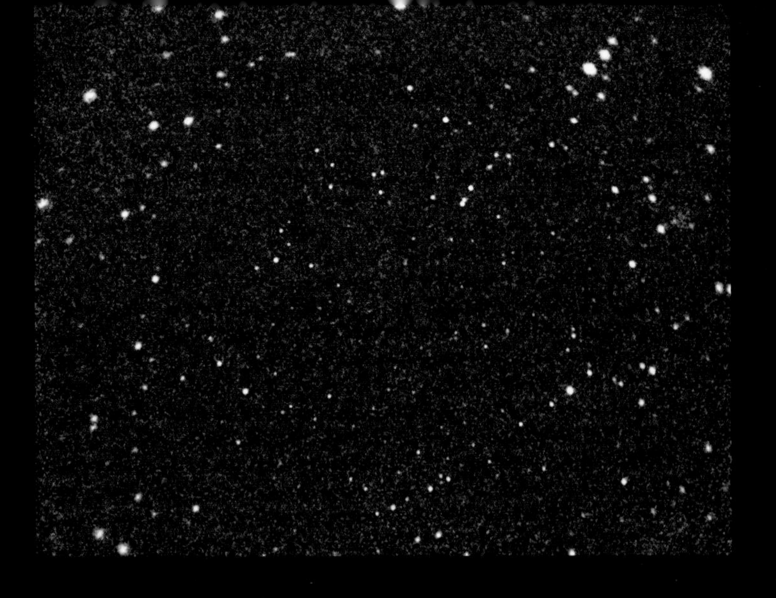

Even now, despite great advances in astronomy and astrophysics, there are many things that astronomers struggle to explain. One of the most baffling problems is that of the origin of galaxies. The most distant quasars we can see are well over 10 billion light-years away, which means they must have formed before 10 billion years ago—within a few billion years or so of the Big Bang. But how? In general, there are two possible ideas. The first is that the galaxies were whittled away from initially much larger gas clouds, much as a new-born star is. Alternatively, a hierarchical formation mechanism suggests that galaxies assembled themselves from smaller building blocks. Our best guess is that perhaps both processes were at work.

BIRTH OF THE MILKY WAY

The Milky Way's creation began at an uncertain time—between one million and one billion years after the Big Bang—when the Universe was little more than a huge fog of hydrogen and helium. Some regions in this early Universe were slightly denser than others. These denser regions pulled in matter from neighboring zones and got heavier and denser, growing larger. Eventually, they detached from the fog and became separate, massive cloudlets: galaxy building blocks. Later, as these cloudlets collided with each other, they merged, and what would one day become the central bulge of the Milky Way took shape. Meanwhile, shockwaves from the collisions compressed the clouds further until they condensed to form stars. For billions of years, cloudlets continued to be sucked into the growing mass. Some joined the bulge, others went into orbit and condensed into globular clusters, and still others settled down and formed the disk, which would later develop the trademark arms of spiral galaxies.

↑ **This shot of deep space**, taken by the Chandra X-ray satellite, shows the Universe when it was very young and its cargo of galaxies had not quite fully formed.

↓ **The Milky Way**, seen here in profile, has been evolving for billions of years. This image reveals the galaxy's dramatic central bulge and the scattering of surrounding globular star clusters.

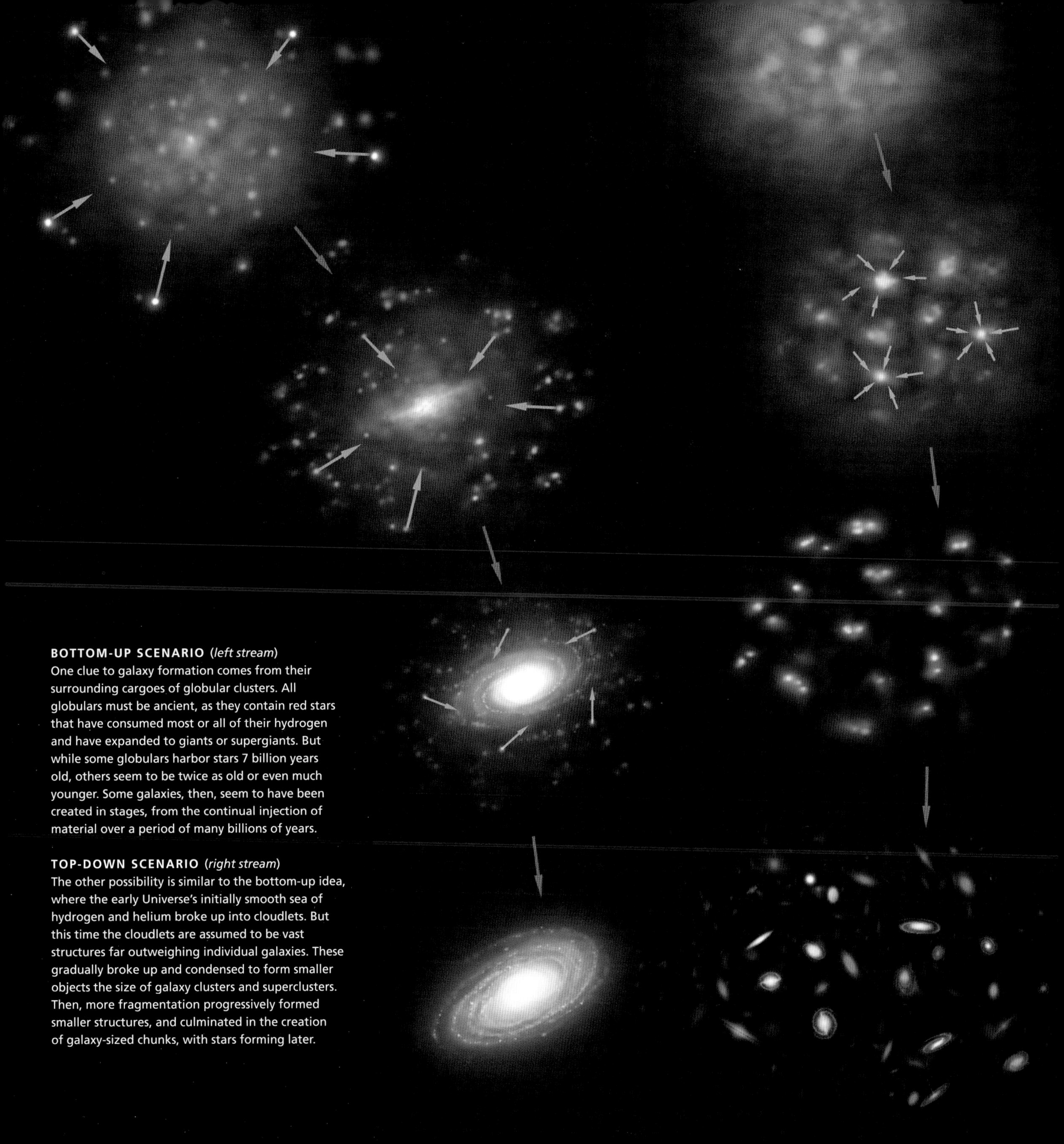

BOTTOM-UP SCENARIO (*left stream*)
One clue to galaxy formation comes from their surrounding cargoes of globular clusters. All globulars must be ancient, as they contain red stars that have consumed most or all of their hydrogen and have expanded to giants or supergiants. But while some globulars harbor stars 7 billion years old, others seem to be twice as old or even much younger. Some galaxies, then, seem to have been created in stages, from the continual injection of material over a period of many billions of years.

TOP-DOWN SCENARIO (*right stream*)
The other possibility is similar to the bottom-up idea, where the early Universe's initially smooth sea of hydrogen and helium broke up into cloudlets. But this time the cloudlets are assumed to be vast structures far outweighing individual galaxies. These gradually broke up and condensed to form smaller objects the size of galaxy clusters and superclusters. Then, more fragmentation progressively formed smaller structures, and culminated in the creation of galaxy-sized chunks, with stars forming later.

Galaxy evolution

The deeper we look into space, the farther into the past we see. This is because it takes more time for the light from the most distant objects to reach us. Curiously, the farther astronomers wind back the cosmic clock, the greater the proportion of spiral galaxies. Spirals were much more common early on. This is perhaps due to the way galaxies evolve: by colliding and merging with their neighbors. Computer simulations show that single elliptical galaxies are frequently the results of galactic collisions. Thus, the merger scenario neatly explains several things. First, spirals are not common closer to home, in the more recent past, as they have been consumed to form ellipticals. Second, the merger picture shows how giant elliptical galaxies grow so huge—they simply swallow their unlucky neighbors. Third, mergers explain why elliptical galaxies have so little gas and only red stars. Collisions send shockwaves through both colliding parties. The shocks compress interstellar gas and cause it to form new, massive stars that become red supergiants within a few million years. Very quickly, the resultant galaxy, an elliptical, is full of red stars and devoid of gas. So the picture that emerges is one of a dynamic, changing Universe in which galaxies compete for space, with the largest ones growing fat at the expense of their smaller cousins.

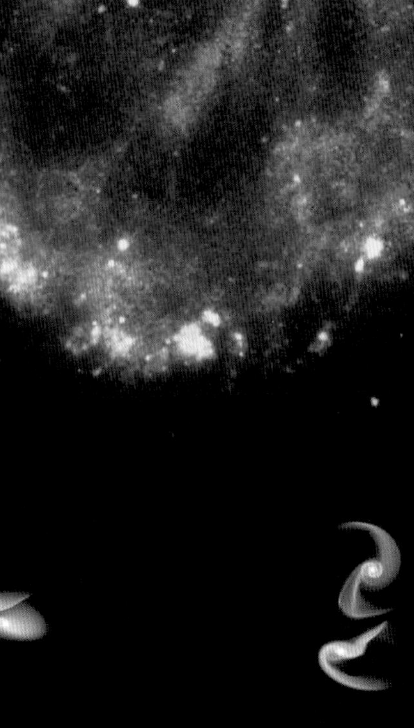

↓ **This simulation shows** a possible future encounter between the Milky Way and our nearest neighbor, the Andromeda Galaxy. Computer simulations such as these have opened up great possibilities in understanding galactic evolution.

← **The Cartwheel,** 500 million light-years away, is a starburst galaxy in which a colossal episode of star birth is currently underway. This galaxy has suffered a recent face-on collision with a dwarf galaxy. When the galaxies collided, shockwaves passed through the spiral, and spread out like ripples on a lake to form the glowing ring at the galaxy's edge.

→ **The Antennae**, NGC 4038 and NGC 4039, are two galaxies in the process of colliding. Located 60 million light-years away, each galaxy has a faint, curved tail extending 360,000 light-years. These filaments are the "antennae" that give these galaxies their name. They are sheets of gas and stars pulled from each galaxy as a result of their mutual gravitational interaction.

↓ **NGC 2207** (left) is colliding with the smaller IC 2163. The stronger gravitational pull of NGC 2207 has distorted the shape of IC 2163. As a result, stars and gas have been ripped from the smaller galaxy and stretched into a long streamer of celestial matter.

The night sky

Previous page Stars appear to move in giant circles through the sky, as seen in this time-lapse photograph.

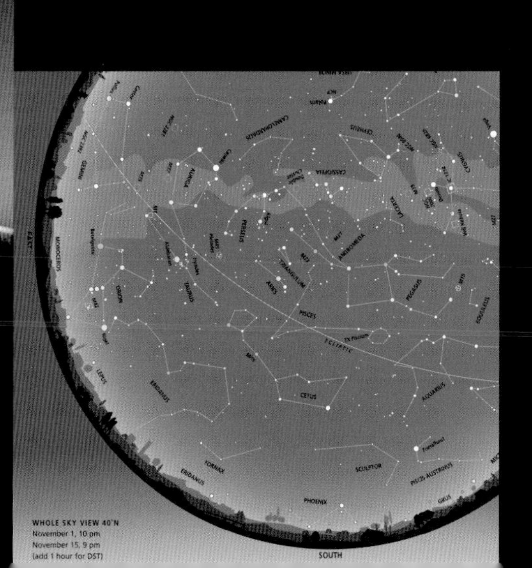

The night sky

The night sky is more than just the Moon and a few dozen constellations of stars. Planets, nebulae, clusters of stars and even a few galaxies are all visible with the unaided eye. And with modest equipment such as a pair of binoculars or a small telescope, the celestial dome takes on a whole new dimension.

Understanding the night sky

We have come a long way in our understanding of the night sky. Some 4000 years ago, the Egyptians pictured Earth and the stars floating on the surface of a giant celestial sea. Centuries later in Greece, the wisdom was that an unmoving Earth lay at the center of the Universe and that the stars were all located at the same distance. They were "pinned" to a gigantic heavenly ball of indeterminate size known as the celestial sphere. As the sphere rotated about us, once in 24 hours, so the stars tracked across the sky. Our ancestors found it hard to reject the idea that Earth was not the hub of the Universe. In the second century AD, Claudius Ptolemaeus (Ptolemy) propounded the notion that the Sun, Moon and planets went around Earth. This held sway for an astonishing 1500 years until, in 1543, Nicolaus Copernicus took a great step toward what we now know to be the truth. Earth goes around the Sun, just like all of the other planets. Since then, with more and more modern instruments, it has become clear that there is no star-studded dome—and there is nothing special about our place in space.

THE EMPTINESS OF SPACE

It is truly difficult to grasp just how widely spaced the stars in a given galaxy are. But to get an idea, let's scale the Universe down to something a little more manageable. Reduce the stars to such a size that you can only just see them with the naked eye—around ¹⁄₂₅₀ inch (0.1 mm) across. That's a reduction factor of about one ten-trillionth (10^{-13}). Even then, many miles would still separate these stars, and the Milky Way would span 45 million miles (70 million km). By contrast, if you scale things down so that the average galaxy is now ¹⁄₂₅₀ inch (0.1 mm) across instead, its nearest neighbors would reside less than half an inch (1 cm) away. Thus there is a very big difference between the density of interstellar space—that which separates the stars—and that of the extragalactic space that separates galaxies.

↑ **Stars fill the sky** in this photograph taken during twilight. Our ancestors imagined that all stars were equidistant from Earth. But we know now that some are much farther than others.

← **The Hubble Space Telescope** took this celebrated image, called the Hubble Deep Field South, in 1998, by staring at the same point on the sky for 10 days, complementing an earlier image taken in the northern hemisphere. These images are the deepest views of space astronomers have ever obtained—perhaps more than 10 billion light-years. The galaxies that fill these images are so distant that we see them as they were when the Universe was just one or two billion years old. These photographs are helping astronomers to understand the early Universe, and galaxy formation.

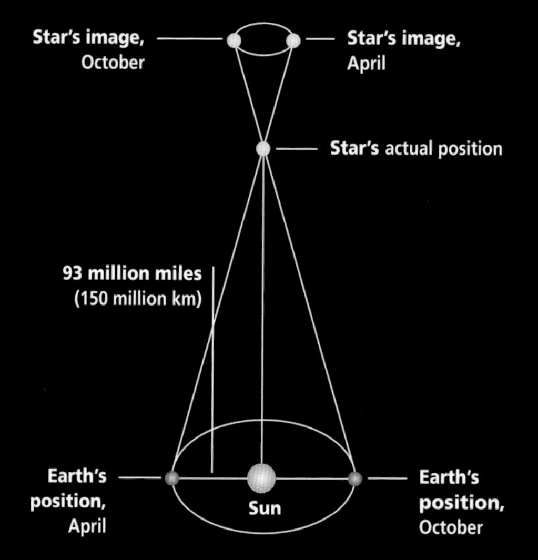

PARALLAX

Hold your finger out in front and move your head side to side. You will see a phenomenon known as parallax—the apparent movement of your finger relative to the more distant objects in the background. Similarly, as Earth orbits the Sun, the nearest stars show a parallax relative to their distant cousins. Parallax enables astronomers to gauge the distance to the closest stars and is the most accurate means we have for doing so.

STAR COLOR AND TEMPERATURE

The color of a star is an indication of its surface temperature. Our Sun, a yellow star, has a surface or photospheric temperature of about 5700 Kelvin (K). The coolest stars have temperatures of about 4000 K or less, and are red in color. At the other end of the scale, the hottest stars have temperatures around 30,000 K to 40,000 K. From the coolest to the hottest stars, the colors follow the sequence red, orange, yellow, white and blue.

STAR COLOR

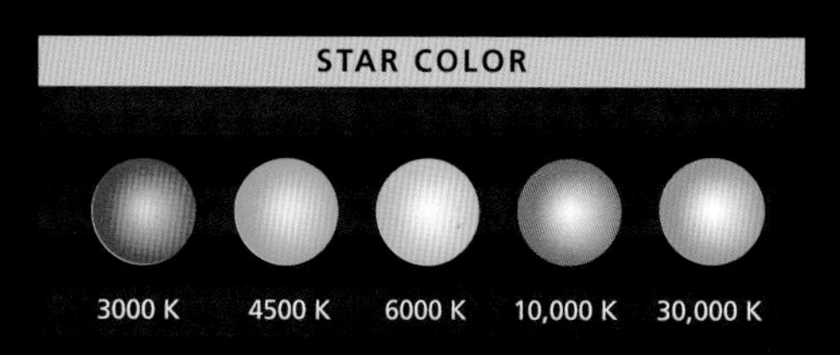

CAPTURING COLOR

Star colors are subtle to the eye. If you look at Antares in Scorpius, Aldebaran in Taurus, or Betelgeuse in Orion, you will see that these stars appear as a muted reddish color—a subtle red rather than the vibrant red of traffic lights. Similarly Vega in Lyra and Rigel in Orion appear a delicate, rather than striking, blue. In photographs, the colors appear more obvious, although they are not a true reflection of star color—they are influenced by the color sensitivity of the film and because bright stars flood the film with light and appear white. Changing camera focus during a star trail exposure overcomes this by spreading out the light to ensure that each star has its color well recorded at some point along its trail. The photograph (*right*) reveals the colors of the Southern Cross and pointers recorded in an exposure of about 30 minutes.

Mapping the heavens

Just as geographers chart the position of objects on the surface of Earth, so astronomers map celestial objects on the sky. Ancient navigators used the night skies to guide them across Earth's oceans and seas. Their star maps were crude but effective, and became invaluable tools. We no longer need star maps to navigate, of course. Instead, modern astronomers chart the positions of astronomical objects—using an equatorial system based on geographical latitude and longitude—merely to keep a record of what's "out there," so that they may better understand the Universe. To help them, they also keep catalogs of stars, clusters, nebulae and galaxies. John Flamsteed (*left*) was one of the first to introduce a star numbering system in his 1725 publication *Coelestis Britannica*. And two of the classic catalogs—Messier's and the *New General Catalog* (NGC)—are still in common usage.

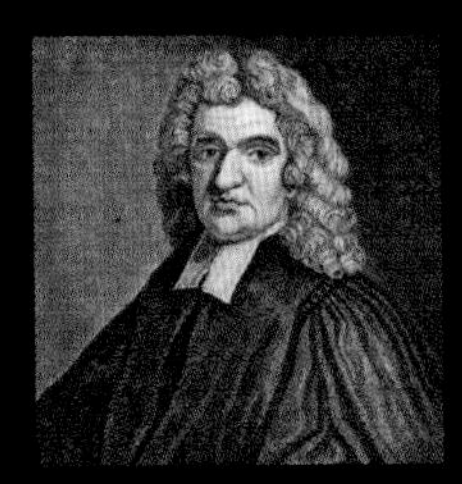

THE MESSIER CATALOG

The Messier Catalog is a list of celestial objects compiled by the French astronomer Charles Joseph Messier (1730–1817). Messier (*below*) assembled the list while he was hunting for comets, and released the first catalog in 1771 with 45 entries. Revised releases followed, and the modern list contains 110 items including nebulae, clusters and galaxies. The first object to appear in the catalog, M1, is the famous supernova remnant, the Crab Nebula.

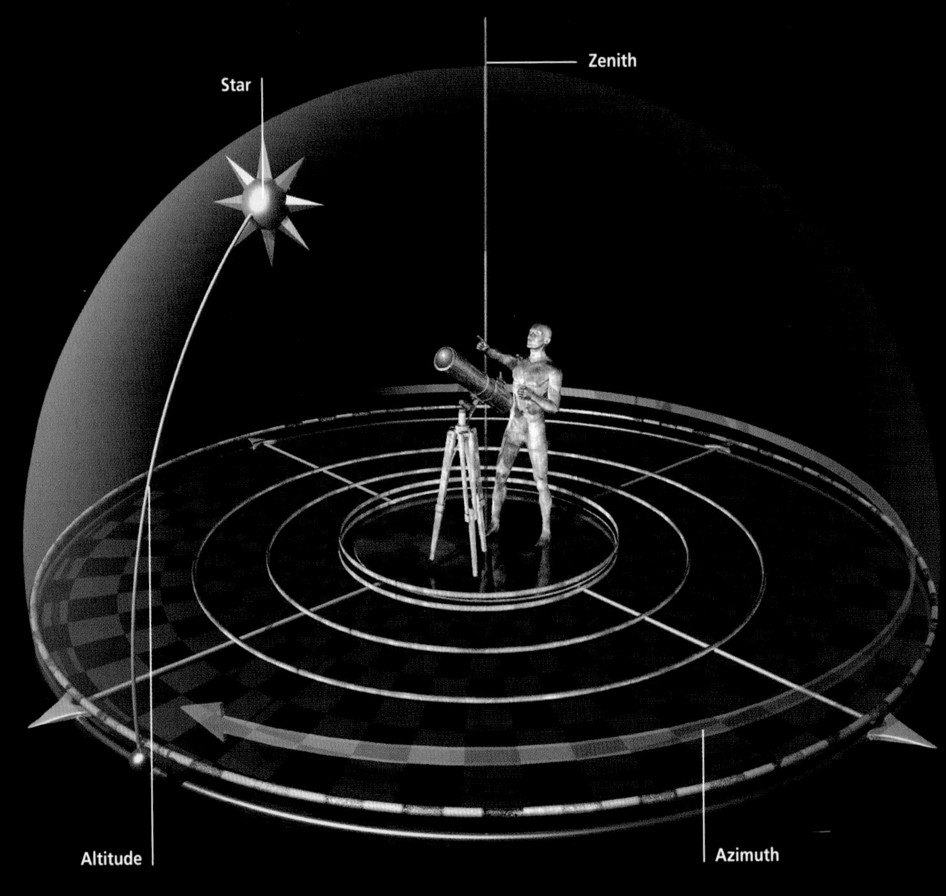

ALTAZIMUTH COORDINATE SYSTEM

The altazimuth system is used in navigation. You specify the height of an object in the sky by its altitude—the angle measured in degrees from the horizon to the object in question. The point directly overhead, at altitude 90 degrees, is the zenith. The azimuth, meanwhile, marks the position of the object measured from north, east along the horizon, to the point on the horizon directly below the object. The meridian is the great circle that passes through the zenith and which splits the sky into two halves.

THE CELESTIAL SPHERE
The celestial sphere and its system of "equatorial" coordinates mimic the surface of Earth—with features equivalent to latitude, longitude, the north and south pole and the equator. The important extra feature is the ecliptic—the apparent path of the Sun across the background of the celestial sphere.

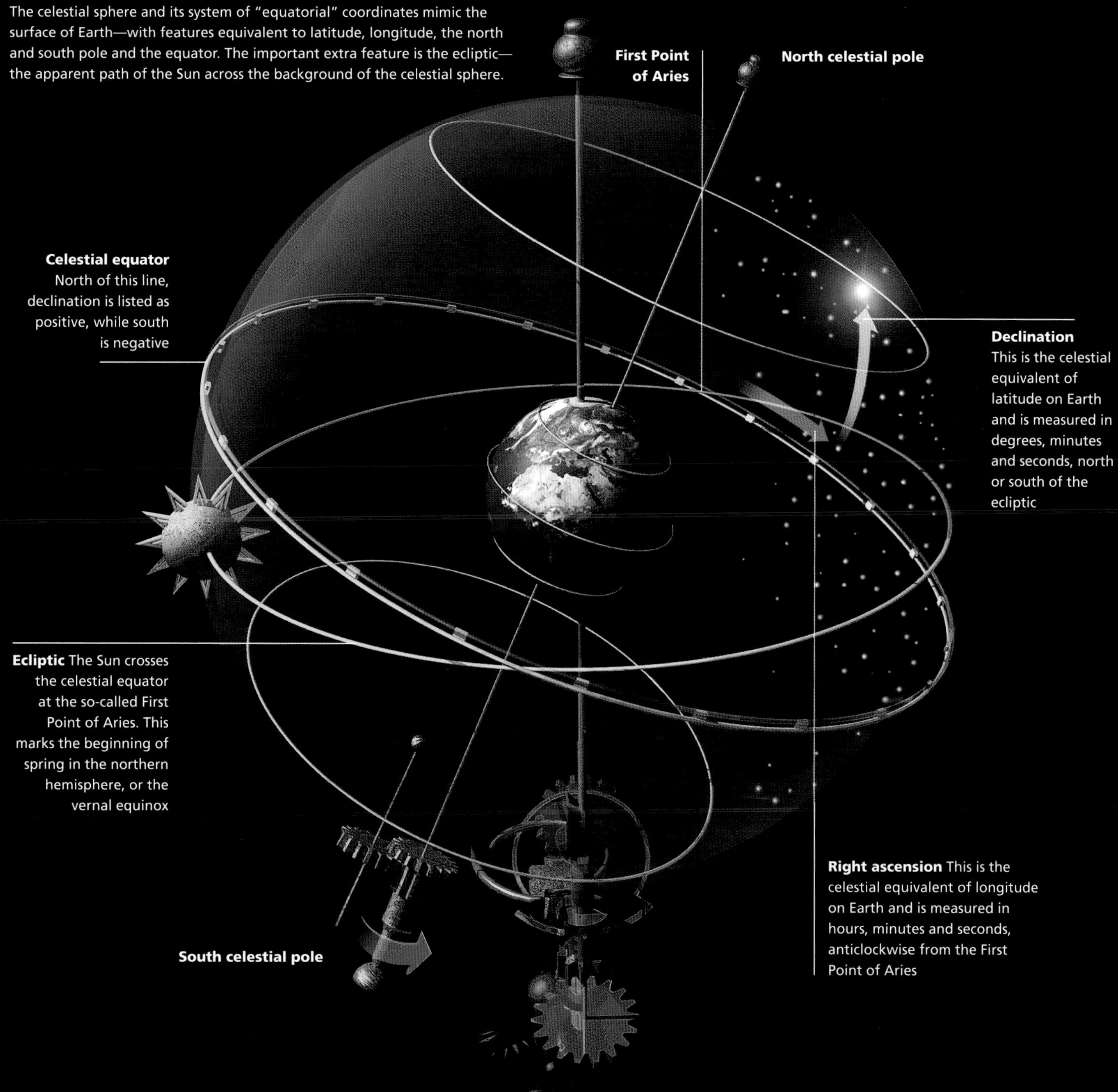

First Point of Aries
North celestial pole
Celestial equator North of this line, declination is listed as positive, while south is negative
Declination This is the celestial equivalent of latitude on Earth and is measured in degrees, minutes and seconds, north or south of the ecliptic
Ecliptic The Sun crosses the celestial equator at the so-called First Point of Aries. This marks the beginning of spring in the northern hemisphere, or the vernal equinox
Right ascension This is the celestial equivalent of longitude on Earth and is measured in hours, minutes and seconds, anticlockwise from the First Point of Aries
South celestial pole

The spinning Earth

Two effects combine to create the seasons: our planet's axial tilt, and its orbit around the Sun. Earth is tipped over by 23.5 degrees relative to the plane in which it orbits. When the planet is in a certain position in its orbit, the northern hemisphere points directly toward the Sun and experiences summer. The Sun is high in the sky, the days last longer, and so the hemisphere gets more heat. The southern hemisphere, at the same time, is pointed away from the Sun and experiences winter. Half an orbit later the reverse ensues, with summer in the south and winter in the north. And in between, both hemispheres receive an equal amount of sunlight, giving rise to the moderate seasons of autumn and spring.

SEASONS ON EARTH
As the Earth orbits the Sun, its axis remains fixed relative to the stars. This means that sometimes one of the hemispheres is pointing more directly toward the Sun than the other, when it will be summer in that hemisphere and winter in the other.

PRECESSION
Earth's spin axis is not fixed. Over a period of around 25,800 years, the axis executes a small circle and sweeps out a conical section of space. This movement is known as precession. Currently, the spin axis points almost directly to Polaris, the pole star, in the north. But this will not be the case for our distant descendants, and likewise our ancestors would have seen a different pole star.

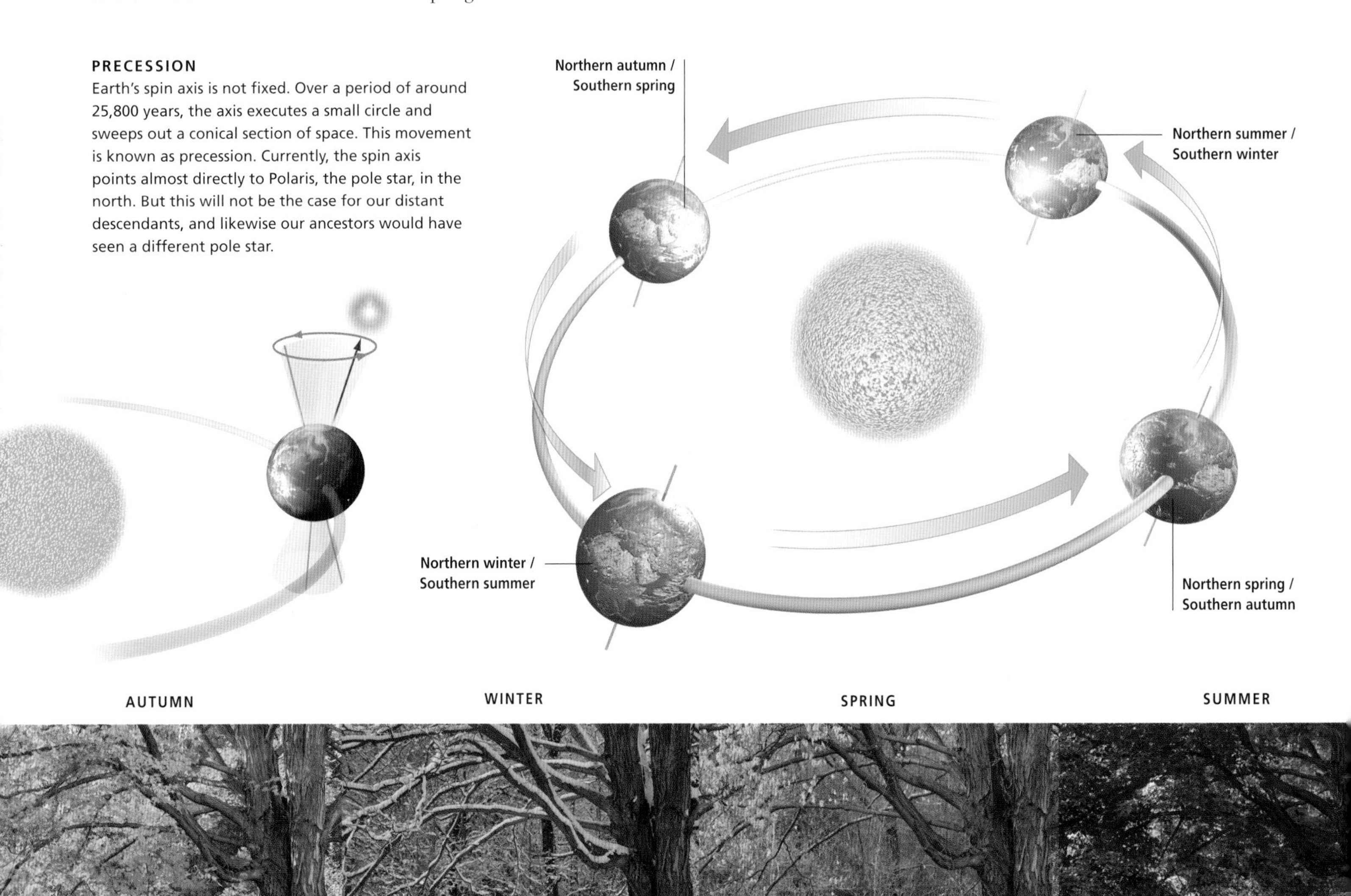

RELATIVE MOVEMENT OF THE STARS

Depending on where you are on Earth, the stars move in different directions across the sky. At the poles, the stars neither set nor rise, but merely circle overhead about the celestial pole. On the equator, the stars rise directly due east, perpendicular to the horizon, and set due west. Between these latitudes, the stars rise and fall at an angle to the horizon—with some stars remaining hidden from view.

↑ **The farther north or south** you go on Earth, the shorter the winter days become, and the longer the summer. During winter, the Sun rises just a little above the horizon and then sets very soon, as seen in this time-lapse photo.

NORTH POLE, 90°N

EQUATOR, 0°

40° NORTH

Constellations and the zodiac

For as long as astronomy has existed, people have seen patterns in the stars. Some groups are so distinctive that they have been recognized since antiquity. The Mesopotamians probably created the earliest constellations about 10,000 years ago—Leo, Scorpius, Taurus and many others feature in carvings—though they may have represented different ideas. As the millennia passed, some constellations were passed on to later civilizations, such as the ancient Greeks, while others vanished, to be replaced by new ones reflecting local tales, deities or animals. Many of the most famous constellations, such as Perseus, Orion and Andromeda, stem from Greek mythology. But there are others, named for scientific instruments, that are much more recent, such as Microscopium and Telescopium. Today, 88 constellations are officially recognized by the Western world. Remember, though, not all civilizations see the same objects.

↓ **This map represents** the constellations of the northern hemisphere and is one of the 29 star maps by the Dutch–German mathematician Andreas Cellarius (1596–1665), from his great work, *Atlas Coelestis seu Harmonia Macrocosmica* (*Celestial Atlas of Harmony*).

STARS IN MOTION
Stars are always moving, but because distances in space are so vast, stars in constellations appear fixed in place. Gradually, however, their movements will make today's constellations unrecognizable. The motion of the Plough, or the Big Dipper, from Ursa Major is illustrated below.

100,000 years ago

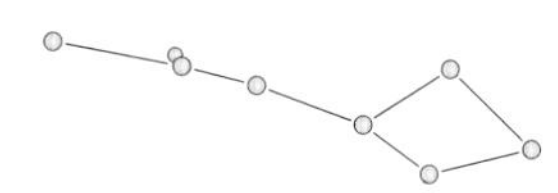

Present day

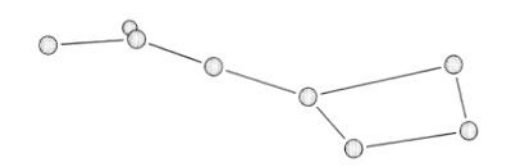

100,000 years hence

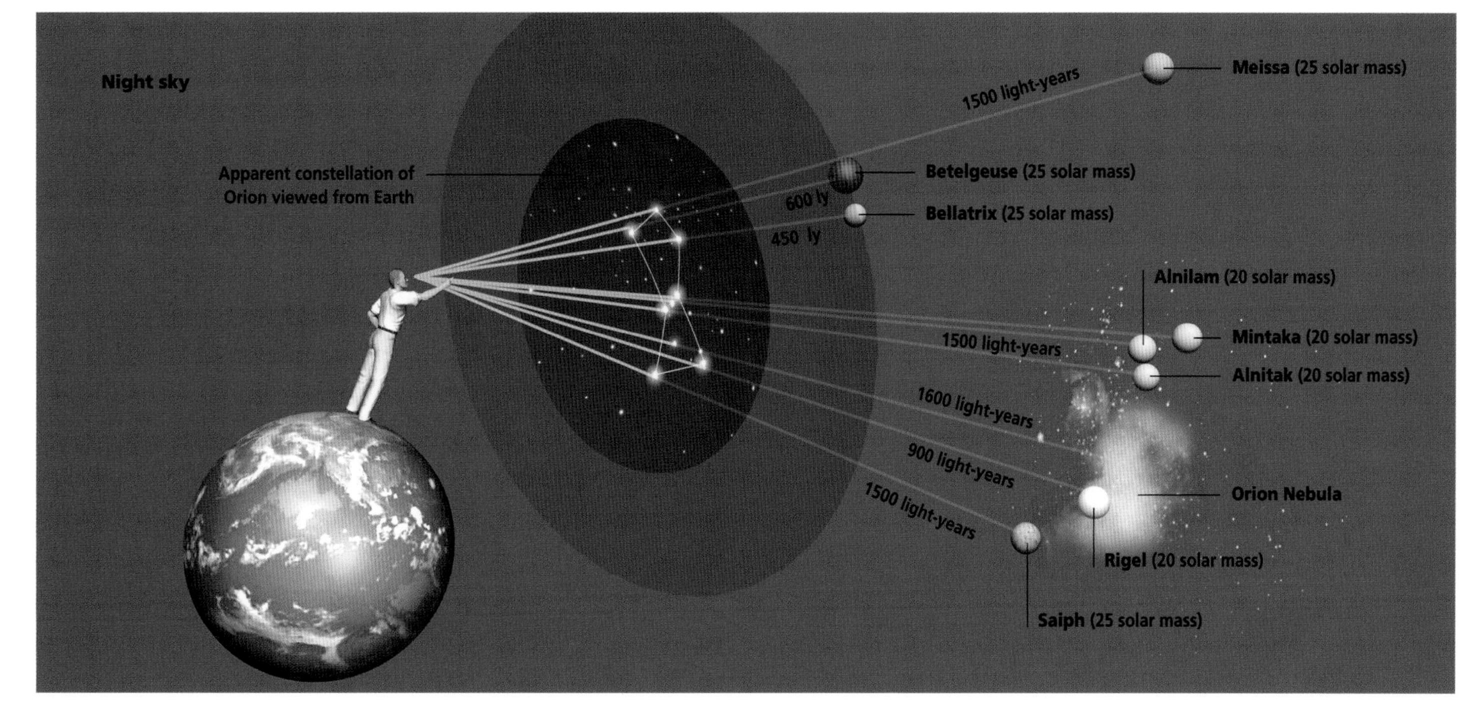

VIEWING THE CONSTELLATIONS

Although the stars appear as though they are equally far, this could not be further from the truth. Take the constellation of Orion, for example. Each of the stars in this constellation lies at a different distance (*see above*). Therefore, if we saw these stars from another vantage point, say from a different star system, the constellation would look very different indeed, as the illustration shows. You can use the constellations as guides to the sky's celestial treasures. For example, follow the stars in Orion's belt toward the east and you will locate Sirius, the brightest star in the sky. Go the other way and you will come to the orange Aldebaran in Taurus, close to the Hyades and Pleiades star clusters. And the zodiacal constellations have their uses too: it is these that will contain the planets.

ORIGINS OF THE ZODIAC

As Earth orbits the Sun, the Sun's path projected onto our skies traces a line known as the ecliptic, which passes through 12 constellations that together comprise the zodiac. These 12 constellations—Aries, Leo, Taurus and so forth—are among the most ancient known, but their modern names come from those applied much later by the Greeks. Some civilizations, such as the ancient Chinese, have entirely different zodiacal creatures. The image (*right*) shows various Asian religious characters surrounded by the 12 signs of the zodiac.

Beginning astronomy

Contrary to what you may believe, you don't need binoculars or a telescope to enjoy the night sky. You can use the oldest of astronomical tools—the naked eye. Many beautiful naked-eye objects have been known since antiquity, before optical aid was invented. You can enjoy them too, without much effort. Simply let your eyes adapt to the darkness—the process whereby the pupils expand to their fullest to let in the most light. After a few minutes, you will be able to see the heavens more clearly. If you must use illumination to consult a star map, for example, use a red torch to preserve your night vision.

So what can you see that the ancients did? Well, from a dark location you will be able to appreciate the majesty of the Milky Way. This stunning band of light and absorbing clouds of dust is a cross-sectional view of our own galaxy, seen from the inside. Even under bright city skies, there is still a great deal to see. Venus is the brightest object in the sky aside from the Moon and Sun, and is particularly attractive, especially if the Moon is nearby. Four other planets are also naked-eye objects: Mercury, Mars, Jupiter and Saturn. Then there are the constellations, the Magellanic Clouds in the southern hemisphere, bright star clusters such as the Pleiades, misty nebulae such as the Orion Nebula, and other spectacles to be had—all visible from urban skies. Plus, at certain times of the year, meteor showers, lunar eclipses, comets (such as Hale-Bopp, *right*) and the beautiful aurorae all make an appearance. Go on and enjoy yourself for free!

MEASURING THE SKY

Astronomers use degrees, minutes and seconds to measure sizes and distances in the sky. An outstretched hand held at arm's length will be about 20 degrees wide from thumb tip to the tip of the little finger—roughly the distance between the first and last stars of the Big Dipper in the constellation Ursa Major. Smaller distances can be measured with your fist at arm's length (about 10 degrees) and your thumb (about 2 degrees).

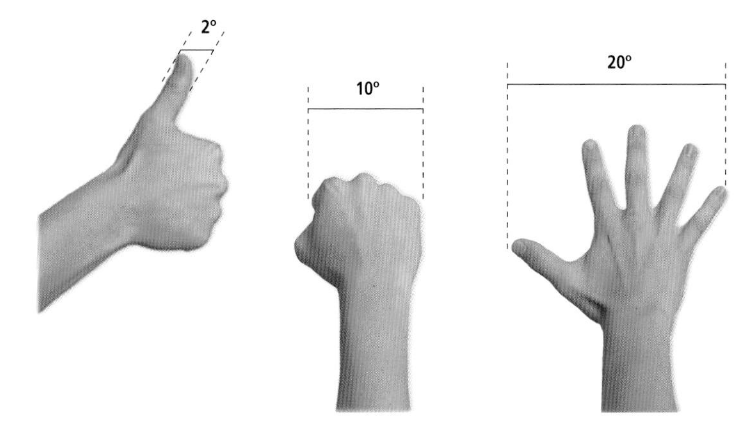

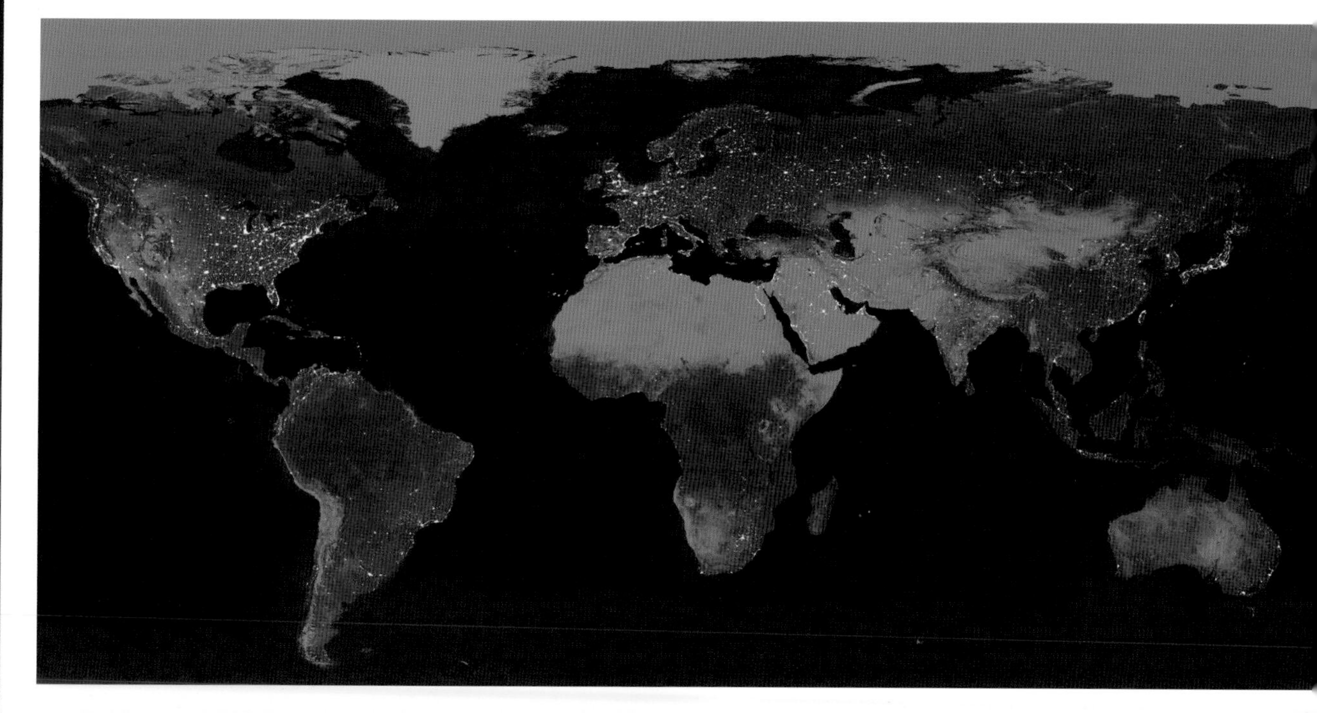

URBAN LIGHT

Your best view of the sky will be had from a dark location, but not everybody has the luxury of living in a rural area. If you are in a city, you will be at the mercy of light pollution. Cities are full of unnecessary or inefficient lights that light up the sky as much as they do the car parks, buildings, roads or signs they are supposed to be illuminating. Don't despair. Try to find a place shadowed from bright streetlights—a park is good. If you are in your own garden, turn off any lights you can see, even those inside the house, or else block them from view.

↑ **The NASA image** above shows the distribution of light polllution across the world—the heaviest concentrations falling in North America and Europe. Searching the cosmos can be more difficult in urban areas, due to man-made light obscuring the heavens.

← **Cities, such as New York,** are saturated in artificial light, making skywatching particularly difficult.

Choosing and using binoculars

Good binoculars are essential to the serious amateur astronomer. Although they lack the superior magnification and large light-gathering capacities of large telescopes, their low power is of benefit. Without binoculars, there are a host of objects that you will miss, simply because they are too large to fit into the field of view of even the most low-power telescope. Binoculars are also lightweight, are more portable than even the smallest good-quality telescope, and require essentially no setting up. Just pick them up and away you go.

7x50

9x63

WHICH BINOCULARS?

Buying binoculars is easier than a telescope, but there are still many things to consider. Most important is the power–aperture specification. Binoculars are specified according to a code such as 7x50. The first number represents the power, and the second is the diameter of each objective lens in millimeters—the aperture. The larger the aperture, the more light the binoculars can take in and the brighter the image will be. For astronomy, you need at least 50 mm or the image is too dark. To help you choose, divide the aperture by the power. This will give you a number called the exit pupil, which is the width of the image that emerges from the eyepieces. Ideally, it should be as close to the size of the dark-adapted pupil (about 7 mm) as possible without being larger. For 7x50 binoculars, the exit pupil is just over 7 mm—so they are ideally suited to most people. 10x50 instruments are also good, especially for older people who have smaller pupils, or if you are viewing from a bright location where the eyes will not dilate more than 5 or 6 mm (⅕ inch). If, as most people do, you want to use your binoculars freehand—not on a tripod—then chose a pair with no more than 10x power and 50 mm of aperture. Higher power means greater image wobble, and larger aperture instruments are heavy, clumsy and uncomfortable to use without a mount. Avoid zoom binoculars, too, as they often have inferior optics.

PORRO-PRISM BINOCULARS

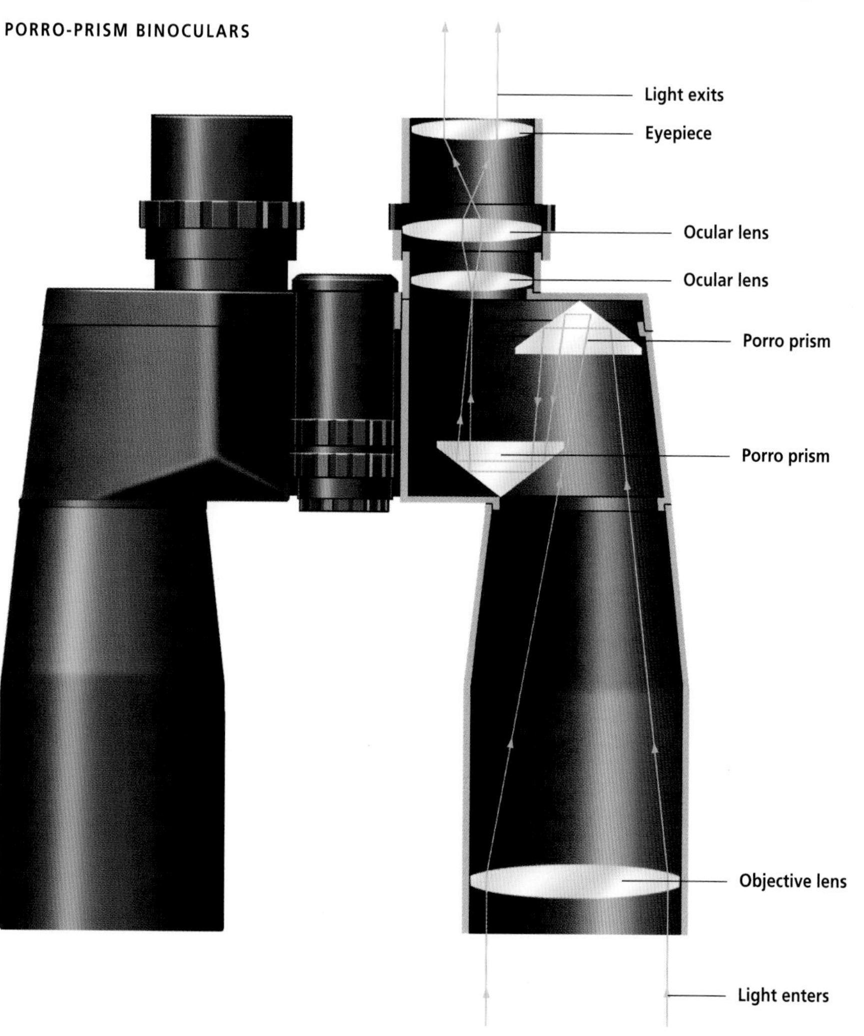

OTHER CONSIDERATIONS

It is a good idea to seek binoculars with multi-coated lenses—a series of films on each optical surface to improve contrast. Cheaper models only have a single coating and not on all optical surfaces, but may still be good. You should also be aware of eye relief—the distance from the eyepiece where you need to put your eyes to see the entire field. If you wear glasses and wish to keep them on when observing, you need binoculars with an extra-long eye relief of about ½ inch (15 mm). Lastly, for more expensive models, you will find that you have a choice of two types of binocular design. The ones with the classic L-shapes have porro prisms inside them—these are best. The other, straight-barrelled design is roof-prism. These binoculars tend to show spikes on bright stars, unless you buy an expensive pair. As with anything else, try before you buy, and go for the most comfortable.

↑ **You don't need a telescope** to make an impression in astronomy. Here, the Japanese amateur astronomer Yuuji Hyakutake is photographed next to his huge comet-hunting binoculars, with which he discovered the famous comet that bears his name.

WHY BINOCULARS?

Look at a large, loose open cluster through a telescope—the Pleiades, say,—and the view will probably disappoint you. The magnification tends to zoom in too far, and many of the cluster members remain outside the field of view. The telescope "unclusters" the cluster. Binoculars have much lower magnifications and therefore a larger field of view. Other targets you can appreciate with binoculars include, but are not limited to, the Eta Carinae Nebula (*below left*), the Large Magellanic Cloud (*below center*) and the Andromeda Galaxy (*below right*).

← **If you must use** a very large pair of binoculars, you will need to mount them on a tripod, such as this one. Otherwise they will soon become very heavy and, if the magnification is high, keeping the image still will prove impossible.

Choosing and using telescopes

While astronomy with the naked eye and binoculars is very rewarding, a telescope opens up so much more that otherwise remains permanently out of view. You can resolve the disks of tiny planetary nebulae, the individual stars in globular clusters and the spiral arms of faraway galaxies—objects too small and faint for binoculars. And many objects that you can see in binoculars, such as the Moon and planets, take on an extra dimension when viewed through a telescope. Even a small telescope will show you craters on the Moon and details on the planets that binoculars cannot reveal.

Altitude motion

Azimuth motion

TELESCOPE TYPES

Once you have decided on budget, portability and aperture, ask yourself what kind of telescope you need. They come in three main types, each with a different optical system. Refractors (1) have a lens at either end. Reflectors (2) have a mirror to collect light. And catadioptric telescopes (3) such as Schmidt-Cassegrains are hybrids of the two. Refractors are the most expensive, costing much more than a reflector with the same aperture, while catadioptrics are somewhere in between. Refractors also tend to be slightly heavier than reflectors, so they are not so easy to move around. The most portable kind of telescope, having the shortest tube, is the catadioptric. You might also consider if you want a telescope with a tracking system that can follow the stars. This will depend on the instrument's mount.

1

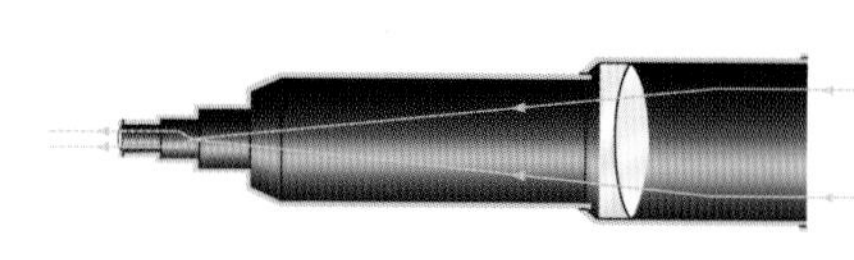

2

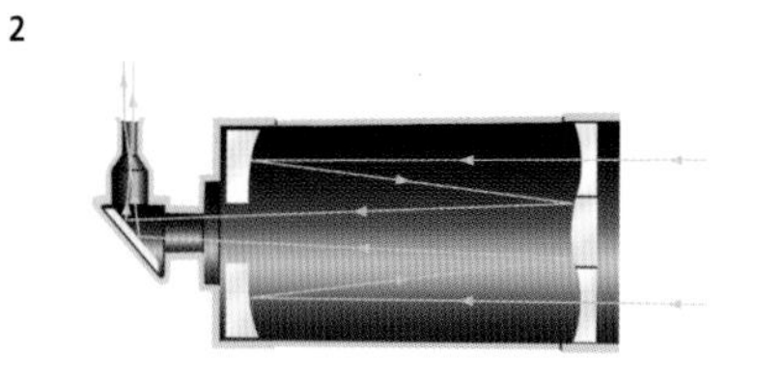

3

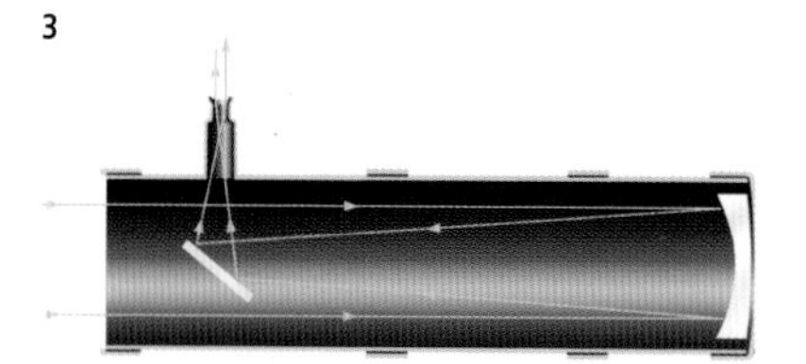

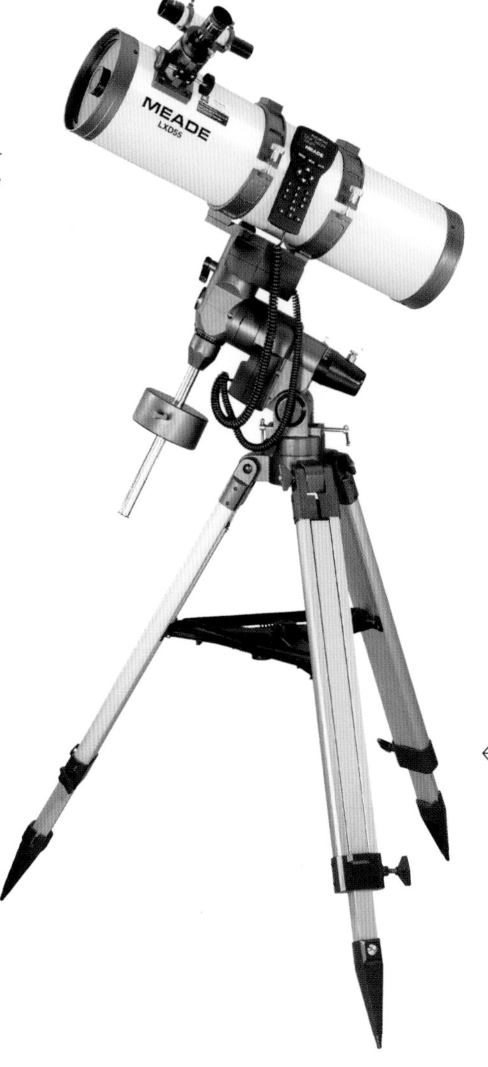

To Celestial Pole

Declination axis

Polar axis

Motion in right ascension

Motion in declination

TELESCOPE MOUNTS

There are two main types of mount, and variations on each. The simplest and most portable, the altazimuth mount (*top*), moves up-down in altitude and left-right in azimuth—hence the name. The other is the equatorial mount (*bottom*). Its two axes are designed so it can move across the sky exactly as the stars do. Although more complex, equatorial mounts with guiding systems are the only choice if you wish to track and photograph celestial objects.

← **This is an 8-inch** (200 mm) reflecting telescope of the Schmidt-Newtonian variety. It is perched on an equatorial mounting, which enables it to track objects as they move across the sky.

CHOOSING A TELESCOPE

Choosing an appropriate telescope is daunting. There are many factors that you must take into account. However, the most essential thing to consider—more important than magnification or the type of telescope—is its aperture. The aperture is the diameter of the telescope's main lens or mirror, usually expressed in either inches or millimeters. You should definitely go for as much aperture as your bank account can stretch to—and if you want it portable, as much as you can comfortably move around. The more aperture, the brighter the image—essential for faint objects. Image brightness is proportional to the square of the aperture. So a 6-inch (150 mm) telescope produces images four times brighter than a 3-inch (80 mm) one. Moreover, resolving power—the ability to separate two close objects—is also proportional to the aperture. The bigger the aperture, the more stars you can resolve in a cluster seen at the same magnification. Magnification, to reiterate, is not important at all in choosing a telescope. Do not be tempted by a telescope that boldly boasts high magnifications factors, because the aperture limits the telescope's practical power. That nice-looking cheap scope in the shopping mall may have a "theoretical maximum" magnification of 400x; but in practice, if the aperture is small and the optics poor, you will be lucky to get it working well at a quarter of that power. Buy from a specialist. The most a telescope can usefully magnify is equal to about 50 times its aperture in inches—so, just 125x for a small (2½-inch / 60 mm) instrument.

FILTERS

To enhance your viewing, you can choose from a range of filters that screw into the eyepiece. Light pollution reduction (LPR) filters do just that, eliminating some of the unwanted colors commonly contributed by street lighting. The narrowband LPR filters are perfect for improving contrast and color in nebulae. They are often called nebula filters. Deep-sky, or broadband, LPR filters are better if you wish to observe a wider range of deep sky objects. And there are other filters that enable you to observe in just one color—great for certain nebula. But if you buy just one filter, make it a narrowband one.

EYEPIECES

The eyepiece is the small interchangeable lens that magnifies and produces the telescope's image. Eyepieces are marked with their focal lengths in millimeters. The smaller it is, the larger the magnification, but the actual power depends on your telescope's focal length too. You should have a selection of two or three of differing powers. Eyepieces come in three main types with differing barrel diameters: 1, 1¼ and 2 inches (24.5, 31.8 and 50.8 mm respectively). The larger two sizes have superior optics, but the 2-inch (50.8 mm) ones will only fit very expensive telescopes.

↑ **This is a refracting telescope,** which uses lenses instead of a lens–mirror combination to gather its images. The mounting is equatorial, providing a simple up-and-down and side-to-side motion which makes it very easy to point and use.

ding the star maps

need a good street plan to save you getting lost in a city, will work wonders for your celestial navigation. Using the ps in this book, plus the two overviews, you should be able r way around the sky at any time of the year, whether you northern or the southern hemisphere. Simply turn to the corresponds to the month and hemisphere in which you serve. Remember that the sky is large, and the size of the ns may surprise you. And don't expect the constellations to ok much like the people, animals or objects they represent— rs who named them had vivid imaginations!

MAGNITUDE

he brightness of celestial objects as seen stronomers use "apparent magnitude." y, the larger the magnitude, the fainter he brightest have magnitudes less than ite magnitude" is the object's apparent seen from a distance of 32 light-years The scale is logarithmic—1 magnitude o a factor of 2.5 in brightness.

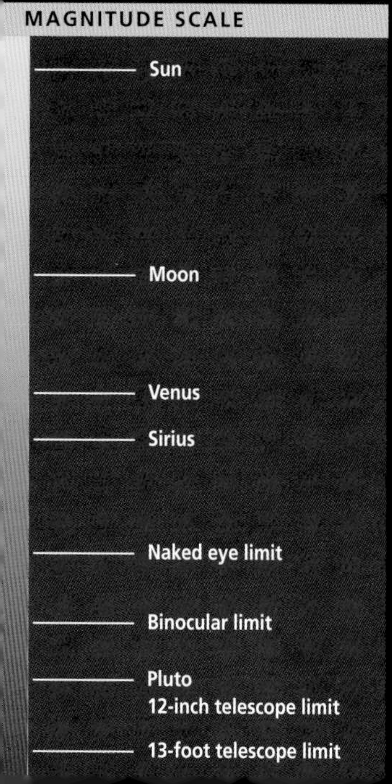

MAGNITUDE SCALE

- Sun
- Moon
- Venus
- Sirius
- Naked eye limit
- Binocular limit
- Pluto
- 12-inch telescope limit
- 13-foot telescope limit

Star sizes correspond to their apparent magnitude—the brightness of stars as seen from Earth

Constellations are indicated by an unbroken line

Stars are visible at latitudes 10 to 15 degrees north and south of this position

Deepsky objects are

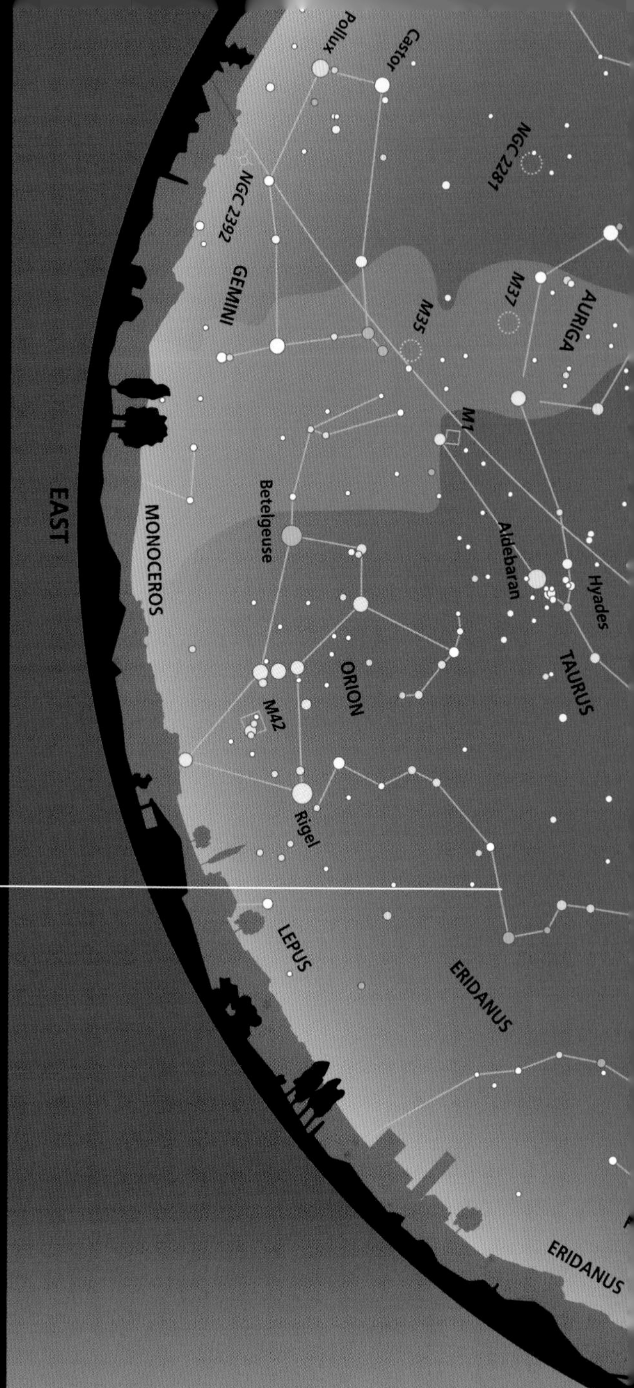

WHOLE SKY VIEW 40°N
November 1, 10 pm
November 15, 9 pm
(add 1 hour for DST)

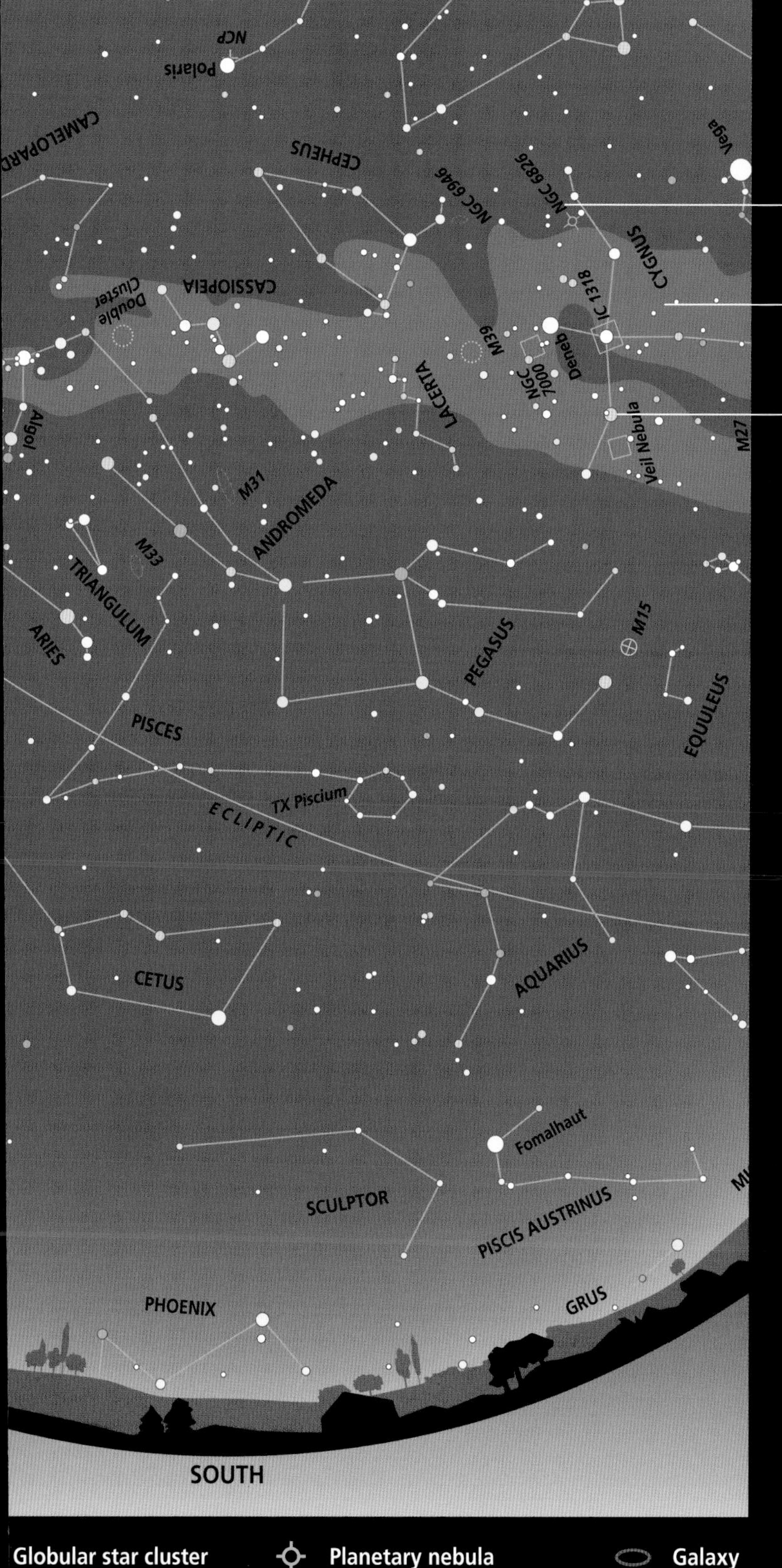

Celestial objects are indicated by their most common name—Messier or NGC catalog number

The Milky Way Galaxy is indicated on the star maps by a strip of light blue streaking across the sky

True star color is indicated

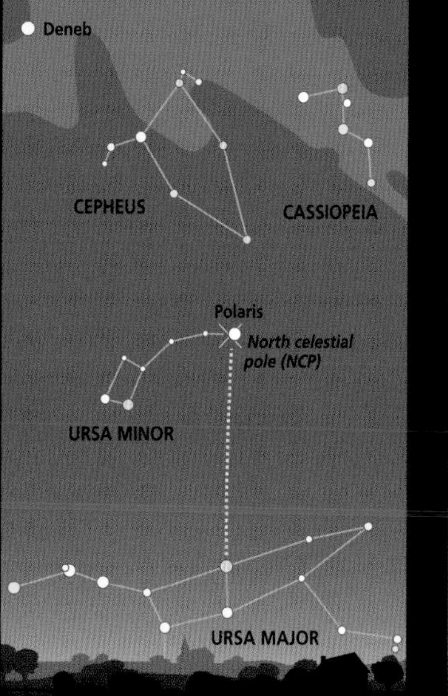

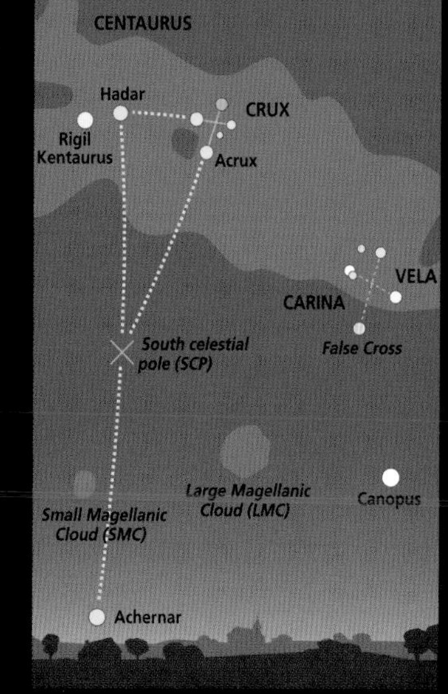

FINDING NORTH IN THE NORTHERN HEMISPHERE

Finding the north polar star, Polaris, is easy. First, locate the famous pattern of stars known as the Plough or the Big Dipper. These are the seven brightest stars in the constellation of Ursa Major, and they form a conspicuous saucepan shape. Depending on where you are and the time of year, the Plough may be upside-down or sideways. Look for it in the north and you should spot it—it spans about 40 degrees of sky. Once you find the Plough, pinpoint the two stars that mark the side of the saucepan's "bowl," farthest from the "handle." These are known as the pointers, as they point almost directly to Polaris. Follow the line they make, in a direction out of the bowl, and the first bright star you come to will be the North Star. Polaris is part of the constellation Ursa Minor.

FINDING SOUTH IN THE SOUTHERN HEMISPHERE

The star closest to the south celestial pole is in the constellation Octans the Octant—it's called Sigma Octantis, or Polaris Australe. However, at magnitude 5.4, Sigma Octantis is only just visible to the naked eye on a dark night. It is not at all as conspicuous as Polaris in the northern hemisphere, and it is not therefore much use in navigation. To locate the general position of the south pole, extend the long axis of Crux, the Southern Cross, four and a half times.

Northern hemisphere—whole sky overview

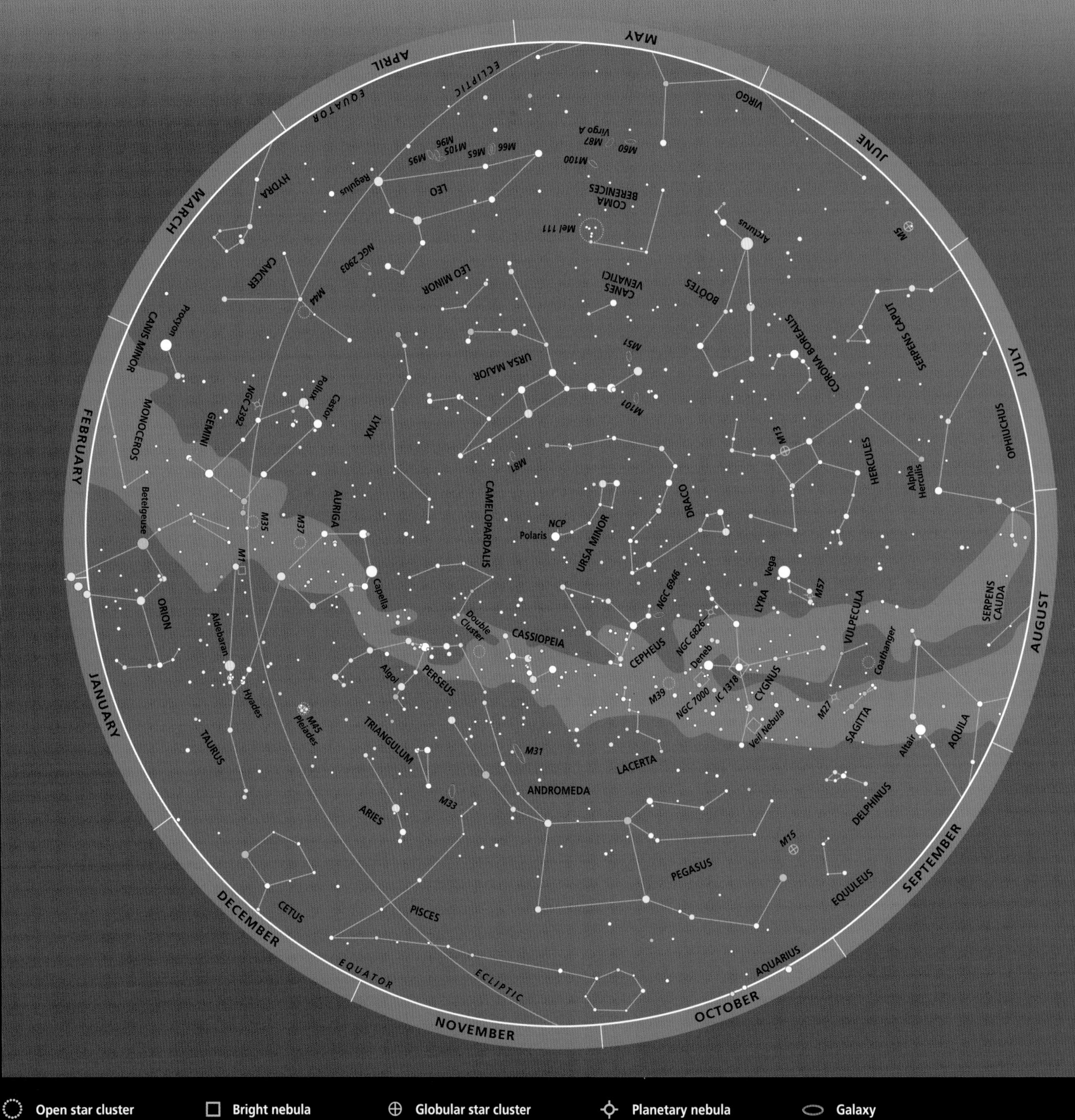

MAGNITUDE SCALE

0.0 +
0.1–0.5
0.6–1.0
1.1–1.5
1.6–2.0
2.1–2.5
2.6–3.0
3.1–3.5
3.6–4.0
4.1–4.5
4.6–5.0

Open star cluster
Bright nebula
Globular star cluster
Planetary nebula
Galaxy

Southern hemisphere—whole sky overview

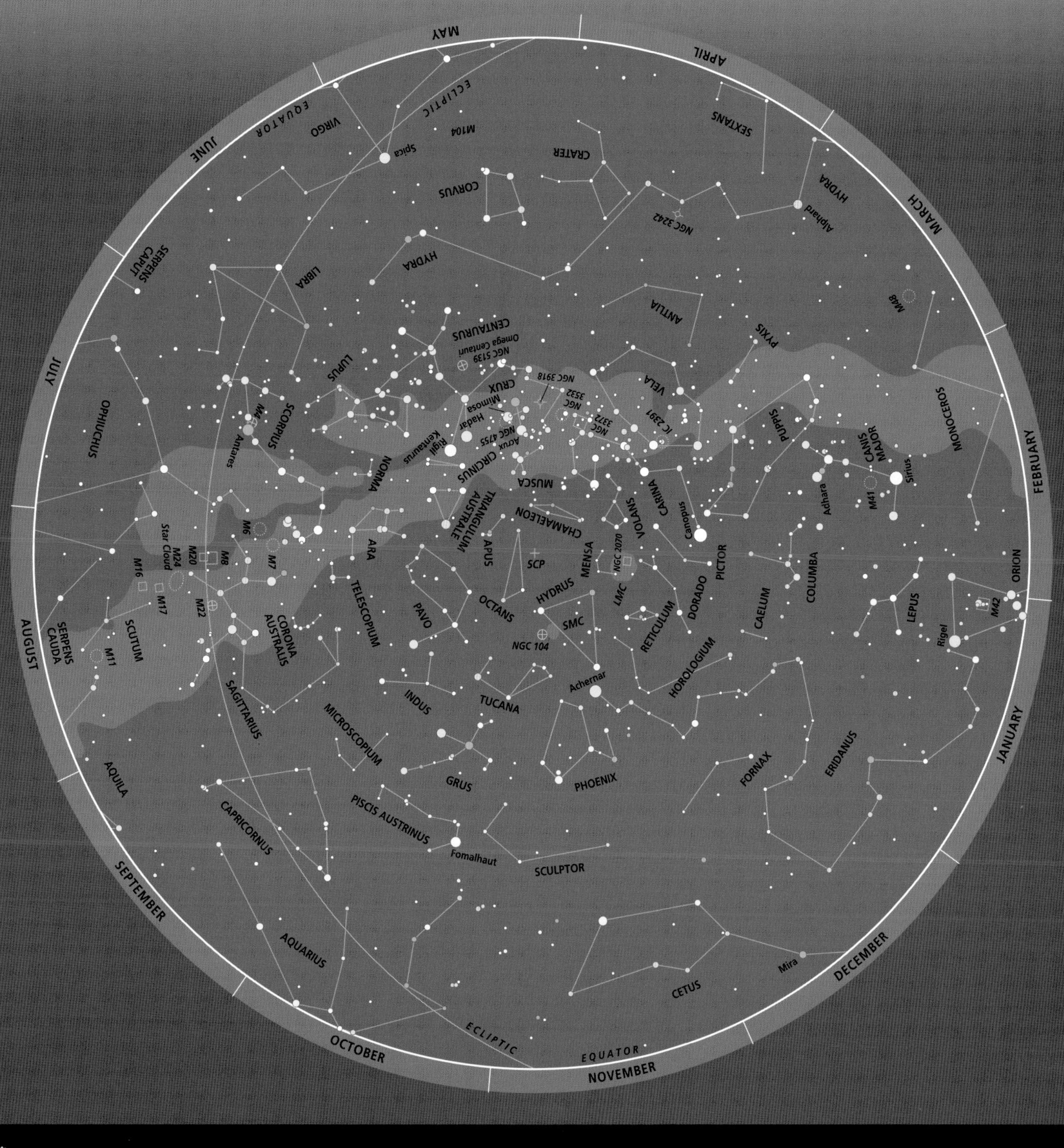

January stars of the northern skies

January is a time of bright winter constellations. As the night unfolds, Orion the Hunter climbs steadily in unending pursuit of Taurus the Bull, chased by his two faithful hunting dogs Canis Major and Canis Minor. Bright Capella in Auriga, Castor and Pollux sweep past the zenith. Meanwhile in the north, Ursa Major emerges perpendicularly from the horizon as it circles Polaris, while the familiar "M" shape of Queen Cassiopeia, along with the constellations of Pegasus, Andromeda and Perseus, slowly descend toward the northwestern horizon.

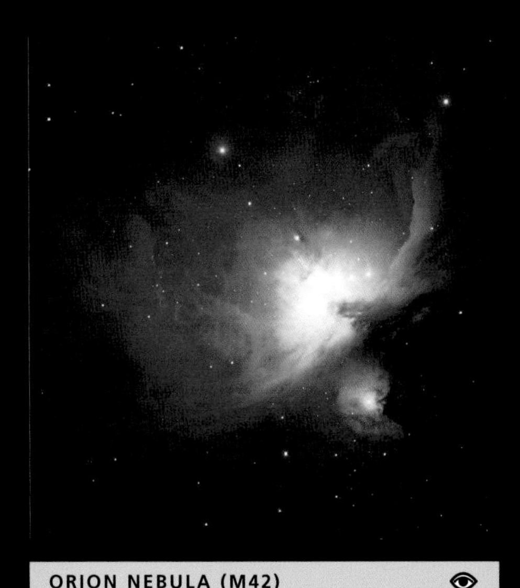

ORION NEBULA (M42)

Object type Diffuse nebula
Constellation Orion
Also known as NGC 1976, Great Nebula in Orion
Description Clearly visible to the naked eye as a bright fuzzy patch even from a city, the Orion Nebula is one of the northern sky's most famous objects. Its location in Orion's sword, just south of the leftmost star in the belt, also makes it an easy object to find. The nebula itself covers as much area as four full Moons, but this is just a small part of a much larger gas cloud that extends over 10 degrees. Although you can see M42 moderately well in binoculars, to truly appreciate it you will need a 6- to 8-inch (150–200 mm) telescope. This size scope will reveal glowing wisps encrusted with numerous stars—the brightest of which are the famous four Trapezium stars at the nebula's striking heart.
Apparent magnitude 4
Apparent size 66 x 60 arcminutes
Actual size 28 light-years
Distance 1600 light-years
Other nebulae in this sky Crab Nebula (M1)
Ghost of Jupiter (NGC 3242)

M37 CLUSTER

Object type Open cluster
Constellation Auriga
Also known as NGC 2099
Description M37 is one of several bright open star clusters in Auriga, forming an elbow shape with nearby M36 and M38. M37 is itself the richest and most striking of the three open clusters. It houses over 200 bright stars spread across an area comparable to the Moon—with up to 200 more fainter stars sprinkled throughout. Compared to its neighboring clusters, M37 is also more evolved, containing a considerable number of red giants and estimated to be 300 million years old. To find it, point your binoculars midway between Theta and Beta Auriga, then scan a couple of degrees toward Gemini. They will reveal a misty patch, while a small telescope will resolve the cluster's many stars.
Apparent magnitude 5.6
Apparent size 24 arcminutes
Actual size 29 light-years
Distance 4100 light-years
Other clusters in this sky Hyades
Beehive Cluster (M44)

PINWHEEL GALAXY (M33)

Object type Spiral galaxy
Constellation Triangulum
Also known as NGC 598
Description The Pinwheel belongs to the Local Group galaxy cluster that includes our own Milky Way. It is a spiral of class Sc, having open spiral arms, and is seen almost face-on from our vantage point. Its angular size is twice that of the full Moon. However, the Pinwheel is a fairly small galaxy. It appears large only because it is relatively close. Its extremely large apparent size does mean, though, that its light is spread thinly, making it difficult to see. It may just be visible to the keenest naked eye, but you will need an 8-inch (200 mm) instrument to resolve any sort of spiral structure. Note that the name "Pinwheel" is often applied to other galaxies, including Ursa Major's M101.
Apparent magnitude 5.7
Apparent size 70 x 40 arcminutes
Actual size 55,000 light-years
Distance 3,000,000 light-years
Other galaxies in this sky M65
Whirlpool Galaxy (M51)

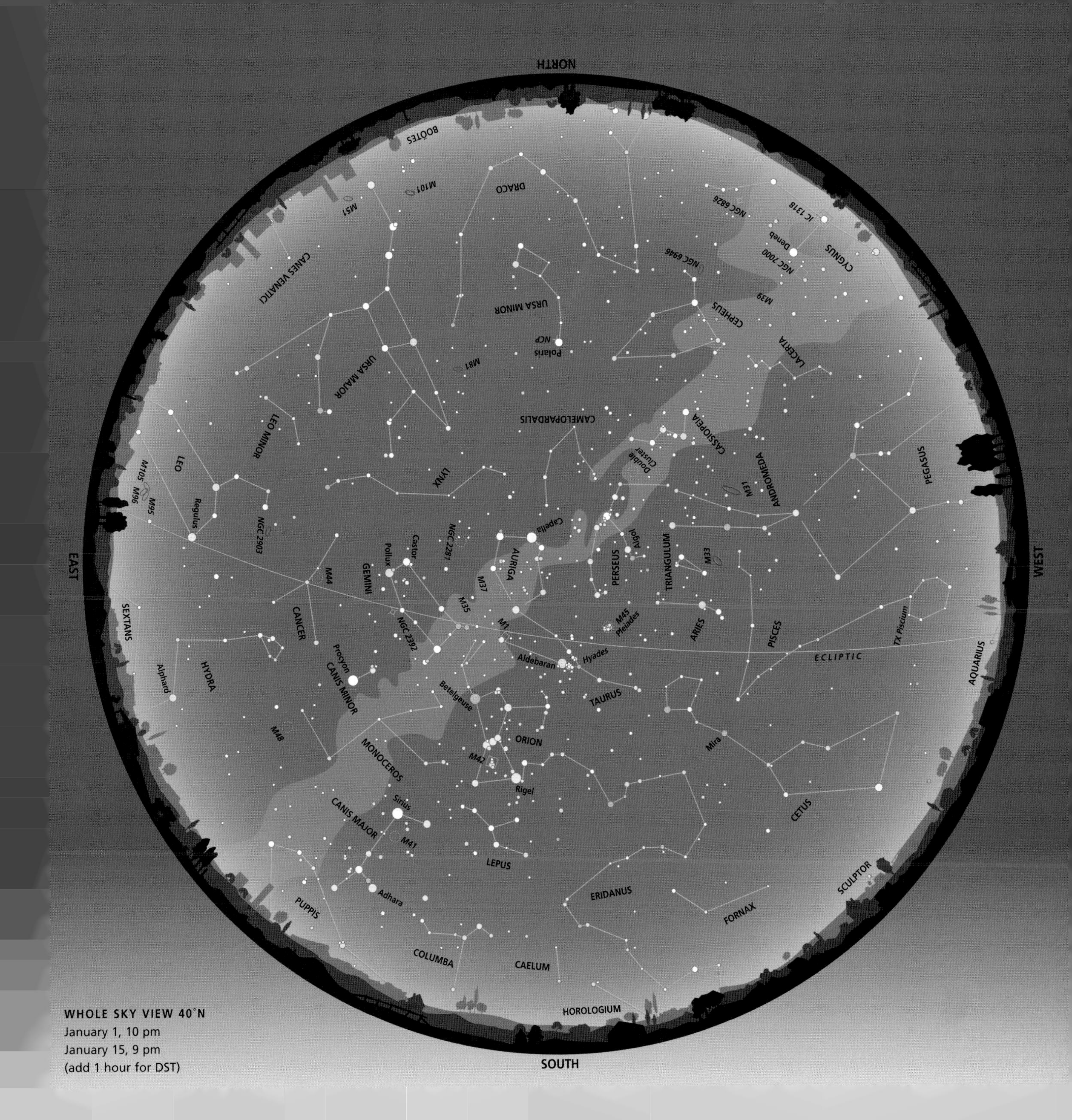

WHOLE SKY VIEW 40°N
January 1, 10 pm
January 15, 9 pm
(add 1 hour for DST)

February stars of the northern skies

Orion, Taurus, Canis Major and Auriga are still the main players in the hours after sunset. From the south, Orion pursues Taurus across the heavens toward the west, with both setting early in the new day. Capella in Auriga and the Gemini Twins Castor and Pollux follow, but remain high and well placed for viewing for most of the night. By 2 am, Ursa Major is high overhead as it leapfrogs upside-down over its smaller northern neighbor, Canis Minor. Meanwhile, in the east, Arcturus in Boötes shines brilliant orange as it ascends into the night.

CRAB NEBULA (M1)

Object type Supernova remnant
Constellation Taurus
Also known as NGC 1952
Description The Crab marks the spot in Taurus where a massive star exploded at the end of its life—an event known as a supernova. It was witnessed and recorded by diligent Chinese astronomers in 1054. Nearly one thousand years later, the shattered remains of the jettisoned star have spread out to form a tangled web of tendrils several light-years across—the furthest arms still racing outward at around 600 miles (1000 km) per second. M1 is located midway between the ecliptic and the star Zeta Tauri. With binoculars it appears as a small fuzzy blob, or as a larger but still fuzzy oval in small telescopes. To see any sort of detail in the Crab requires a good 10-inch (250 mm) instrument.
Apparent magnitude 8.2
Apparent size 6 x 4 arcminutes
Actual size 11 x 7 light-years
Distance 6300 light-years
Other nebulae in this sky Eskimo Nebula (NGC 2392) Orion Nebula (M42)

THE HYADES

Object type Open cluster
Constellation Taurus
Also known as Melotte 25
Description The Hyades is a large, V-shaped grouping of stars spanning about 10 lunar diameters and situated near the brilliant orange Aldebaran that marks the single, unblinking eye of Taurus the Bull (seen clearly in the image above). Altogether, this cluster contains about 200 stars, including many doubles and binaries, and at around 700 million years old it is much younger than the Solar System. The Hyades is an important yardstick for astronomers, who use it to calibrate astronomical distances. You can easily pick out the brightest members with the naked eye, but a pair of binoculars, or a low-power, wide-angle telescope, provides the best view.
Apparent magnitude 0.5
Apparent size 5 degrees
Actual size 15 light-years
Distance 150 light-years
Other clusters in this sky NGC 2548 Beehive Cluster (M44)

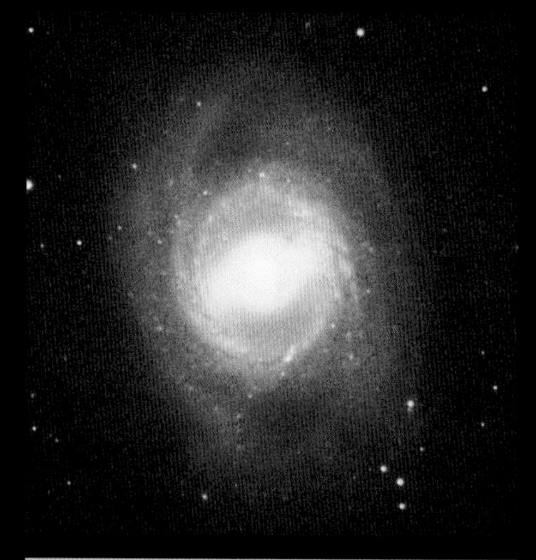

M95 GALAXY

Object type Barred spiral galaxy
Constellation Leo
Also known as NGC 3351
Description This galaxy has an extremely bright central core, which is surrounded by wisps of nebulosity. It is seen almost face-on, and has almost perfectly circular, spiral arms. It is a member of the Leo I galaxy group, which includes M96, M105 and several fainter NGC galaxies. M95 was one of the galaxies used in an experiment carried out using the Hubble Space Telescope, to estimate the so-called Hubble Constant which is related to the age of the Universe. Observationally, M95 is faint and even a 10-inch (250 mm) telescope will be hard pressed to reveal much in the way of its obvious spiral structure. However, the intensity of the core is evident in a 4-inch (100 mm) instrument.
Apparent magnitude 9.7
Apparent size 6.1 x 3.9 arcminutes
Actual size 63,000 light-years
Distance 36,000,000 light-years
Other galaxies in this sky M65 Whirlpool Galaxy (M51)

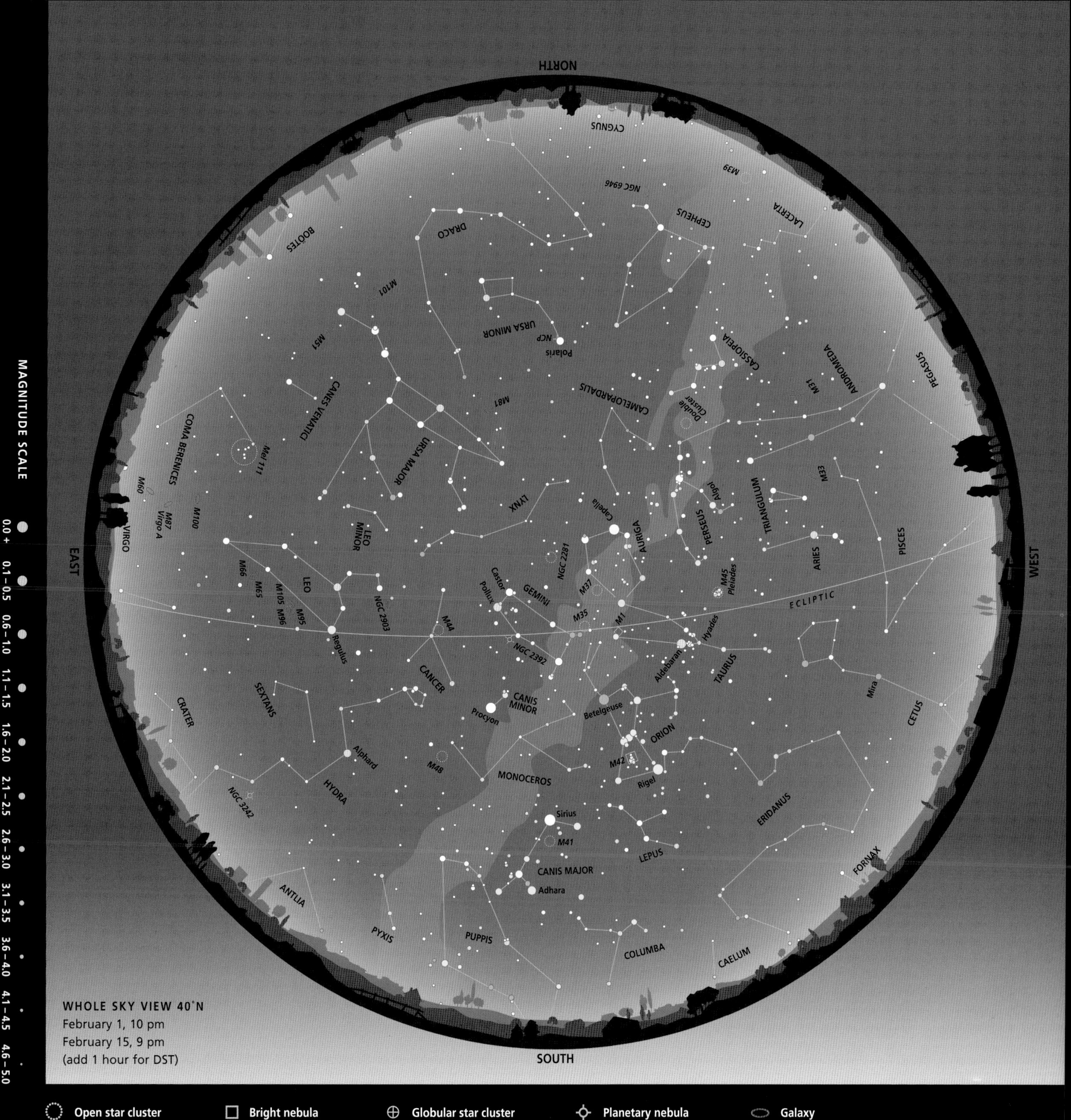
NORTH
SOUTH
EAST
WEST
WHOLE SKY VIEW 40°N
February 1, 10 pm
February 15, 9 pm
(add 1 hour for DST)
MAGNITUDE SCALE
0.0 +
0.1–0.5
0.6–1.0
1.1–1.5
1.6–2.0
2.1–2.5
2.6–3.0
3.1–3.5
3.6–4.0
4.1–4.5
4.6–5.0
Open star cluster
Bright nebula
Globular star cluster
Planetary nebula
Galaxy
CYGNUS
CEPHEUS
LACERTA
DRACO
BOÖTES
URSA MINOR
Polaris
CASSIOPEIA
ANDROMEDA
PEGASUS
CAMELOPARDALIS
CANES VENATICI
URSA MAJOR
COMA BERENICES
LYNX
PERSEUS
TRIANGULUM
AURIGA
Capella
Algol
LEO MINOR
VIRGO
PISCES
ARIES
LEO
GEMINI
Castor
Pollux
ECLIPTIC
Regulus
CANCER
TAURUS
Aldebaran
Hyades
SEXTANS
CANIS MINOR
Procyon
Betelgeuse
ORION
CETUS
Mira
CRATER
Alphard
HYDRA
MONOCEROS
Rigel
ERIDANUS
Sirius
LEPUS
CANIS MAJOR
Adhara
FORNAX
ANTLIA
PYXIS
PUPPIS
COLUMBA
CAELUM
M39
NGC 6946
M101
M51
M81
M31
Double Cluster
M33
Mel 111
M60
M87
Virgo A
M100
NGC 2281
M45
Pleiades
M66
M65
M105
M96
M95
NGC 2903
M37
M35
M1
M44
NGC 2392
M48
M42
NGC 3242
M41

March stars of the northern skies

With spring in the air, Orion and Taurus still corner the southern skies, but appear lower as the month progresses. Following the stars in Orion's belt downward finds Sirius the Dog Star, the sky's brightest star, which reaches its maximum height above the horizon this month. Ursa Major is well placed, the familiar saucepan shape dominating the north. Meanwhile, orange Arcturus climbs higher as March plays out, followed by Hercules. High in the southwest, Leo's stars, dominated by Regulus, form an urgent celestial question-mark.

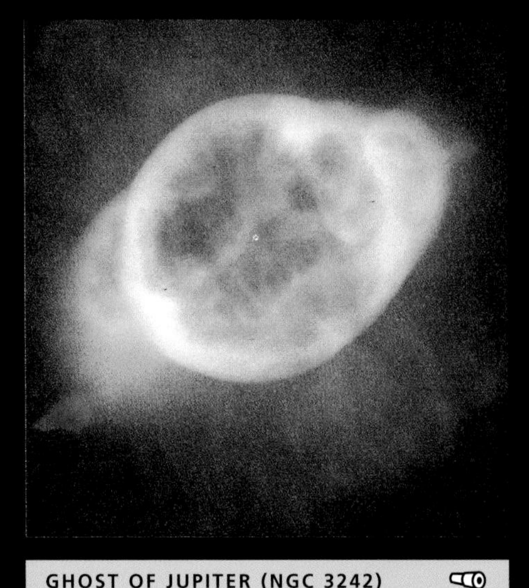

GHOST OF JUPITER (NGC 3242)

Object type Planetary nebula
Constellation Hydra
Description Through a low-power telescope this nebula is similar in size and appearance to the planet Jupiter, hence its name. It can be found near the middle of the constellation of Hydra, 2 degrees away from Mu Hydrae. A small instrument exposes little more than an elongated planet-like disk with a slight blue–green tint. A 6-inch (150 mm) telescope shows the object's unusual brightness and reveals its sharp boundary. Meanwhile, for those with 10-inch (250 mm) apertures or more, a greater level of detail awaits—the center of the nebula contains an elliptical ring that bears some resemblance to a human eye. This illusion is completed by the twelfth-magnitude central star of the nebula—a white dwarf, seen in the above image.
Apparent magnitude 8.6
Apparent size 20 arcminutes
Actual size 16 light-years
Distance 2600 light-years
Other nebulae in this sky Crab Nebula (M1) Orion Nebula (M42)

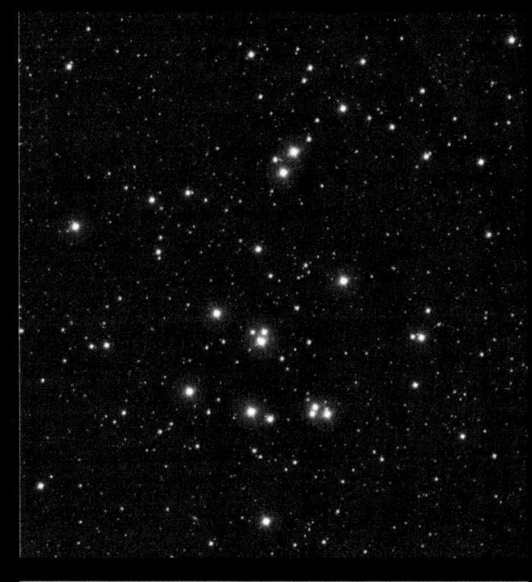

BEEHIVE CLUSTER (M44)

Object type Open cluster
Constellation Cancer
Also known as Praesepe, NGC 2632
Description The Beehive Cluster is a third-magnitude grouping of about 200 stars spread over a huge patch of sky—the equivalent size of nine full Moons. Because of its brightness and proximity—it is one of the nearest open clusters to the Solar System—it is a naked-eye object and has been known since antiquity. The Greek astronomer Hipparchus of Nicaea catalogued it as a "cloudy star" in 130 BC, but it wasn't until 1609 that Galileo turned his attention to it and found it to be a cluster. Although a naked-eye object, it requires a dark viewing location far from urban light pollution. Either binoculars or a low-power telescope with a wide field are the best tools for viewing this object.
Apparent magnitude 3.1
Apparent size 1.5 degrees
Actual size 15 light-years
Distance 580 light-years
Other clusters in this sky M37 Coma Star Cluster (Melotte 111)

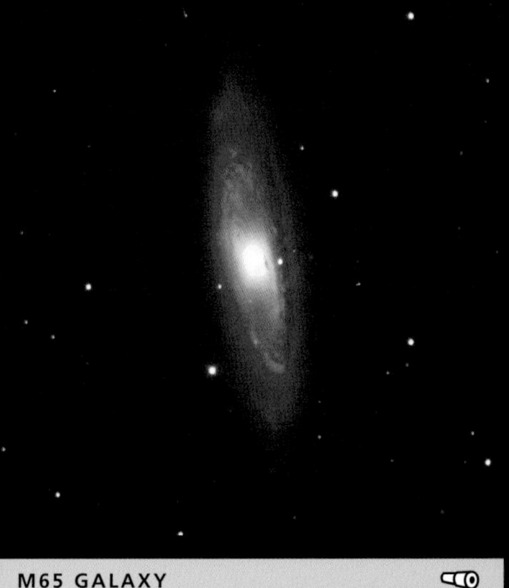

M65 GALAXY

Object type Spiral galaxy
Constellation Leo
Also known as NGC 3263
Description Technically, M65 is classed as a "peculiar" galaxy, because it has a slight distortion owing to its gravitational interaction with M66 and NGC 3268, two neighboring spiral galaxies. Together, these three galaxies form the so-called Leo Triplet. However, despite its purported peculiarity, M65 looks like a normal spiral of class Sa, with tightly wound arms and a dark lane of dust on the facing edge. You can see M65 in binoculars, but a low-power telescope provides the best view of M65 and M66, which form a lovely pair. To find them, aim a couple of degrees northwest of Theta Leonis. M66 is the smaller of the pair, and a 4-inch (100 mm) telescope will begin to reveal lanes of dust.
Apparent magnitude 9.3
Apparent size 8 x 2.5 arcminutes
Actual size 82,000 light-years
Distance 35,000,000 light-years
Other galaxies in this sky M95 Sombrero Galaxy (M104)

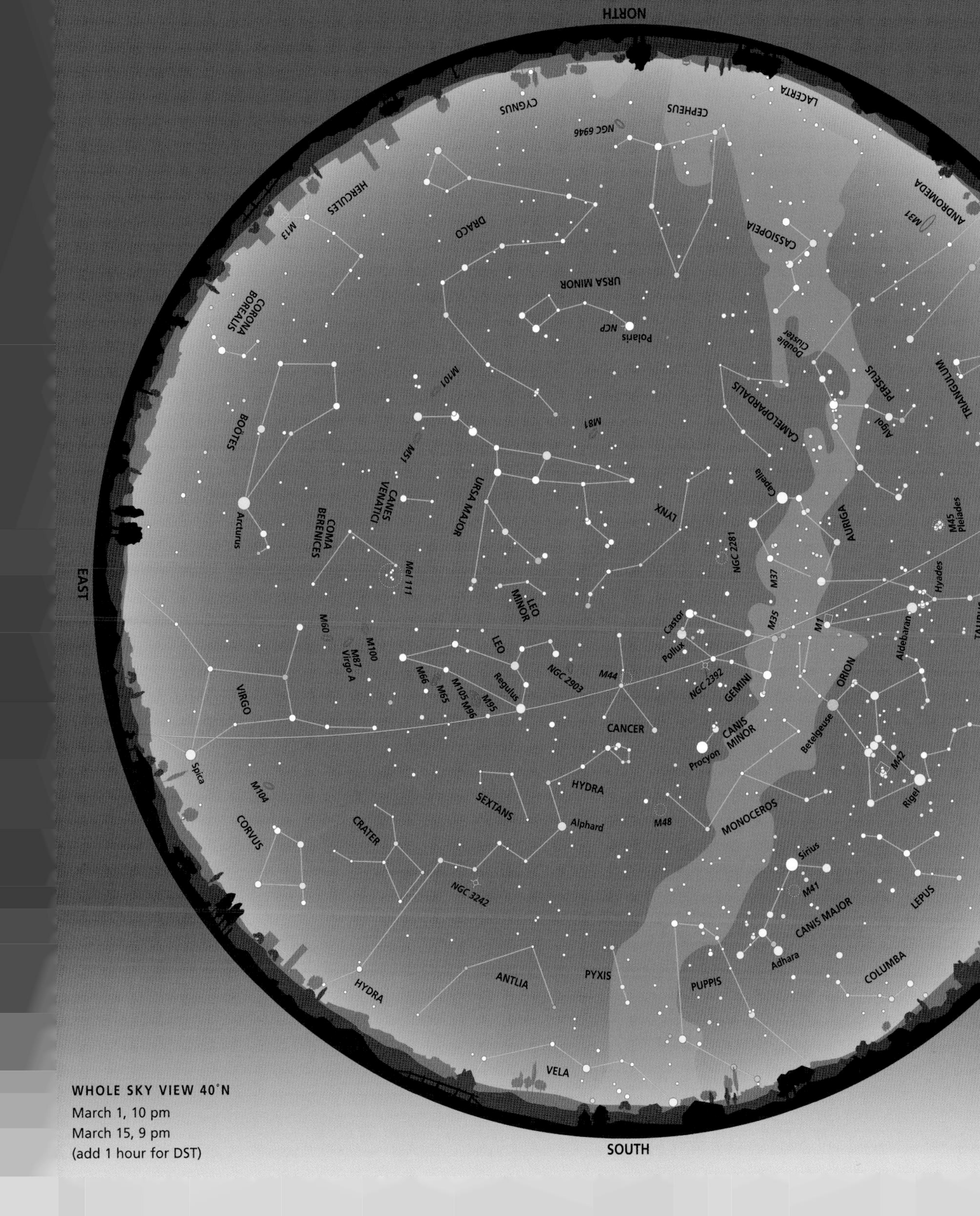

WHOLE SKY VIEW 40°N
March 1, 10 pm
March 15, 9 pm
(add 1 hour for DST)

April stars of the northern skies

The bright winter stars have all but left the scene. The dominant constellation is now Ursa Major, moving high overhead. Arcturus in Boötes is now unmistakable, similarly high in the southeastern sky. To find it, follow the curve of the handle of Ursa Major's saucepan. Extend the curve a similar distance to reach the first-magnitude Spica, in Virgo. Meanwhile, Leo dives head-first toward the western horizon, leaving the east for Lyra and Cygnus, the bright summer constellations, which become more prominent as the night progresses.

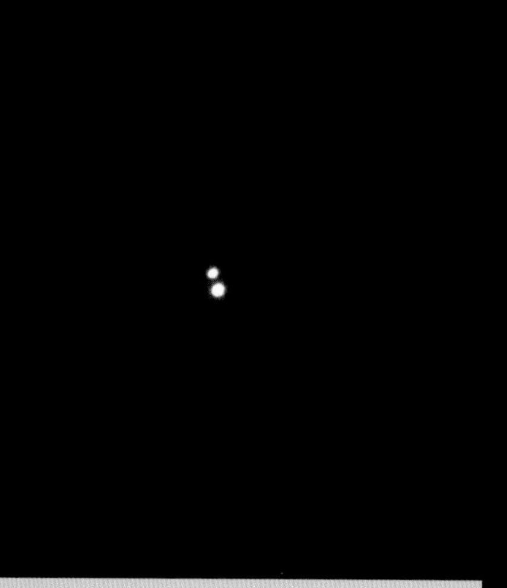

MIZAR

Object type Multiple star
Constellation Ursa Major
Description Mizar is the second star from the end of the Plough's handle in Ursa Major. Even with the naked eye, you can see that Mizar is a double star, paired with nearby Alcor, 12 arcminutes away. Mizar and Alcor are not physically associated—they merely appear next to each other in the sky. However, in 1883, Italian astronomer Giovanni Riccioli discovered that Mizar itself is a spectroscopic binary—the first star found to have an orbiting companion, known as fourth-magnitude partner Mizar B. Like a fractal, Mizar reveals more detail the closer you look, for Mizar B is also a spectroscopic binary, making this a complex three-star system. Binoculars reveal the Mizar–Alcor pair well, while a small telescope will also reveal Mizar's binary companion Mizar B.
Apparent magnitude 2.3
Apparent separation 14 arcseconds
Actual separation 0.005 light-year
Distance 74 light-years
Other significant stars in this sky Polaris
Regulus

M48 CLUSTER

Object type Open cluster
Constellation Hydra
Also known as NGC 2548
Description M48 is Charles Messier's "lost" object. He included it on his famous catalog in 1771, but because he reported its position incorrectly, it disappeared and was only "discovered" again officially in 1959. This open cluster lies just inside Hydra, near the border with Monoceros, away from any bright stars. A keen eye combined with a dark sky will reveal M48 without any optical aid. However, to get the best view, use binoculars or a small telescope. This will reveal about 50 stars brighter than magnitude 13. Most of these are concentrated in a dense core that extends for about 30 arcminutes or the apparent diameter of the Moon—half the cluster's total span.
Apparent magnitude 5.8
Apparent size 54 arcminutes
Actual size 28 light-years
Distance 1700 light-years
Other clusters in this sky M37
Coma Star Cluster (Melotte 111)

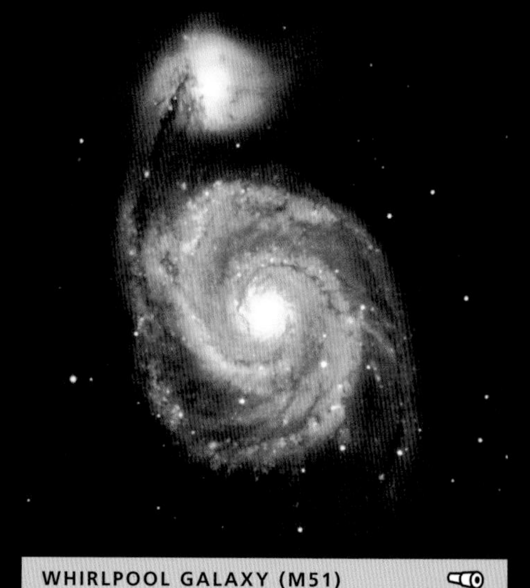

WHIRLPOOL GALAXY (M51)

Object type Spiral galaxy
Constellation Canes Venatici
Also known as NGC 5194
Description M51 is a famous face-on spiral galaxy with a distinctive hurricane-like appearance from which it derives its name. It is an Sc-type galaxy, having open spiral arms, and was the first spiral galaxy in which spiral arms were clearly identified. 7x50 binoculars reveal little more than a fuzzy blob, while in larger binoculars the object appears more elongated. To begin to see the spiral arms well, a telescope with an 8-inch (200 mm) aperture is needed, while a 12-inch (300 mm) scope will show the object in its true glory. At the end of one of the spiral arms is a smaller, background galaxy called NGC 5195 (seen in the above image), with which the Whirlpool is engaged in a gravitational tug-of-war.
Apparent magnitude 8.4
Apparent size 10 x 5.5 arcminutes
Actual size 100,000 light-years
Distance 34,000,000 light-years
Other galaxies in this sky M65
Sombrero Galaxy (M104)

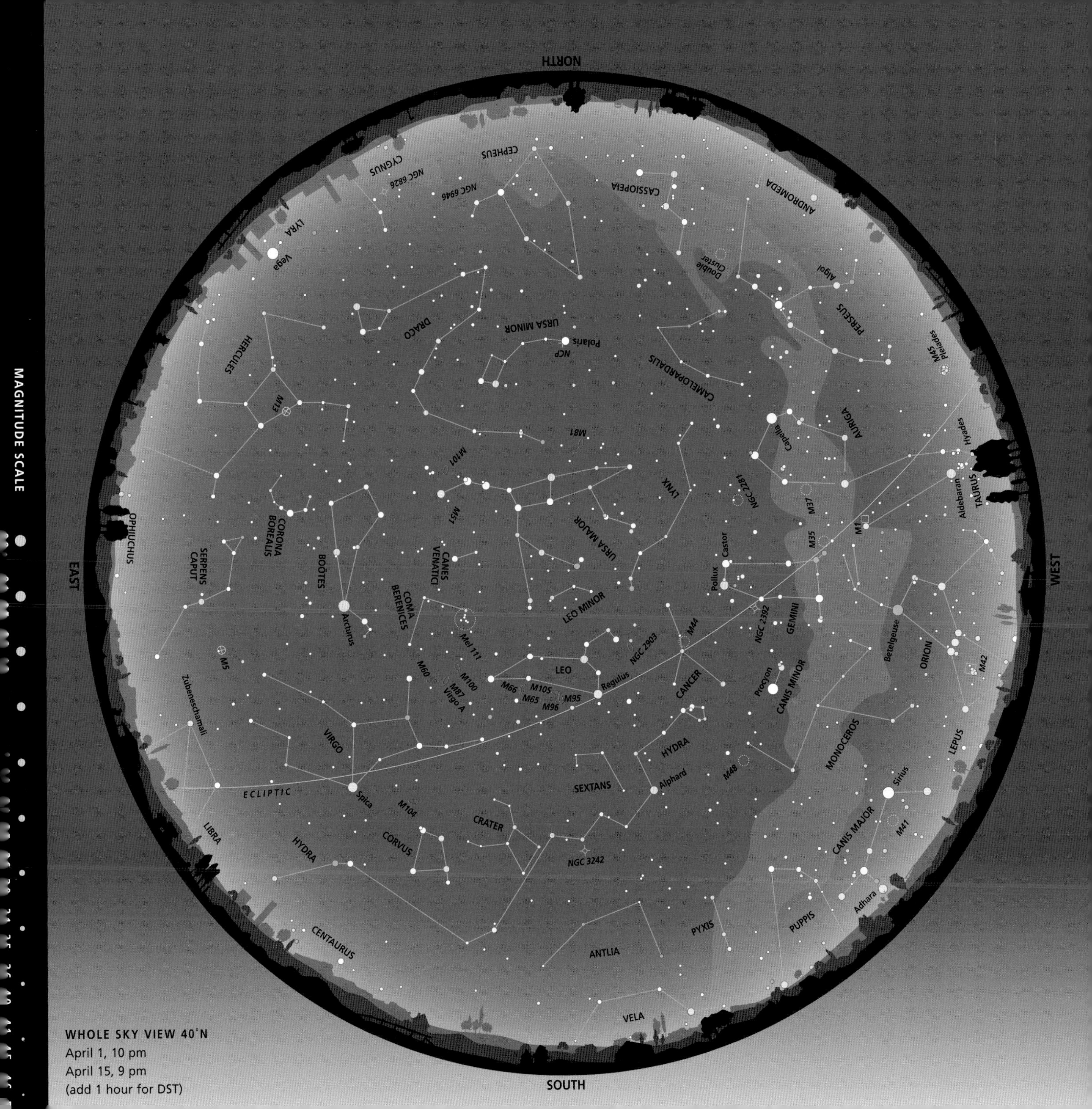
MAGNITUDE SCALE
NORTH
EAST
WEST
SOUTH
CYGNUS
NGC 6826
CEPHEUS
NGC 6946
CASSIOPEIA
ANDROMEDA
LYRA
Vega
Double Cluster
Algol
PERSEUS
DRACO
URSA MINOR
Polaris
NCP
CAMELOPARDALIS
M45 Pleiades
HERCULES
M13
AURIGA
Capella
M81
Hyades
M101
LYNX
NGC 2281
Aldebaran
TAURUS
M37
M51
URSA MAJOR
M1
CORONA BOREALIS
OPHIUCHUS
SERPENS CAPUT
BOÖTES
CANES VENATICI
M35
Castor
Pollux
COMA BERENICES
LEO MINOR
GEMINI
NGC 2392
Arcturus
Mel 111
M44
Betelgeuse
ORION
M5
M42
NGC 2903
M60
M100
LEO
Regulus
CANCER
Procyon
CANIS MINOR
M87
Virgo A
M66
M105
M65
M95
M96
Zubeneschamali
VIRGO
HYDRA
MONOCEROS
LEPUS
M48
SEXTANS
Alphard
ECLIPTIC
Spica
M104
Sirius
CRATER
CANIS MAJOR
M41
LIBRA
HYDRA
CORVUS
NGC 3242
Adhara
PUPPIS
PYXIS
CENTAURUS
ANTLIA
VELA
WHOLE SKY VIEW 40°N
April 1, 10 pm
April 15, 9 pm
(add 1 hour for DST)

May stars of the northern skies

Winter's bright constellations are first lost in the haze of the western horizon, then sink out of site within an hour or two after sunset. Leo is still high but has edged farther toward the western horizon. Once again, the Great Bear's Plough stars dominate the north, and Arcturus and Spica reign in the east while Cassiopeia's celestial "W" skirts the northern horizon and rises into the night. As the night progresses, the constellations that will mark the coming summer skies are already becoming prominent in the east: Cygnus, Lyra and Aquila.

IC 1318 NEBULA

Object type Diffuse nebula
Constellation Cygnus
Also known as Butterfly Nebula
Description IC 1318 is sometimes known as the Butterfly Nebula. But confusingly, that name is also applied to three other nebulae: IC 2200 in Carina, the Little Dumbbell in Perseus, and the Bug Nebula in Scorpius. IC 1318 is a diffuse tangle of a nebula surrounding the star Gamma Cygni, just east of which the nebula is mottled and broken up by several small patches of dark gas and dust. Its size is larger than the Moon, which spreads its light so thinly that it is difficult to detect unaided, even at magnitude 2.3. To see this nebula, you will need at least a 10-inch (250 mm) telescope. But don't expect the dazzling array of color seen in the photograph above. It will appear as little more than a faint wisp.
Apparent magnitude 2.3
Apparent size 45 x 25 arcminutes
Actual size 240 light-years
Distance 3 light-years
Other nebulae in this sky Ring Nebula (M57) Blinking Planetary (NGC 6826)

COMA STAR CLUSTER

Object type Open cluster
Constellation Coma Berenices
Also known as Melotte 111
Description This very conspicuous cluster spans an area of around 80 full Moons in its entirety—slightly less than 5 degrees across. Despite its size, it was not included in either the New General Catalog or Charles Messier's catalog, because it was not known until 1938 that the stars within it truly form a physically bound cluster. However, in 1915 the Coma Star Cluster was included as number 111 on the Melotte Catalog—a catalog of clusters compiled by French astronomer Philibert Melotte. It contains about 40 to 50 stars, the brightest being fourth magnitude. The best way to see this cluster is with a pair of low-power binoculars. The cluster then fills the field of view and is a dazzling sight.
Apparent magnitude 1.8
Apparent size 4.5 degrees
Actual size 22 light-years
Distance 288 light-years
Other clusters in this sky The Coathanger Hercules Cluster (M13)

SOMBRERO GALAXY (M104)

Object type Spiral galaxy
Constellation Virgo
Also known as NGC 4594
Description The Sombrero is an Sa- or Sb-class spiral seen edge-on in Virgo, the brightest of the galaxies in that constellation and part of the giant Coma–Virgo cluster of galaxies. There is a conspicuous dark lane of dust girdling the galaxy's equator that cuts the image in two and, from Earth, gives it the appearance of the Mexican hat from which its name is derived. M104 was the first galaxy shown to be moving away from us, as a consequence of an expanding Universe. You can find it by first locating bright Spica, then moving 11 degrees to the west. A small telescope shows the Sombrero as a small blurry blob of gray light. With an 8-inch (200 mm) telescope or more you will easily pick out the dust lane.
Apparent magnitude 8.3
Apparent size 9 x 4 arcminutes
Actual size 130,000 light-years
Distance 50,000,000 light-years
Other galaxies in this sky M81 Whirlpool Galaxy (M51)

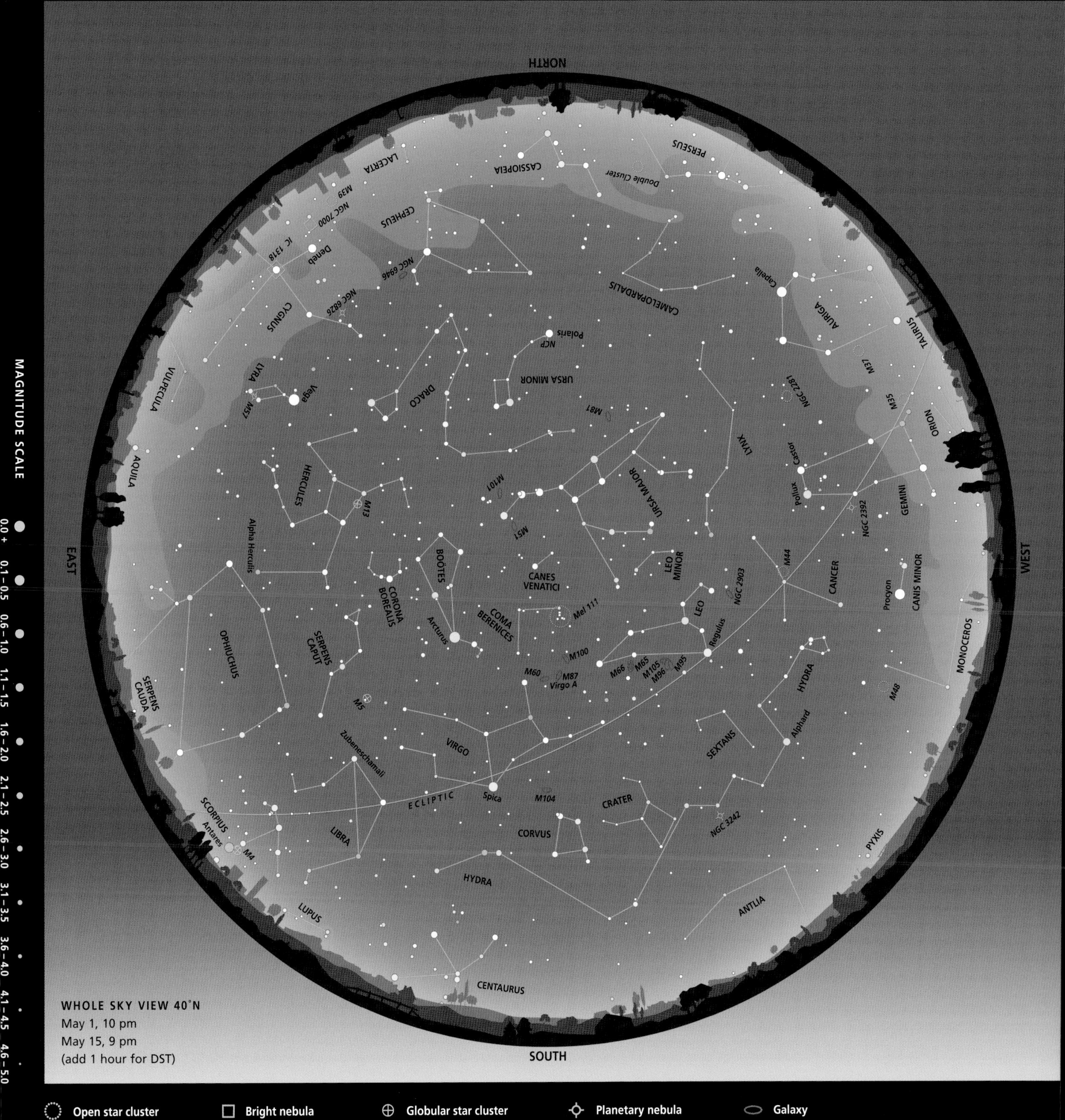
NORTH
SOUTH
EAST
WEST
MAGNITUDE SCALE
0.0 +
0.1 – 0.5
0.6 – 1.0
1.1 – 1.5
1.6 – 2.0
2.1 – 2.5
2.6 – 3.0
3.1 – 3.5
3.6 – 4.0
4.1 – 4.5
4.6 – 5.0
WHOLE SKY VIEW 40°N
May 1, 10 pm
May 15, 9 pm
(add 1 hour for DST)
Open star cluster
Bright nebula
Globular star cluster
Planetary nebula
Galaxy
CASSIOPEIA
PERSEUS
Double Cluster
LACERTA
M39
NGC 7000
CEPHEUS
IC 1318
Deneb
NGC 6946
CAMELOPARDALIS
Capella
AURIGA
TAURUS
CYGNUS
NGC 6826
Polaris
NCP
URSA MINOR
M37
NGC 2281
LYRA
Vega
M57
VULPECULA
DRACO
M81
M35
ORION
AQUILA
HERCULES
M101
URSA MAJOR
LYNX
Castor
Pollux
GEMINI
NGC 2392
M13
M51
Alpha Herculis
BOÖTES
CANES VENATICI
LEO MINOR
M44
CANCER
CANIS MINOR
Procyon
CORONA BOREALIS
COMA BERENICES
Mel 111
NGC 2903
LEO
OPHIUCHUS
SERPENS CAPUT
Arcturus
Regulus
MONOCEROS
M100
M60
M87
Virgo A
M66
M65
M105
M96
M95
HYDRA
M48
SERPENS CAUDA
M5
Zubeneschamali
VIRGO
SEXTANS
Alphard
ECLIPTIC
Spica
M104
CRATER
SCORPIUS
Antares
M4
LIBRA
CORVUS
NGC 3242
PYXIS
HYDRA
ANTLIA
LUPUS
CENTAURUS

June stars of the northern skies

Leo deserts the sky for hunting in more southerly latitudes. The Great Bear is now standing on its front paws as it rounds Polaris on its way to the northwestern horizon. Arcturus is close behind, becoming less prominent as spring gives way to summer. It is the high-latitude eastern skies that now dominate. Around midnight, Lyra passes high overhead, its brilliant blue Vega clearly defining the zenith by 1 am, with Cygnus the Swan and Aquila the Eagle always close by. As the night draws to a close, the Square of Pegasus is visible in the east.

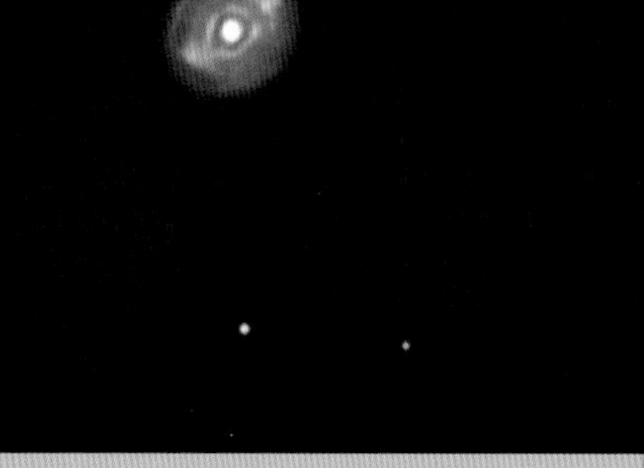

BLINKING PLANETARY (NGC 6826)

Object type Planetary nebula
Constellation Cygnus
Description This planetary nebula gets its name from an optical illusion. When you move your vision back and forth between the surrounding nebula and its tenth-magnitude central star, the nebulosity seems to appear and then vanish, giving the appearance of a blinking eye. You can find this nebula 1.3 degrees east of Theta Cygni, one of the two stars that form the wing tips of Cygnus. However, it is small, at around half an arcminute, and with a small telescope you will see little more than a faint star-like object. To see any sort of detail you will need an 8-inch (200 mm) telescope, which will show the object's bright neon green disk. Larger apertures and powers reveal two conspicuous red specks called "fliers"—which astronomers cannot yet explain.
Apparent magnitude 8.8
Apparent size 27 x 24 arcseconds
Actual size 0.4 light-year
Distance 3000 light-years
Other nebulae in this sky IC 1318

THE COATHANGER

Object type Open cluster
Constellation Vulpecula
Also known as Brocchi's Cluster, Collinder 399
Description The Coathanger is an open cluster of six bright stars, and possibly as many as 40 altogether—possibly, because it is not known for sure if all of the Coathanger stars form a real cluster. There is evidence that only the brightest half-dozen have a common distance, meaning that the rest may be just more distant background stars. Its brightest members, of fifth, sixth and seventh magnitude, trace an east–west arc, while another four stars form a hook shape. Overall, this gives the cluster a distinctive coathanger-like appearance. The Coathanger is best seen through binoculars or a finderscope. Larger instruments have too limiting a field of view to capture the extent of the cluster.
Apparent magnitude 3.6
Apparent size 60 arcminutes
Actual size 7 light-years
Distance 420 light-years
Other clusters in this sky Hercules Cluster

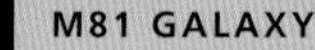

M81 GALAXY

Object type Spiral galaxy
Constellation Ursa Major
Also known as NGC 3031
Description M81 is one of several galaxies forming a small, nearby galaxy cluster called the M81 group. It has a bright central core with relatively faint spiral arms because of their lower stellar concentrations. It also has a remarkably symmetrical shape. This object is a treat—a bright galaxy easily visible in binoculars even from a city. To find it, trace your eye from Gamma to Alpha Ursae Majoris, then continue in the same direction for a similar distance. In a low-power telescope or binoculars you should be able to make out M81 and neighboring M82. With larger telescopes, the bright nucleus of M81 becomes more conspicuous, while an 8-inch (200 mm) instrument begins to show traces of the delicate spiral arms.
Apparent magnitude 6.9
Apparent size 21 x 10 arcminutes
Actual size 66,000 light-years
Distance 11,000,000 light-years
Other galaxies in this sky M101

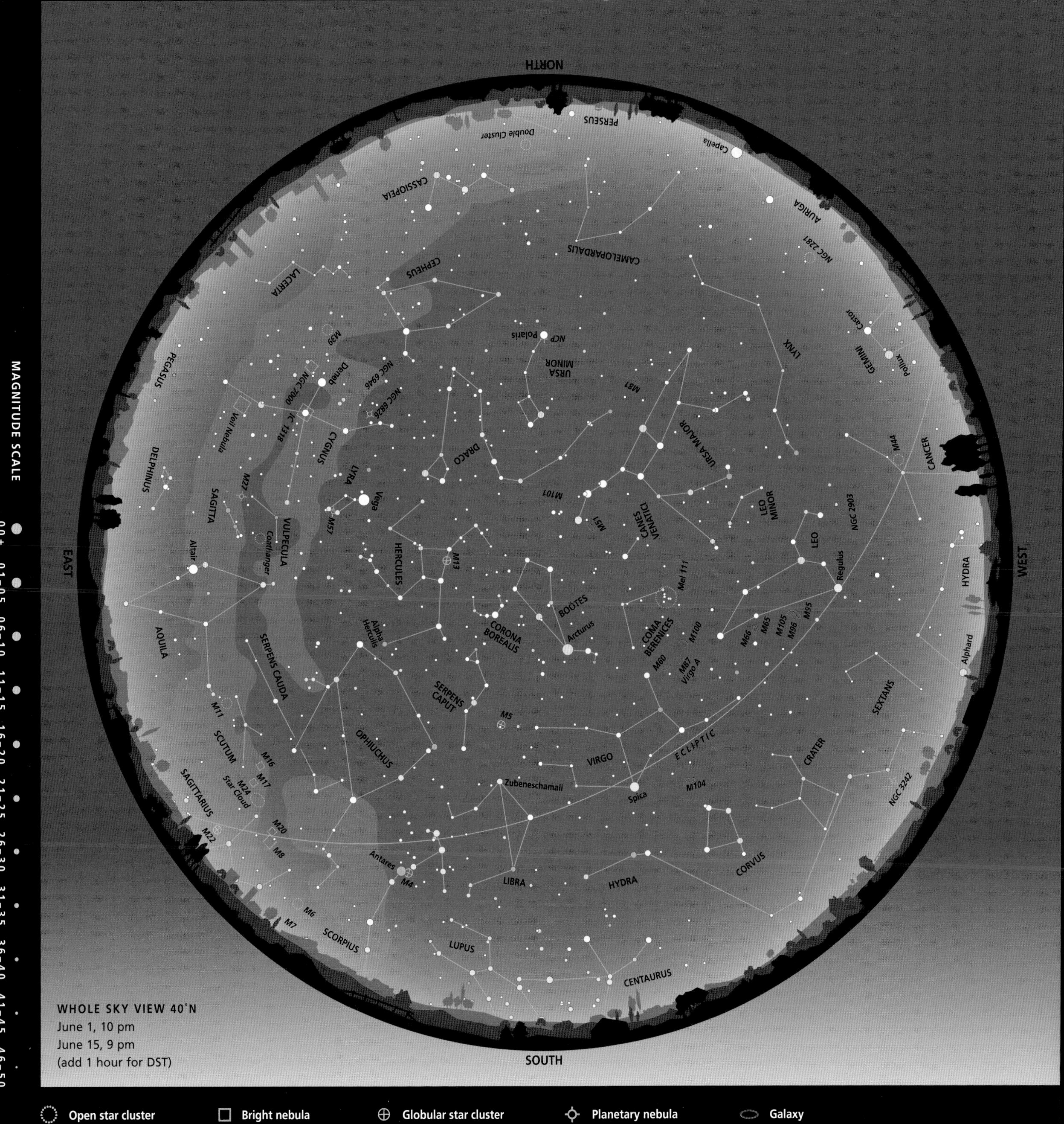

NORTH
SOUTH
EAST
WEST
MAGNITUDE SCALE
WHOLE SKY VIEW 40˚N
June 1, 10 pm
June 15, 9 pm
(add 1 hour for DST)
Open star cluster
Bright nebula
Globular star cluster
Planetary nebula
Galaxy

July stars of the northern skies

Summer is in full swing in the northern hemisphere and to mark the occasion, the bright stars Deneb in Cygnus, Vega in Lyra and Altair in Aquila, gather high overhead to form the famous Summer Triangle. If you are fortunate enough to have a dark sky, look directly upward and you will see Cygnus the Swan and Aquila the Eagle soaring through the brilliant band of stars that marks a spiral arm of our galaxy, the Milky Way. The bright stars of Ursa Major are lower down in the north now, giving way in the east to Pegasus and Perseus.

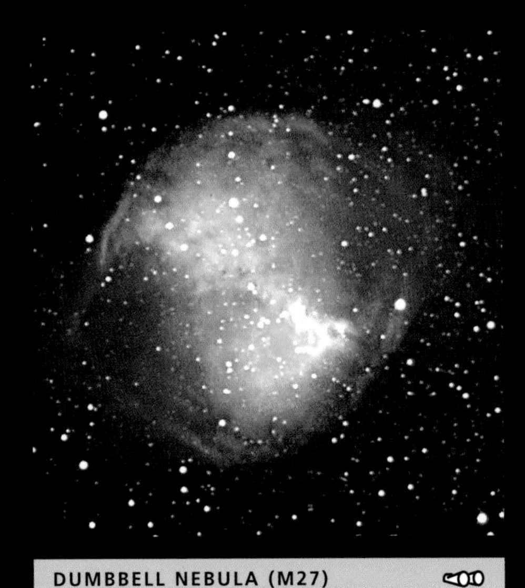

DUMBBELL NEBULA (M27)

Object type Planetary nebula
Constellation Vulpecula
Also known as NGC 6853
Description With no bright stars, Vulpecula is not a particularly exciting constellation. However, it does boast the Dumbbell Nebula. At greater than 2 light-years across, this famous object is one of the largest known planetary nebulae, and one of the best to pick out with a small telescope. You can locate it just 3 degrees north of Gamma Sagittae in the neighboring constellation. At seventh magnitude, the Dumbbell is visible in binoculars, though as little more than a blurry spot. But in a 4-inch (100 mm) telescope on a dark night, the object's shape becomes apparent. Larger apertures reveal still greater detail, but nothing less than 10 inches (250 mm) will reveal its central star.
Apparent magnitude 7.6
Apparent size 8 x 6 arcminutes
Actual size 2.7 light-years
Distance 1150 light-years
Other nebulae in this sky Blinking Planetary (NGC 6826) North American Nebula (NGC 7000)

HERCULES CLUSTER (M13)

Object type Globular cluster
Constellation Hercules
Also known as NGC 6205
Description M13, discovered by Edmond Halley in 1714, is without doubt the most spectacular globular cluster in the entire northern sky—a smooth globe of one million stars squeezed into a volume of space occupying 100 light-years. You can see M13 with the naked eye on a dark night using averted vision, when it appears as a faint smudge of light. The view is not much improved in binoculars. To really appreciate this object you need a 4-inch (100 mm) telescope. With this scope, you will begin to resolve individual stars glistening on the cluster's comet-like periphery. With an even-larger instrument, say 10 inches (250 mm), M13 is simply breathtaking—a city of stars radiating from a dense central core.
Apparent magnitude 5.9
Apparent size 15 arcminutes
Actual size 100 light-years
Distance 24,000 light-years
Other clusters in this sky M15 The Coathanger

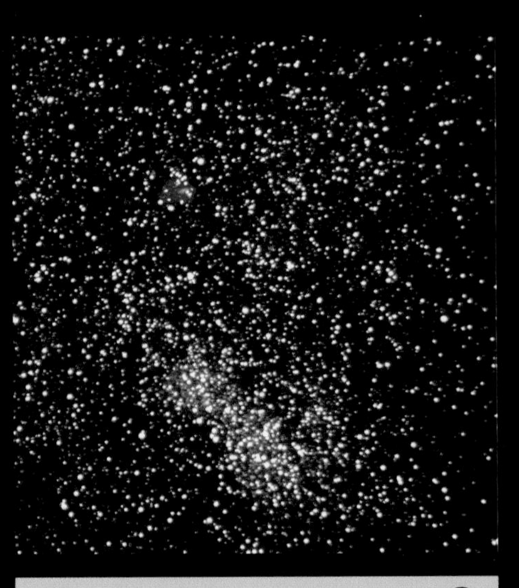

M24 STAR CLOUD

Object type Star cloud
Constellation Sagittarius
Also known as Sagittarius Star Cloud
Description M24 is a unique Messier object. Neither a nebula, a cluster nor a galaxy, it is a vast patch of stars that forms part of a spiral arm of the Milky Way. However, the star cloud does itself contain an eleventh-magnitude open star cluster, NGC 6603, which is sometimes referred to as M24 in error. The Sagittarius Star Cloud can be seen through a comparatively clear "window" of the Milky Way's dark blobs of obscuring dust. It is visible to the naked eye on dark nights—a faint patch of light the size of several Moons. In a small telescope the cloud resolves into countless individual stars. Meanwhile, you will need at least a 6- to 8-inch (150-200 mm) telescope to see the star cluster NGC 6603.
Apparent magnitude 4.6
Apparent size 90 arcminutes
Actual size 260 light-years
Distance 10,000 light-years
Other features in this sky Hercules Cluster (M13) Wild Duck Cluster (M11)

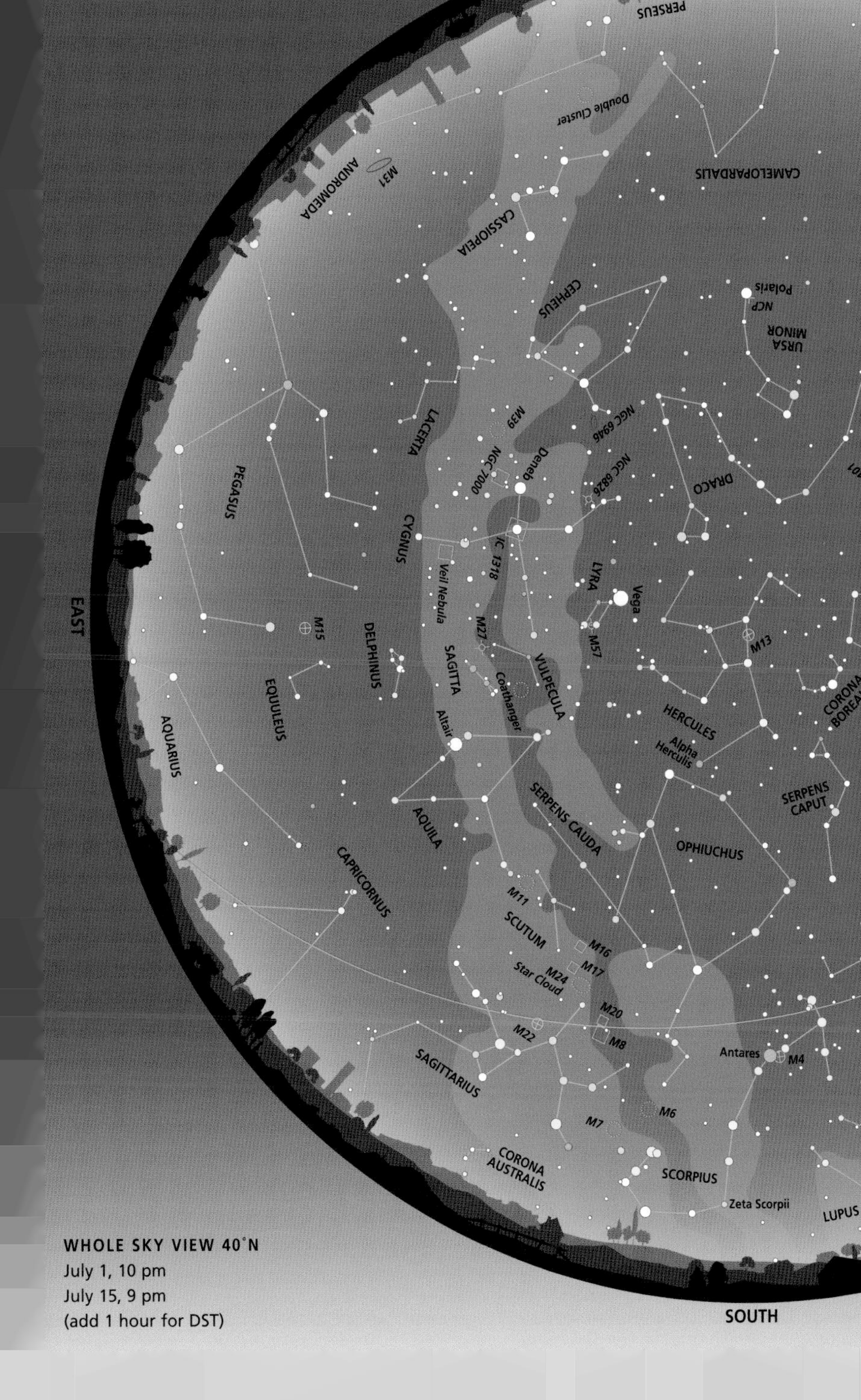

WHOLE SKY VIEW 40°N
July 1, 10 pm
July 15, 9 pm
(add 1 hour for DST)

August stars of the northern skies

As in July, it is still the Summer Triangle that dominates the heavens, threaded by the patchy band of light that is the Milky Way Galaxy. Look for the brightest regions of the Milky Way in Scutum the Shield and in Sagittarius the Archer—the latter marking the very center of our galaxy. Draco and Hercules are still high in the west and northwest, while Ursa Major now skims the northern horizon on all fours as it orbits Polaris. In the east, Cassiopeia and Andromeda reign over Perseus, while Pegasus passes near the zenith as the nights progress.

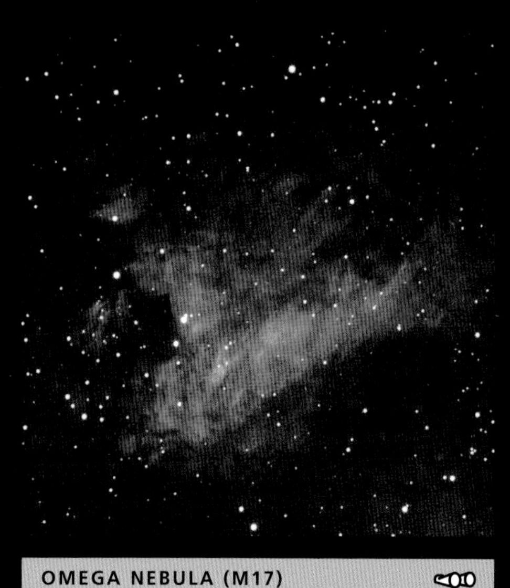

OMEGA NEBULA (M17)

Object type Diffuse nebula
Constellation Scutum
Also known as NGC 6618
Description This diffuse nebula in Sagittarius is variously known as the Swan Nebula or the Horseshoe Nebula, and to some eyes it resembles a celestial number two or the Greek letter omega. The brightest part of the Omega Nebula, the baseline of its number two shape, is readily visible in binoculars as an extended bar of light several arcminutes long. With a small telescope on a dark night, more nebulosity becomes visible on either side of this bar, including an arc-shaped part that forms the curve of the number two or the neck, if you like, of the swan. Still larger instruments will reveal even more of the nebula's structure—faint dark streaks across the bright bar caused by interstellar dust.
Apparent magnitude 6
Apparent size 11 arcminutes
Actual size 17 light-years
Distance 5500 light-years
Other nebulae in this sky Blinking Planetary (NGC 6826) North American Nebula (NGC 7000)

WILD DUCK CLUSTER (M11)

Object type Open cluster
Constellation Scutum
Also known as NGC 6705
Description The Wild Duck Cluster is a superb open cluster situated in a particularly rich portion of the sky, on the edge of a dense star cloud in Scutum. Containing about 3000 tightly arranged stars, it is one of the densest known open clusters. The Wild Duck Cluster gets its name from a V shape, like a flock of ducks, formed by its brightest members. The sharp-eyed ought to be able to spot M11 on a dark night away from the city, while binoculars and small telescopes will reveal a slightly hazy starlike object. In larger telescopes, from about 6 inches (150 mm), the V-shape pattern of stars begins to emerge, while a 10-inch (250 mm) scope will resolve several hundred fainter stars.
Apparent magnitude 5.8
Apparent size 14 arcminutes
Actual size 23 light-years
Distance 5800 light-years
Other clusters in this sky M39 The Coathanger

M101 GALAXY

Object type Spiral galaxy
Constellation Ursa Major
Also known as Pinwheel Galaxy, NGC 5457
Description Located about 5 degrees east of Ursa Major's multiple star Mizar, M101 is one of the largest and brightest spiral galaxies, and also one of the closest to the Solar System. It was the first known "spiral nebula"—the name given to these objects before astronomers realized they were galaxies outside our own. As it is so large, at nearly half a degree across, its light is spread thinly, making it difficult to spot in binoculars or small telescopes. Ideally, you need an aperture of 10 inches (250 mm) and a wide field of view (low magnification) to make out the galaxy's spiral arms. Note that M101 is often referred to as the Pinwheel Galaxy, a name also used, confusingly, to describe Triangulum's M33.
Apparent magnitude 7.7
Apparent size 22 arcminutes
Actual size 150,000 light-years
Distance 24,000,000 light-years
Other galaxies in this sky M81 Andromeda Galaxy (M31)

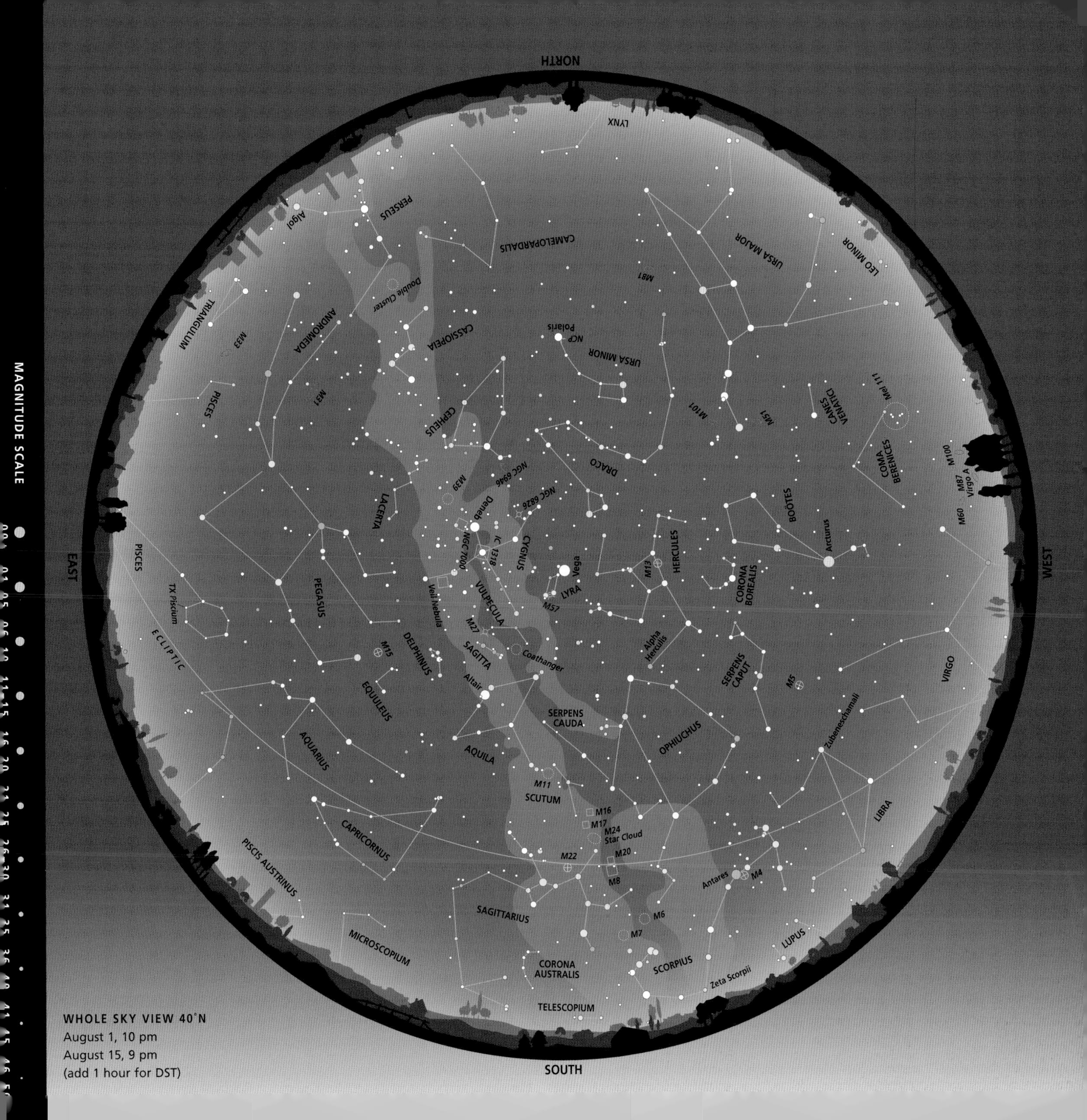

WHOLE SKY VIEW 40°N
August 1, 10 pm
August 15, 9 pm
(add 1 hour for DST)

September stars of the northern skies

The Milky Way now seems to divide the sky into two equal halves, north and south. The Summer Triangle still beckons overhead after sunset, but moves to the west to make way for the increasing dominance of Andromeda, Pegasus, Cassiopeia and Perseus in the east. Autumn is in the air. Ursa Major still surveys the northern horizon—its legs vanishing farther south. Aldebaran in Taurus and Capella in Auriga are now becoming prominent in the east, where Orion also makes its first appearance for months at around 2–3 am.

NTH. AMERICAN NEBULA (NGC 7000)

Object type Diffuse nebula
Constellation Cygnus
Description This is an intriguing diffuse nebula that bears a striking resemblance to the outline of North America. There is even a dark patch of obscuring dust hiding just the right parts of the cloud to form a suitable Gulf of Mexico! You can just about spot this nebula with the unaided eye on a dark night as a fuzzy patch, as it is four times the apparent width of the Moon. Because of its size, use only low-power binoculars or a very low-power telescope with a nebula filter, or you will not get the whole object in the field. However, the spectacular North American shape is visible only in photographs (such as above). Various other patches of nebulosity surround NGC 7000, including IC 5070 in the south, and the Pelican Nebula and the Skull Nebula in the west.
Apparent magnitude 6
Apparent size 2 x 1.5 degrees
Actual size 54 light-years
Distance 1500 light-years
Other nebulae in this sky Veil Nebula Dumbbell Nebula (M27)

M39 CLUSTER

Object type Open cluster
Constellation Cygnus
Also known as NGC 7092
Description This is a loose scattering of around 30 stars, which you can just make out with the naked eye, superimposed on a rich background of Milky Way star fields. On a clear night, look 8 degrees east of bright Deneb, the tail star of Cygnus, to find Rho and Pi_2 Cygni. The M39 cluster forms an elbow shape between these two stars, each about 2 degrees away. M39 is visible in small binoculars, which reveal it as a nebulous glow. Larger binoculars will begin to resolve stars. Through a small telescope with low magnification, you will be able to discern the equilateral triangle of three stars that defines M39's boundaries. The triangle contains most of the cluster's stars, with many of them grouped into pairs.
Apparent magnitude 4.6
Apparent size 32 arcminutes
Actual size 8 light-years
Distance 825 light-years
Other clusters in this sky Wild Duck Cluster (M11) Double Cluster (NGC 869/NGC 884)

TX PISCIUM (19 PISCIUM)

Object type Variable star
Constellation Pisces
Also known as 19 Piscium
Description The "TX" in TX Piscium is a variable star designation, though this star also goes under its Flamsteed name. TX Piscium is a red giant—a highly evolved, inflated star much cooler than the Sun—and like many red giants its brightness varies irregularly. In TX Piscium's case, the magnitude goes from 5 to 6 over a period of around 18 months. TX is distinctly red to the eye or through binoculars. To find it, first locate the Pisces Circlet—the ring of stars, including Gamma and Iota Piscium, at the western end of Pisces. TX is about midway between Gamma and Iota, a little to the east, outside the circlet. The photograph above shows TX as the distinctly red star to the left in the Pisces Circlet.
Apparent magnitude 5 to 6
Star type Red giant variable
Period Irregular, about 18 months
Distance 1000 light-years
Other significant stars in this sky Vega Deneb

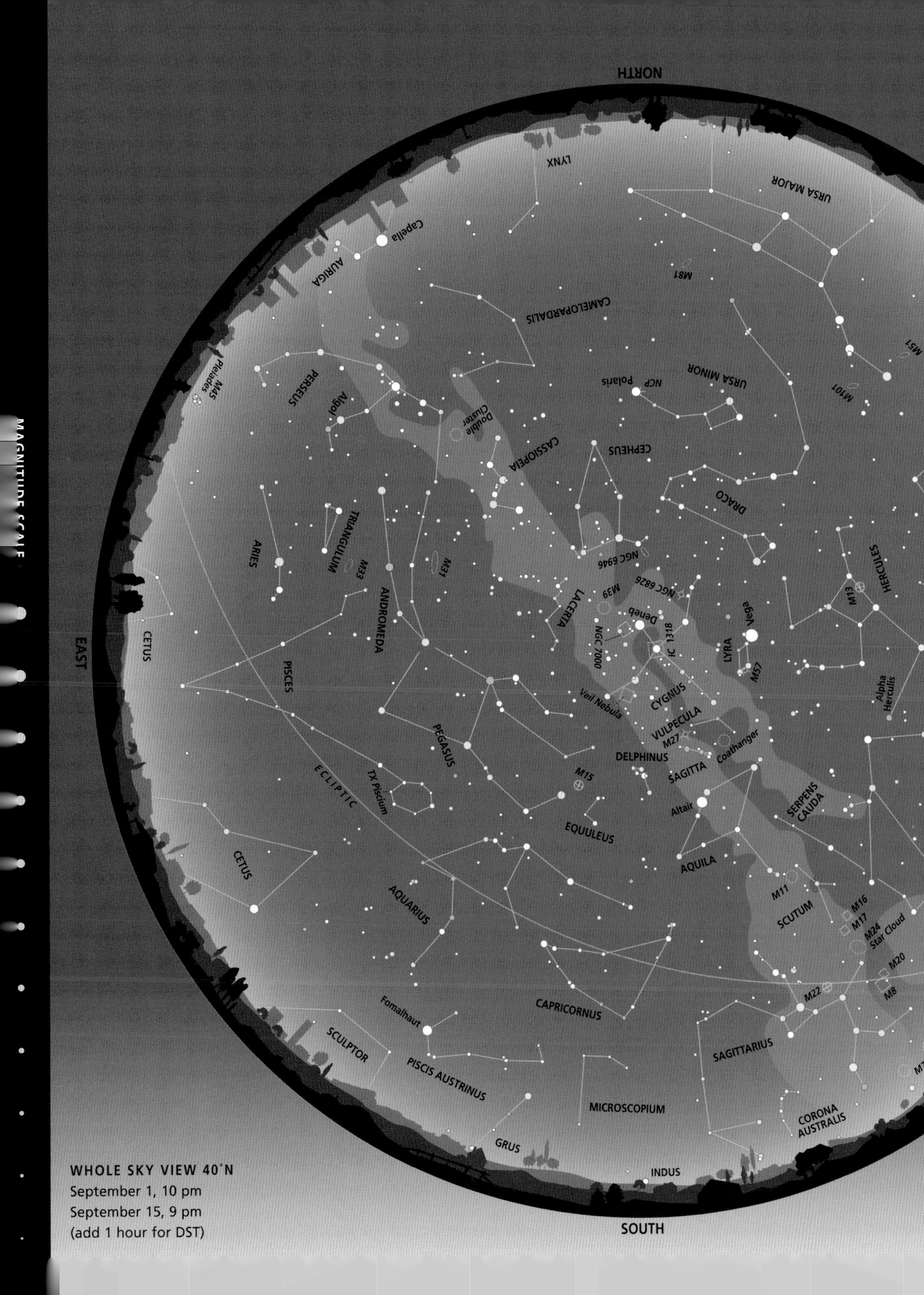
NORTH
SOUTH
EAST
WEST
URSA MAJOR
LYNX
CAMELOPARDALIS
URSA MINOR
CEPHEUS
CASSIOPEIA
PERSEUS
AURIGA
Capella
Algol
Double Cluster
M45 Pleiades
M81
M51
M101
NCP Polaris
CANES VENATICI
COMA BERENICES
BOÖTES
Arcturus
DRACO
HERCULES
CORONA BOREALIS
M13
TRIANGULUM
ARIES
M33
M31
ANDROMEDA
LACERTA
NGC 6946
NGC 6826
M39
Deneb
NGC 7000
IC 1318
Vega
LYRA
M57
CETUS
PISCES
Veil Nebula
CYGNUS
VULPECULA
M27
Coathanger
DELPHINUS
SAGITTA
Alpha Herculis
SERPENS CAPUT
VIRGO
M5
PEGASUS
M15
TX Piscium
ECLIPTIC
Altair
SERPENS CAUDA
OPHIUCHUS
EQUULEUS
AQUILA
LIBRA
AQUARIUS
M11
SCUTUM
M16
M17
M24 Star Cloud
M20
M8
SCORPIUS
M22
CAPRICORNUS
Fomalhaut
SCULPTOR
PISCIS AUSTRINUS
SAGITTARIUS
M7
M6
MICROSCOPIUM
CORONA AUSTRALIS
GRUS
INDUS
MAGNITUDE SCALE
WHOLE SKY VIEW 40°N
September 1, 10 pm
September 15, 9 pm
(add 1 hour for DST)

October stars of the northern skies

Summer has given way to autumn. While the Summer Triangle still dominates the western skies—with Cygnus' Deneb passing the zenith as the sun sets and Altair setting around 2 am—it is Pegasus, Perseus, Cassiopeia and Andromeda who now take the scene. Cassiopeia's "M" shape is directly opposite Ursa Major, with Polaris between them in the north, and Pegasus is a prominent square in the southwest. Meanwhile in the east, Auriga, Taurus and, later, Orion, climb ever higher in the early hours, signaling the winter that is yet to come.

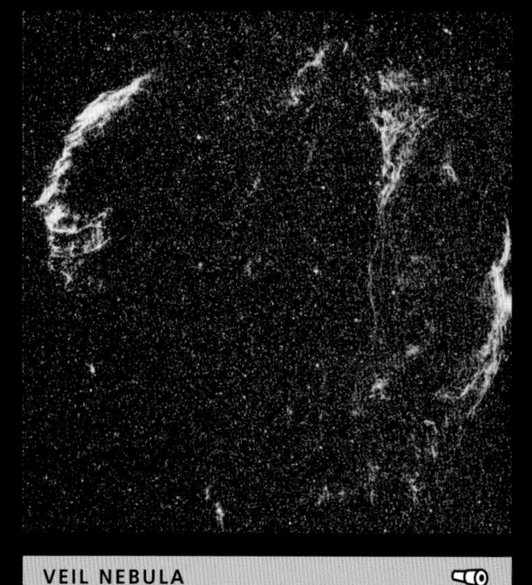

VEIL NEBULA

Object type Supernova remnant
Constellation Cygnus
Also known as NGC 6960, NGC 6992 and NGC 6995
Description These three long arc-shaped filaments are actually all part of the same nebular complex—a huge ring of gas known as the Cygnus Loop. The Cygnus Loop is a supernova remnant, the puffed-out shell of star that exploded perhaps 30,000 years ago. NGC 6960 forms the westernmost part of the loop, while NGC 6992 and NGC 6995 mark the eastern portion. Altogether, the Veil Nebula spans several degrees. This means it has quite a low surface brightness and requires a minimum of a 6-inch (150 mm) telescope with a high-contrast filter to see the intricacies of its delicate structure well. However, you ought to be able to pick out both halves of the Veil Nebula with the aid of binoculars.
Apparent magnitude 7
Apparent size 3 degrees
Actual size 140 light-years
Distance 2600 light-years
Other nebulae in this sky Ring Nebula (M57) North American Nebula (NGC 7000),

DOUBLE CLUSTER (NGC 869/NGC 884)

Object type Open cluster pair
Constellation Perseus
Description This striking close pair of bright star clusters lies in the Perseus spiral arm of our galaxy. Of the two, NGC 869 is the richer, with about 200 stars compared to 150 for NGC 884. Although they appear side by side, NGC 884 is several hundred light-years deeper into space, but astronomers think that both clusters evolved from a common interstellar cloud. With the unaided eye, you will easily see the double cluster as a lengthy smudge between Perseus and Cassiopeia. In binoculars, the clusters begin to resolve into a figure-of-eight scattering of tiny stars. But the object's true beauty can only be fully appreciated with a 4-inch (100 mm) telescope and a wide-angle field of view.
Apparent magnitude 4
Apparent size 30 arcminutes each cluster
Actual size 63 light-years each cluster
Distance 7100 light-years (NGC 869), 7400 light-years (NGC 884)
Other clusters in this sky M15 Pleiades (M45)

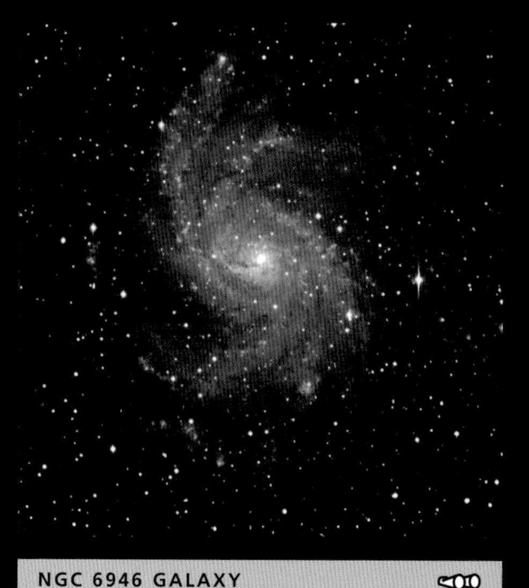

NGC 6946 GALAXY

Object type Spiral galaxy
Constellation Cepheus
Description Galaxies are relatively rare in this part of the sky. Still, NGC 6946 is a fairly dark spiral galaxy seen face-on, situated just on the western border of Cepheus, with Cygnus. It was discovered by the German-born British astronomer Sir William Herschel in 1798, and astronomers have since witnessed a total of six supernovae in this galaxy. You ought to be able to make out this galaxy with binoculars, but it will appear as little more than a small patch. Interestingly, it shares the field of view with nearby open cluster NGC 6938, just 40 arcminutes away. NGC 6938 has the same apparent size as the galaxy, but it is similarly dark, and appears nebulous through binoculars. However, you will need a large telescope to make out any sort of detail in either object.
Apparent magnitude 9.6
Apparent size 10 arcminutes
Actual size 30,000 light-years
Distance 10,000,000 light years
Other galaxies in this sky M101 Andromeda Galaxy (M31)

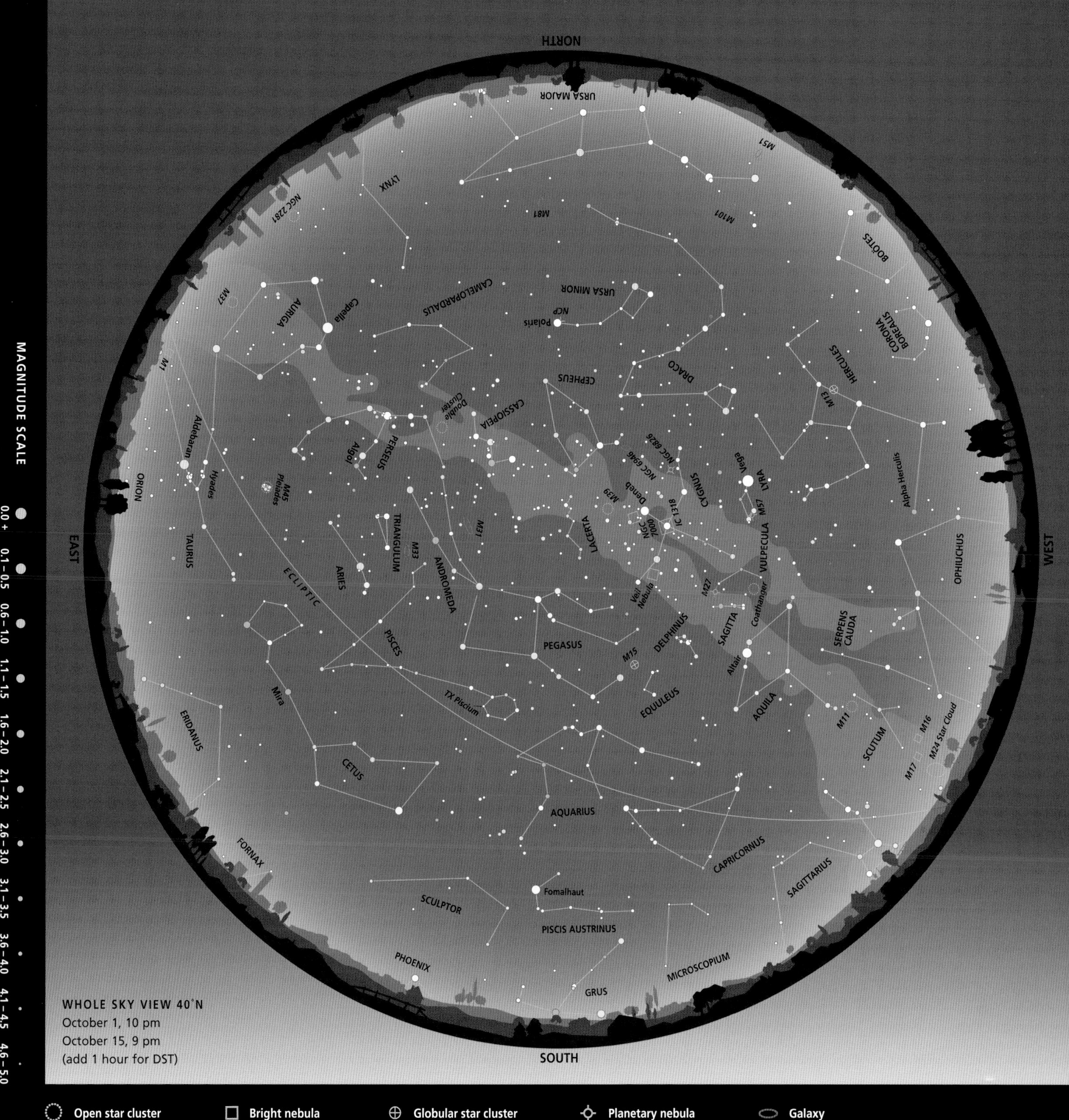
NORTH
SOUTH
EAST
WEST
MAGNITUDE SCALE
0.0 +
0.1–0.5
0.6–1.0
1.1–1.5
1.6–2.0
2.1–2.5
2.6–3.0
3.1–3.5
3.6–4.0
4.1–4.5
4.6–5.0
URSA MAJOR
LYNX
M51
M81
M101
NGC 2281
BOÖTES
CAMELOPARDALIS
URSA MINOR
NCP
Polaris
M37
AURIGA
Capella
CORONA BOREALIS
HERCULES
DRACO
CEPHEUS
M1
M13
CASSIOPEIA
Double Cluster
Aldebaran
Algol
PERSEUS
NGC 6826
NGC 6946
Vega
LYRA
CYGNUS
ORION
Hyades
M45 Pleiades
Alpha Herculis
Deneb
M39
IC1318
NGC 7000
M57
TRIANGULUM
M31
TAURUS
M33
LACERTA
VULPECULA
OPHIUCHUS
ARIES
ANDROMEDA
ECLIPTIC
M27
Veil Nebula
Coathanger
PISCES
DELPHINUS
SAGITTA
SERPENS CAUDA
PEGASUS
M15
Altair
Mira
EQUULEUS
TX Piscium
AQUILA
M11
ERIDANUS
M16
M24 Star Cloud
SCUTUM
CETUS
M17
AQUARIUS
FORNAX
CAPRICORNUS
SAGITTARIUS
Fomalhaut
SCULPTOR
PISCIS AUSTRINUS
PHOENIX
MICROSCOPIUM
GRUS
WHOLE SKY VIEW 40°N
October 1, 10 pm
October 15, 9 pm
(add 1 hour for DST)
Open star cluster
Bright nebula
Globular star cluster
Planetary nebula
Galaxy

November stars of the northern skies

November's skies are replete with bright stars. Even this late in the year, the Summer Triangle still holds supreme for several hours after sunset. Directly opposite, in the east, the brilliant stars of Aldebaran in Taurus, Capella in Auriga, the Twins Castor and Pollux, the Pleiades star cluster, and of course Orion, are now well placed, high above the horizon and getting higher as the night deepens. Just as the Summer Triangle sets for the night, Sirius appears over the eastern horizon—the most brilliant star among the dozens already there.

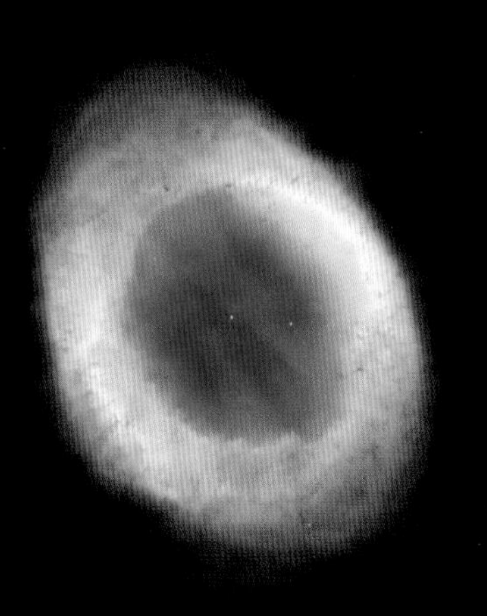

RING NEBULA (M57)

Object type Planetary nebula
Constellation Lyra
Also known as NGC 6720
Description This interstellar smoke ring lying between Gamma and Beta Lyrae is one of the best-known planetary nebulae. Despite its name, the Ring Nebula is not ring-shaped at all, but is rather a cylinder which we see essentially end-on. As with all planetary nebulae, this one is the cast-off remains of a dead star—a white dwarf, tiny and hot, resides at its heart. However, you will need a 13- to 15-inch (325–375 mm) telescope to see the star, and a very steady, dark sky. Still, observers equipped with a much smaller, 3- to 4-inch (75–100 mm) instrument will be able to make out this object, looking a little like an out-of-focus star, while higher powers will reveal its oval outline, and also the central hole.
Apparent magnitude 9.1
Apparent size 1.1 x 1.4 arcminutes
Actual size 1.3 light-years
Distance 3200 light-years
Other nebulae in this sky Crab Nebula (M1)
Eskimo Nebula (NGC 2392)

M15 CLUSTER

Object type Globular cluster
Constellation Pegasus
Also known as NGC 7078
Description M15 is probably the densest known globular cluster in the Milky Way Galaxy, and one of the more spectacular for northern observers, easily competing with M13 in Hercules. It is a massive swarm of a million stars, spanning 120 light-years, with an extremely compact core. You will probably not be able to see M15 with the naked eye, as at magnitude 6.4 it is beyond the range of most people's vision. Binoculars will expose a circular, fuzzy object. With a 4-inch (100 mm) telescope, M15 still appears quite nebulous, with no stars resolved, but with a slightly patchy appearance. In larger telescopes, you will start to discern the outermost stars, but the dense core remains unresolved.
Apparent magnitude 6.4
Apparent size 12 arcminutes
Actual size 120 light-years
Distance 34,000 light-years
Other clusters in this sky Pleiades (M45)
Double Cluster (NGC 869/NGC 884)

ANDROMEDA GALAXY (M31)

Object type Spiral galaxy
Constellation Andromeda
Also known as NGC 224
Description The Andromeda Galaxy has the reputation of being the farthest object the naked eye can see. In fact, M33 in Triangulum is farther and may just be visible if you have 20/20 vision. Still, the Andromeda Galaxy is rightly famous, known to Persian astronomers since at least AD 905, and now a subject of intense research. It is our nearest large galaxy, bigger than the Milky Way and probably containing more stars. It appears without optical aid on a dark night as a faint smudge—you will be hard-pressed to see it from a city. To get a better view, use a large pair of binoculars or a small telescope. You will then spot M31's "Magellanic Clouds"—its satellite galaxies M32 and M110, in close attendance.
Apparent magnitude 3.4
Apparent size 180 x 60 arcminutes
Actual size 150,000 light-years
Distance 2,900,000 light-years
Other galaxies in this sky NGC 6946
Pinwheel Galaxy (M33)

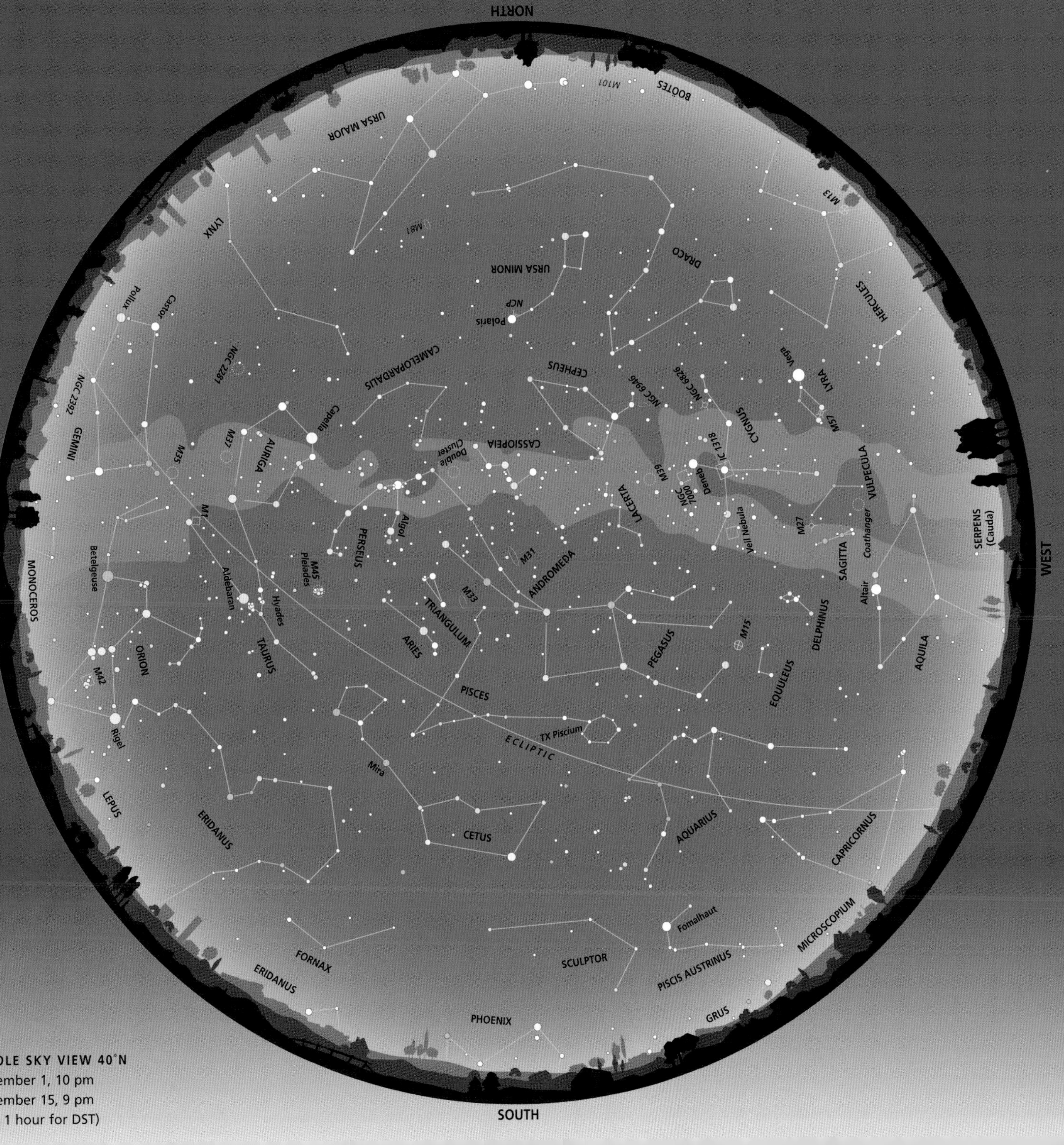

NORTH

EAST

WEST

SOUTH

WHOLE SKY VIEW 40°N
November 1, 10 pm
November 15, 9 pm
(add 1 hour for DST)

December stars of the northern skies

It is winter in the north and as the sun sets, the southern skies are filled with faint constellations. However, the rest of the heavens more than compensate. Around midnight, Capella passes the zenith, and the skies are filled with the sparkling constellations of Auriga, Orion, Taurus, Gemini, Canis Major and Canis Minor. In the east, Leo pounces onto the scene around 10 o'clock, while Pegasus descends the western sky, along with the other familiar autumn constellations. Meanwhile, Ursa Major rises to prominence again in the north.

ESKIMO NEBULA (NGC 2392)

Object type Planetary nebula
Constellation Gemini
Also known as Clown Face Nebula
Description This colorful nebula was discovered in 1787 by British astronomer Sir William Herschel. In 2000, the Hubble Space Telescope took a photograph of the nebula that made it world famous. The photo reveals in detail the furrowed fringe, like the lining of an Inuit's furry hood, from which this object's name is derived—and astronomers are struggling to explain how this complexity arises. This is a comparatively bright nebula, but you will still need a large telescope with high magnification, and possibly a nebula filter, to make out any details. Still, don't expect to see much evidence of the "face." That is only visible in high-resolution photographs taken with very large, professional instruments.
Apparent magnitude 8.3
Apparent size 47 x 43 arcseconds
Actual size 0.7 light-year
Distance 3000 light-years
Other nebulae in this sky Crab Nebula (M1)
North American Nebula (NGC 7000)

THE PLEIADES (M45)

Object type Open cluster
Constellation Taurus
Also known as Seven Sisters
Description The famous Pleiades is without contest the best open cluster in the whole sky. Despite its alternative name, there are considerably more than seven stars in this group—in fact there are several hundred very young, brilliant blue stars. M45 is readily visible even from a city, spanning an area four times the diameter of the Moon. With a keen naked eye and a dark night, you may make out the seven brightest stars that gave the cluster its name, but most people only spot six. They glisten beautifully while the fainter members add a lovely nebulous background glow. To get an even better view, use a pair of low-power binoculars. The view is simply awe-inspiring.
Apparent magnitude 1.2
Apparent size 2 degrees
Actual size 13 light-years
Distance 400 light-years
Other clusters in this sky Hyades
Beehive Cluster (M44)

NGC 2903 GALAXY

Object type Barred spiral galaxy
Constellation Leo
Description NGC 2903 is a typical, fairly large spiral galaxy among several to be found in Leo the Lion. Charles Messier missed this oblique, ninth-magnitude galaxy when he was compiling his famous catalog, and its discovery was left to the astronomer Sir William Herschel. Recently, astronomers studied the Hubble data and discovered how NGC 2903's bar funnels material to the galaxy's core, so new stars form there. You can find NGC 2903 about 1.5 degrees below and 2 degrees to the right of Epsilon Leonis, the star that forms the sharp end of Leo's sickle. This galaxy will show up in small telescopes of around 2 inches (50 mm), while a slightly larger 4-inch (100 mm) scope will reward you with detail including some of the spiral-arm structure.
Apparent magnitude 9.0
Apparent size 12 x 6 arcminutes
Actual size 73,000 light-years
Distance 20,000,000 light-years
Other galaxies in this sky M65
Pinwheel Galaxy (M33)

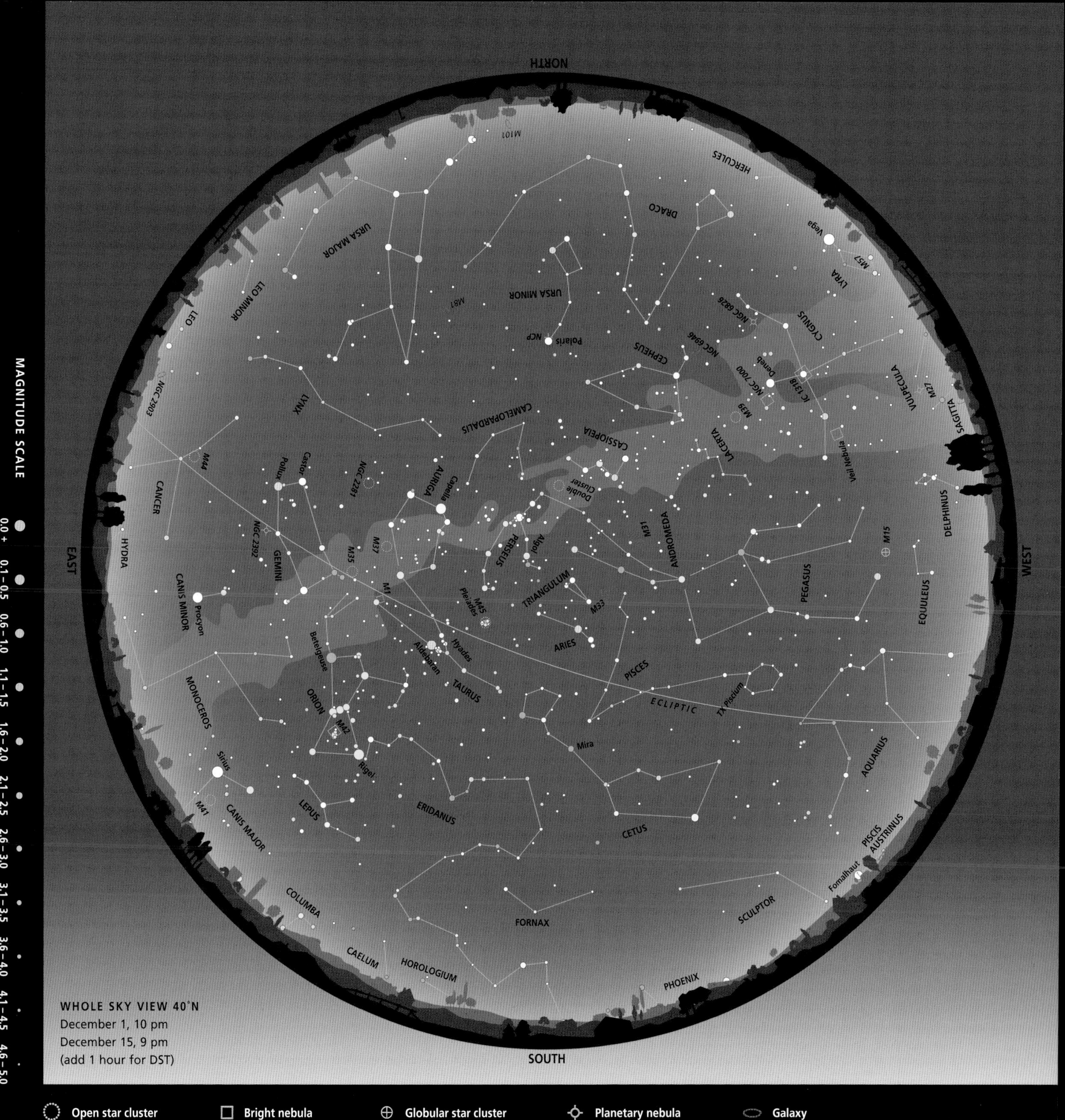

NORTH
SOUTH
EAST
WEST
MAGNITUDE SCALE
0.0 +
0.1–0.5
0.6–1.0
1.1–1.5
1.6–2.0
2.1–2.5
2.6–3.0
3.1–3.5
3.6–4.0
4.1–4.5
4.6–5.0
WHOLE SKY VIEW 40°N
December 1, 10 pm
December 15, 9 pm
(add 1 hour for DST)
Open star cluster
Bright nebula
Globular star cluster
Planetary nebula
Galaxy
M101
HERCULES
DRACO
Vega
LYRA
M57
URSA MAJOR
URSA MINOR
M81
LEO MINOR
LEO
NGC 6826
CYGNUS
Polaris
NCP
NGC 6946
CEPHEUS
Deneb
NGC 7000
IC 1318
VULPECULA
M27
SAGITTA
NGC 2903
LYNX
CAMELOPARDALIS
M39
LACERTA
Veil Nebula
M44
CASSIOPEIA
Castor
Pollux
NGC 2281
AURIGA
Capella
Double Cluster
CANCER
DELPHINUS
HYDRA
NGC 2392
GEMINI
M37
M35
PERSEUS
Algol
M31
ANDROMEDA
M15
CANIS MINOR
Procyon
M1
M45
Pleiades
TRIANGULUM
M33
PEGASUS
EQUULEUS
Betelgeuse
Aldebaran
Hyades
ARIES
PISCES
MONOCEROS
ORION
TAURUS
ECLIPTIC
TX Piscium
M42
Mira
AQUARIUS
Sirius
Rigel
M41
CANIS MAJOR
LEPUS
ERIDANUS
CETUS
PISCIS AUSTRINUS
Fomalhaut
COLUMBA
FORNAX
SCULPTOR
CAELUM
HOROLOGIUM
PHOENIX

January stars of the southern skies

Summer in the southern hemisphere brings a wealth of bright stars to the sky, including the three most brilliant of all—Sirius, Canopus and Rigil Kentaurus—appearing after midnight in the southeast. Orion, who dominates winter in the northern hemisphere, is also a southern constellation—albeit inverted. In January he chases Taurus high overhead while Sirius passes close to the zenith. As the night deepens, the Southern Cross climbs higher, circling the south celestial pole, and Leo—another inverted constellation—traverses the sky east to west.

TARANTULA NEBULA (NGC 2070)

Object type Diffuse nebula
Constellation Dorado
Also known as 30 Doradus
Description This is a vast star factory in the Large Magellanic Cloud (LMC)—one of the two small galaxies that orbit our own galaxy. The bulk of this nebula, 30 times larger than the northern hemisphere's Orion Nebula, is around 1000 light-years across, but there are traces of fainter parts that extend much farther. If the Tarantula were as close as the Orion Nebula, it would cover a massive 30 degrees of sky. You can see this nebula with the naked eye in the eastern half of the LMC. A 3-inch (75 mm) telescope will reveal some of the sprawling, spider-like structure from which the nebula's name is derived. However, with an 8-inch (200 mm) telescope, the Tarantula will truly impress you with its detail.
Apparent magnitude 5
Apparent size 30 arcminutes
Actual size 1000 light-years
Distance 160,000 light-years
Other nebulae in this sky Coalsack Nebula Eta Carinae Nebula (NGC 3372)

M41 CLUSTER

Object type Open cluster
Constellation Canis Major
Also known as NGC 2287
Description This naked-eye open cluster is situated in a rich star field about 4 degrees south of brilliant Sirius—the sky's brightest star. M41 is bright but quite sparse, with about 100 stars spanning 25 light-years. There are several red giants, with the brightest one a vibrant orange diamond dominating the cluster's heart. At only 200 million years old, M41 dates to around the time of the earliest dinosaurs and is certainly much younger than the Solar System. The conspicuous blue star in the cluster's southeast corner (bottom left in the above image), however, is a foreground star, not related to the other stars. M41 is best seen with binoculars or a low-power telescope or finderscope.
Apparent magnitude 4.6
Apparent size 38 arcminutes
Actual size 25 light-years
Distance 2300 light-years
Other clusters in this sky Omicron Velorum Cluster Omega Centauri (NGC 5139)

LARGE MAGELLANIC CLOUD (LMC)

Object type Irregular galaxy
Constellation Dorado and Mensa
Also known as The Nebecula Major
Description You could be forgiven for dismissing the LMC as little more than a wisp of cloud. But this is no Earthly phenomenon. The LMC is a "cloud" of stars—an entire galaxy, which is in orbit around our own in a plane perpendicular to the Milky Way's disk. It is classified as an irregular galaxy, has traces of spiral structure, and is richer than our own galaxy in clouds of interstellar gas. Subsequently, the LMC harbors many areas of active star formation. The LMC fills the field of view in 7x50 binoculars, but you can also explore this object using a low-power telescope. The eastern half is particularly satisfying, as it contains the giant Tarantula Nebula—a huge stellar factory illuminated by a nest of new stars.
Apparent magnitude 1
Apparent size 7 degrees
Actual size 20,000 light-years
Distance 160,000 light-years
Other galaxies in this sky M105 Small Magellanic Cloud

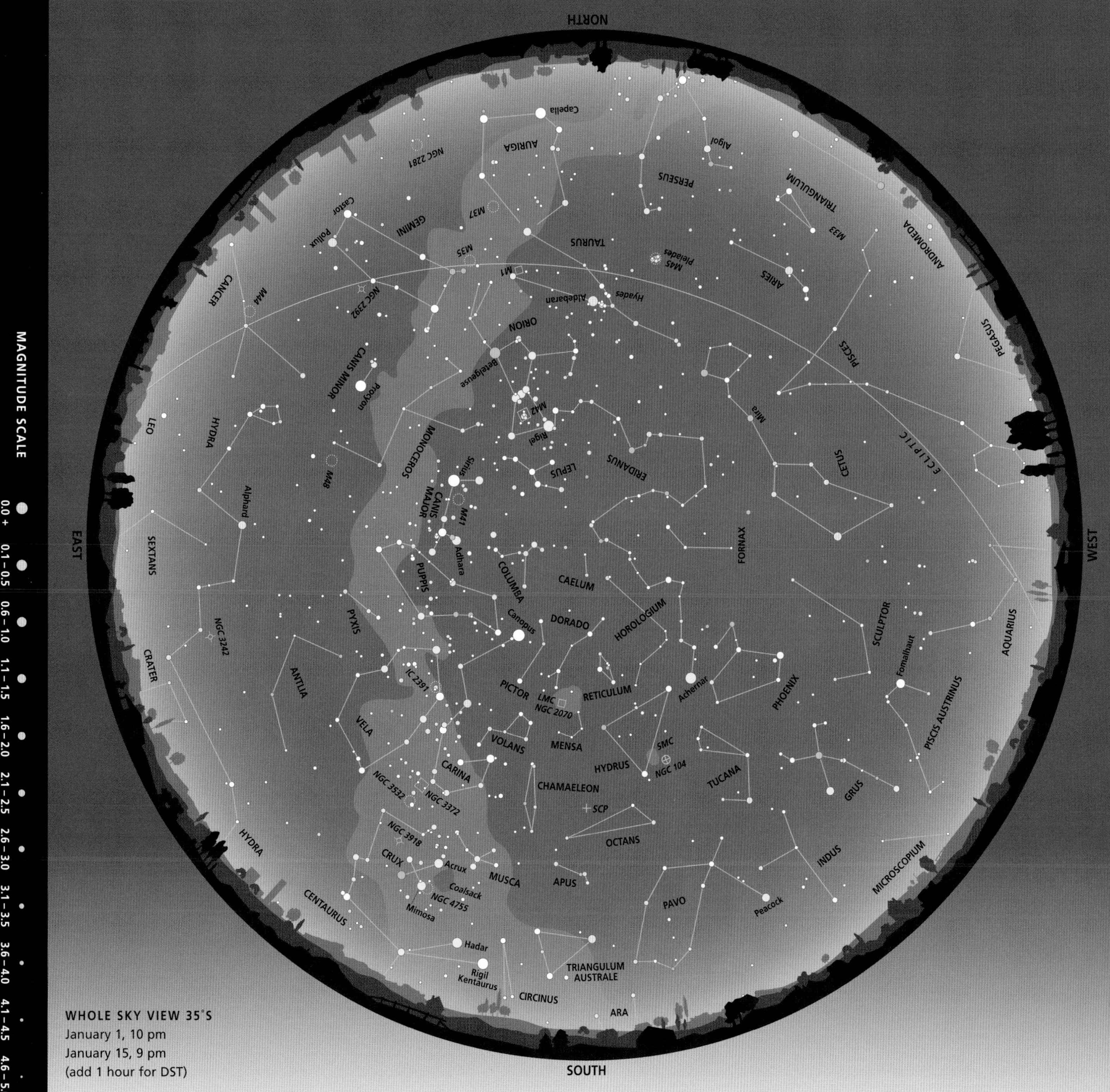

WHOLE SKY VIEW 35°S
January 1, 10 pm
January 15, 9 pm
(add 1 hour for DST)

February stars of the southern skies

As the February nights commence, Canis Major and Orion are still high in the northwest. The sky there is ablaze with bright stars like Sirius, Procyon, Aldebaran, Rigel and Betelgeuse, and the three brilliant points that mark Orion's belt. On the other side of the celestial dome, the Milky Way rises out of the southern horizon and spills upward as if in defiance of gravity, carrying with it the Southern Cross and Centaurus the Centaur. To the right of this, like detached portions of the Milky Way, fly the Magellanic Clouds.

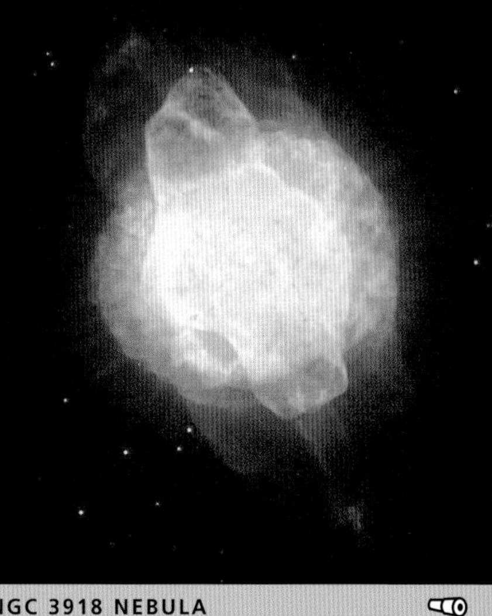

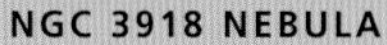

NGC 3918 NEBULA

Object type Planetary nebula
Constellation Centaurus
Description A bright planetary nebula not far from the Southern Cross, NGC 3918 was the subject of the striking photograph above, taken by the Hubble Space Telescope, which exposed complex balloons of inflated gas billowing away from the central star. However, at only 10 arcseconds across, NGC 3918 is very small, and despite its brightness you will not see any of the detail of the Hubble photo above, even with a large telescope. With a finderscope or binoculars, you might be able to spot NGC 3918, but as little more than a faint, blue star. A 15-inch (375 mm) telescope will begin to show the object's oval disk, but you will need even more aperture than this to see the central star, lost in the relatively bright glare of the nebula.
Apparent magnitude 8
Apparent size 10 arcseconds
Actual size 0.3 light-year
Distance 3000 light-years
Other nebulae in this sky Coalsack Nebula
Eta Carinae Nebula (NGC 3372)

NGC 3532 CLUSTER

Object type Open cluster
Constellation Carina
Description Just three degrees east and a little north of Eta Carinae lies this lovely open cluster comprising stars of magnitude 6 and fainter. NGC 3532 is breathtaking in a small telescope and is one of the sky's best clusters. The naked eye will make it out as a fuzzy blob, oval in shape, while binoculars will resolve it into a few dozen stars. With a 3-inch (75 mm) telescope under low magnification, NGC 3532 will appear as a grouping of many more fairly bright stars, strewn tightly in an east–west direction. Still larger instruments will resolve even more stars and reveal a denser clustering near the center. The cluster is all the more attractive for the presence of a yellow supergiant—left of center in the image. But it is closer than the cluster, not an actual member.
Apparent magnitude 3
Apparent size 1 degree
Actual size 23 light-years
Distance 1300 light-years
Other clusters in this sky Omega Centauri (NGC 5139) Omicron Velorum Cluster

SMALL MAGELLANIC CLOUD (SMC)

Object type Irregular galaxy
Constellation Tucana
Also known as The Nebecula Minor
Description This is the second and more distant of the two galaxies in orbit around the Milky Way, the other being the Large Magellanic Cloud (LMC). Both the LMC and the SMC—described by the Portuguese explorer, Ferdinand Magellan, whose name they now bear—orbit the Milky Way in a plane perpendicular to its spiral disk. The SMC contains less gas but more dust than the LMC, and a smaller complement of clusters and nebulae. It is an easy naked-eye object, appearing as a hazy patch spanning six or seven full Moons in the sky, some 20 degrees west of its larger cousin. It is a rewarding sight in 10x50 binoculars, where it occupies the majority of the field of view, but the LMC still provides the more spectacular sight.
Apparent magnitude 2.3
Apparent size 3.5 degrees
Actual size 9000 light-years
Distance 190,000 light-years
Other galaxies in this sky M105
Large Magellanic Cloud

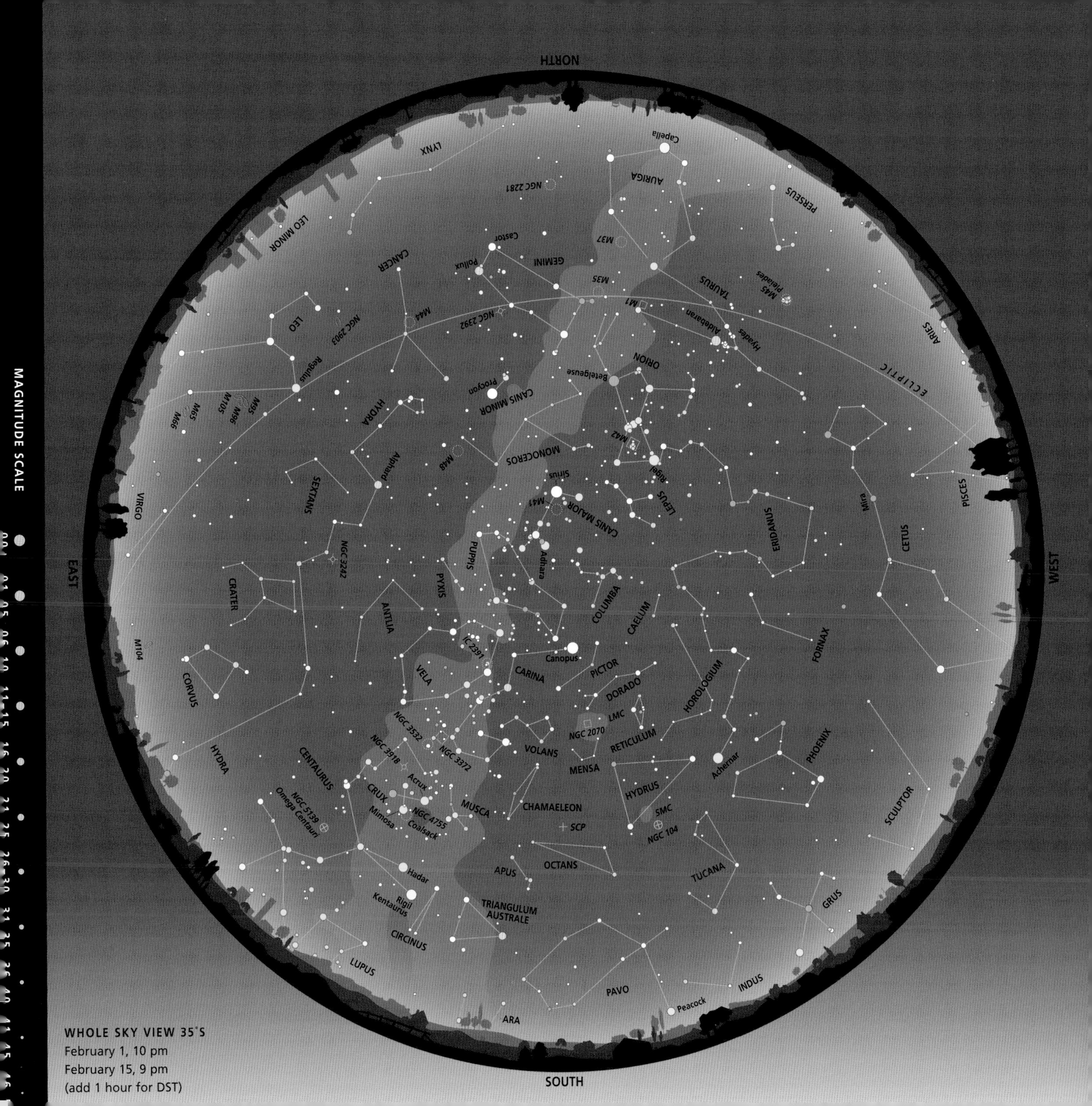
MAGNITUDE SCALE
NORTH
Capella
AURIGA
LYNX
NGC 2281
PERSEUS
LEO MINOR
M37
Castor
GEMINI
CANCER
Pollux
M35
TAURUS
M45
Pleiades
LEO
NGC 2903
M44
NGC 2392
M1
Aldebaran
Hyades
ARIES
Regulus
ORION
Betelgeuse
ECLIPTIC
M95
M96
M105
M65
M66
HYDRA
Procyon
CANIS MINOR
M42
MONOCEROS
M48
Alphard
Rigel
SEXTANS
Sirius
PISCES
VIRGO
LEPUS
M41
CANIS MAJOR
Mira
ERIDANUS
CETUS
PUPPIS
NGC 3242
Adhara
EAST
WEST
CRATER
PYXIS
COLUMBA
ANTLIA
CAELUM
FORNAX
M104
IC 2391
Canopus
VELA
CARINA
PICTOR
CORVUS
HOROLOGIUM
DORADO
NGC 3532
LMC
NGC 2070
RETICULUM
NGC 3918
NGC 3372
VOLANS
HYDRA
CENTAURUS
MENSA
Achernar
PHOENIX
Acrux
CRUX
NGC 5139
Omega Centauri
Mimosa
NGC 4755
Coalsack
MUSCA
CHAMAELEON
HYDRUS
SMC
SCP
NGC 104
SCULPTOR
APUS
OCTANS
TUCANA
Hadar
Rigil Kentaurus
TRIANGULUM AUSTRALE
GRUS
CIRCINUS
LUPUS
PAVO
INDUS
Peacock
ARA
SOUTH
WHOLE SKY VIEW 35°S
February 1, 10 pm
February 15, 9 pm
(add 1 hour for DST)

March stars of the southern skies

The hunter Orion and his two dogs, Canis Major and Canis Minor, still dominate in the west in the first few hours after sunset. A glance due south at midnight reveals the glorious band of light that marks the Milky Way, spanning the heavens in an east–west arc. It is encrusted with brilliant stars such as Canopus in the southwest and Alpha and Beta Centauri in the southeast, while in the middle, Crux the Southern Cross points toward the horizon. There, the Small Magellanic Cloud remains hidden for most of the rest of the night.

ETA CARINAE NEBULA (NGC 3372)

Object type Diffuse nebula
Constellation Carina
Description Without doubt, the Eta Carinae Nebula is the sky's finest. Easily visible to the naked eye, it spans 2 degrees of sky—an area of eight Moons—making it even larger in angular size than the great Orion Nebula. Binoculars are great for viewing this nebula. They will show you how it appears split into several bright fragments by a V-shaped rift of dust. A telescope of any size will also serve you well. If you focus on the star Eta itself with moderate magnification, you will see that it is embedded in its own cloud of nebulosity, the Homunculus Nebula. This is the blister of gas that Eta Carinae jettisoned in the nineteenth century in a mysterious explosion. Eta Carinae is a hypergiant variable and one of the most massive known stars at about 100 solar masses.
Apparent magnitude 1
Apparent size 2 degrees
Actual size 320 light-years
Distance 9000 light-years
Other nebulae in this sky Coalsack Nebula
Tarantula Nebula (NGC 2070)

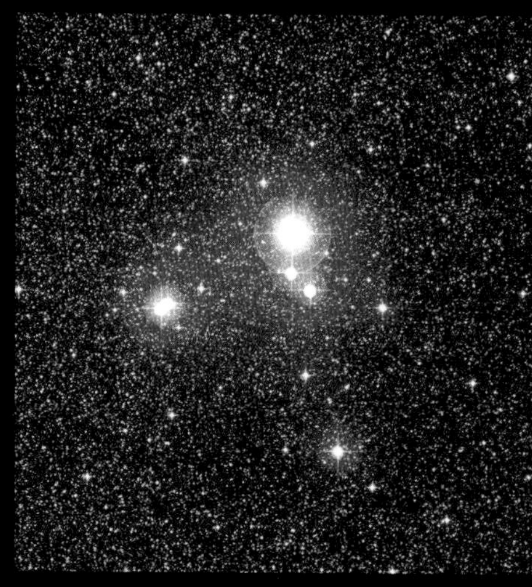

OMICRON VELORUM CLUSTER

Object type Open cluster
Constellation Vela
Also known as IC 2391
Description Known familiarly as the Omicron Velorum Cluster, IC 2391 is a loose and small scattering of about 20 to 30 stars dotted around a fourth-magnitude star called, unsurprisingly, Omicron Velorum. Persian astronomer Al Sufi, who now has a Moon crater named after him, first described IC 2391 in AD 964. The brightest star in the cluster shines at magnitude 3.6, a blue giant. However, the cluster's overall magnitude is 2.5. Thus, the Omicron Velorum Cluster is easily bright enough to be picked out with the naked eye, a patch of light about 2 degrees north of Delta Velorum. But the best views are obtained using binoculars or a small telescope with a wide-angle eyepiece.
Apparent magnitude 2.5
Apparent size 50 arcminutes
Actual size 9 light-years
Distance 580 light-years
Other clusters in this sky NGC 3532
Omega Centauri (NGC 5139)

M105 GALAXY

Object type Elliptical galaxy
Constellation Leo
Also known as NGC 3379
Description M105 is often considered the typical prototype elliptical galaxy, classed as E1 according to the Hubble scheme. It is one of a collection of stellar islands that populate the constellation of Leo and belongs to the Leo I group of galaxies—a small clustering around 38 million light-years away. You will probably not spot M105 through binoculars. Through a moderate telescope, however, it forms a 0.1-degree triangle with NGC 3389—also a member of the Leo I group—and NGC 3384, which is a more distant background object. Because M105 is an elliptical galaxy, not a spiral, you will find that no matter how much you try to focus, it always appears as a fuzzy, circular blob.
Apparent magnitude 9.2
Apparent size 2 arcminutes
Actual size 22,000 light-years
Distance 38,000,000 light-years
Other galaxies in this sky Virgo A (M87)
Large Magellanic Cloud

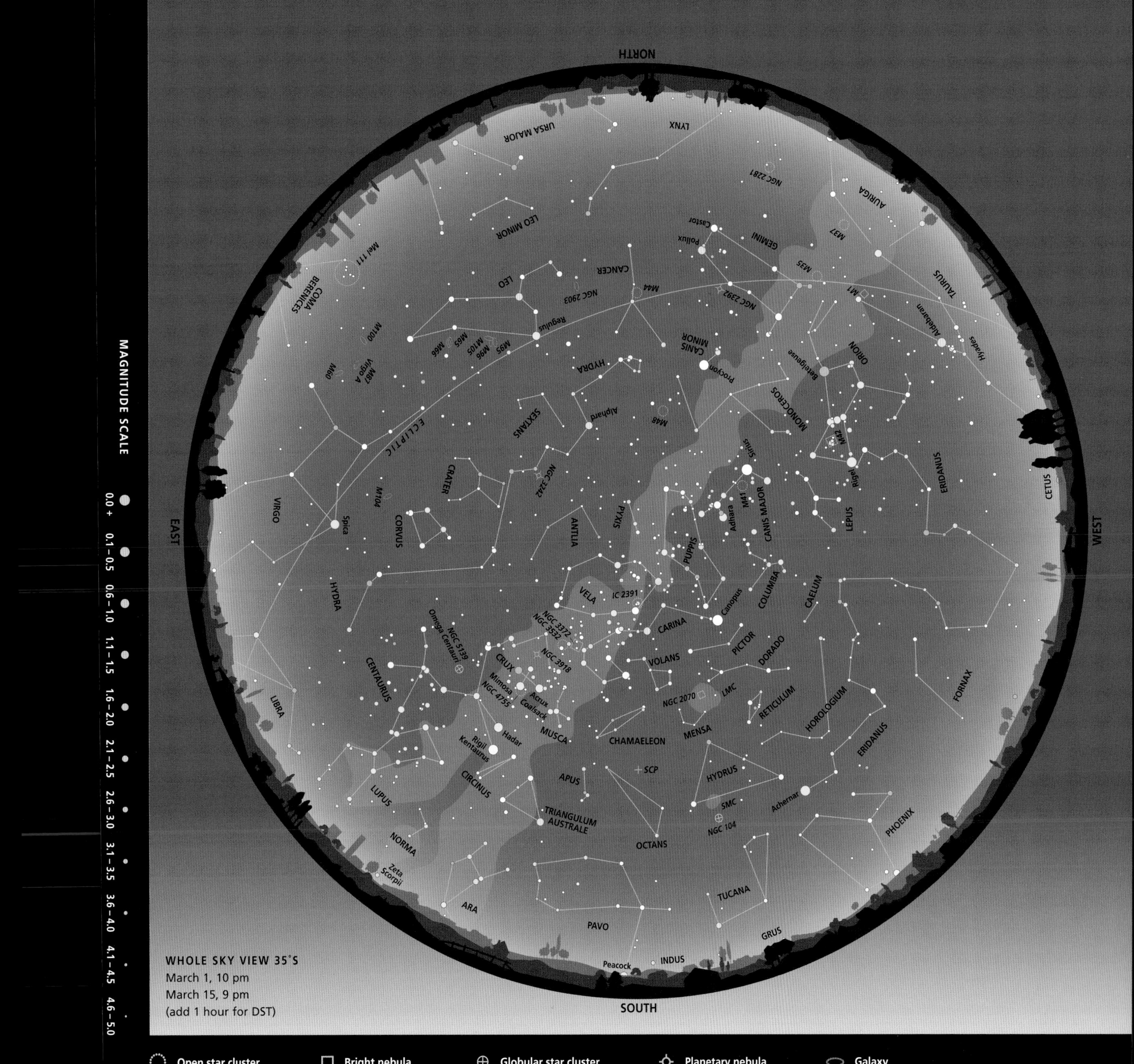

WHOLE SKY VIEW 35˚S
March 1, 10 pm
March 15, 9 pm
(add 1 hour for DST)
NORTH
SOUTH
EAST
WEST
MAGNITUDE SCALE
0.0 +
0.1 – 0.5
0.6 – 1.0
1.1 – 1.5
1.6 – 2.0
2.1 – 2.5
2.6 – 3.0
3.1 – 3.5
3.6 – 4.0
4.1 – 4.5
4.6 – 5.0
Open star cluster
Bright nebula
Globular star cluster
Planetary nebula
Galaxy
URSA MAJOR
LYNX
NGC 2281
AURIGA
M37
LEO MINOR
Castor
Pollux
GEMINI
M35
TAURUS
M1
Mel 111
COMA BERENICES
CANCER
LEO
NGC 2903
M44
NGC 2392
Aldebaran
Hyades
Regulus
CANIS MINOR
Procyon
ORION
Betelgeuse
M100
M66
M65
M105
M96
M95
HYDRA
M60
M87
Virgo A
ECLIPTIC
SEXTANS
Alphard
M48
MONOCEROS
M42
Sirius
Rigel
ERIDANUS
CETUS
CRATER
NGC 3242
M104
VIRGO
Spica
CORVUS
ANTLIA
PYXIS
PUPPIS
Adhara
M41
CANIS MAJOR
LEPUS
HYDRA
VELA
IC 2391
COLUMBA
CAELUM
Canopus
NGC 3372
NGC 3532
Omega Centauri
NGC 5139
CARINA
NGC 3918
CRUX
CENTAURUS
Mimosa
NGC 4755
Acrux
Coalsack
VOLANS
PICTOR
DORADO
NGC 2070
LMC
RETICULUM
HOROLOGIUM
FORNAX
LIBRA
Hadar
Rigil Kentaurus
MUSCA
CHAMAELEON
MENSA
ERIDANUS
SCP
APUS
CIRCINUS
HYDRUS
LUPUS
TRIANGULUM AUSTRALE
SMC
NGC 104
Achernar
PHOENIX
NORMA
OCTANS
Zeta Scorpii
TUCANA
ARA
PAVO
GRUS
Peacock
INDUS

April stars of the southern skies

As autumn takes a firm hold, Orion sinks even closer to the western horizon. Sirius still shines nearby, but it too is fast disappearing. The south and east are still ablaze with the Milky Way and its brilliant cargo of stars such as Alpha and Beta Centauri, Canopus, and the conspicuous four stars of the Southern Cross, Crux. In the north, Regulus in Leo and Spica in Virgo pace steadily from right to left, bringing with them another northern constellation now making a dominant appearance in the southern hemisphere—Boötes.

COALSACK NEBULA

Object type Dark nebula
Constellation Crux
Description The Coalsack is the sky's largest dark interstellar cloud, containing as much mass as 3500 Sunlike stars and stretching for dozens of light-years. In fact, the Coalsack does emit light—but with only 10 percent of the luminosity of the surrounding star clouds, it appears dark by comparison. It lies mainly in Crux, the Southern Cross, but also spreads into Musca in the south and Circinus in the east.
You need no optical aid to see this seeming void, just the unaided eye and a dark night. Under those conditions, the Coalsack will show up beautifully as a black emptiness, spanning almost as much of the sky as the Southern Cross, against the brighter star clouds of the Milky Way. Just above the Coalsack, below Beta Crucis, is the famous Jewel Box Cluster.
Apparent magnitude N/A
Apparent size 7 x 5 degrees
Actual size 68 light-years
Distance 560 light-years
Other nebulae in this sky NGC 3918
Tarantula Nebula (NGC 2070)

OMEGA CENTAURI (NGC 5139)

Object type Globular cluster
Constellation Centaurus
Description This globular cluster, a spherical island of stars orbiting the Milky Way, is the sky's finest, lying in the constellation Centaurus. It has 10 times the mass of other large globulars—comparable to a small galaxy—and at third magnitude, it is so bright that it was mistaken for a star and given the star designation Omega Centauri. Although you can see this cluster with the naked eye, you will certainly benefit from any type of optical aid. In binoculars, the cluster is fuzzy with a mottled appearance. In a small telescope of about 4 inches (100 mm), you will begin to resolve individual stars. In a larger instrument, the object is simply stunning, showing countless points of light that sparkle on the cluster's edge against the blackness of the surrounding space.
Apparent magnitude 3.7
Apparent size 36 arcminutes
Actual size 200 light-years
Distance 17,000 light-years
Other clusters in this sky Jewel Box Cluster (NGC 4755) Ptolemy's Cluster (M7)

M100 GALAXY

Object type Spiral galaxy
Constellation Coma Berenices
Also known as NGC 4321
Description M100 is a large spiral galaxy with open spiral arms viewed face-on. Like M101, it was one of the first known spiral galaxies. M100 is one of several galaxies in the Coma–Virgo Cluster of galaxies and the one with the largest apparent size. You can find M100 about 8 degrees to the left of Denobola in Leo. However, despite its size, this object is very difficult to see well unless you have at least an 8-inch (200 mm) telescope. In smaller instruments it looks a little like a globular cluster. With 8 to 10 inches (200–250 mm) of aperture, you will be able to spot the dense nucleus using averted vision, but the spiral arms may still evade you unless the night is exceptionally clear.
Apparent magnitude 9.4
Apparent size 7 x 6 arcminutes
Actual size 114,000 light-years
Distance 56,000,000 light-years
Other galaxies in this sky Virgo A (M87)
Large Magellanic Cloud

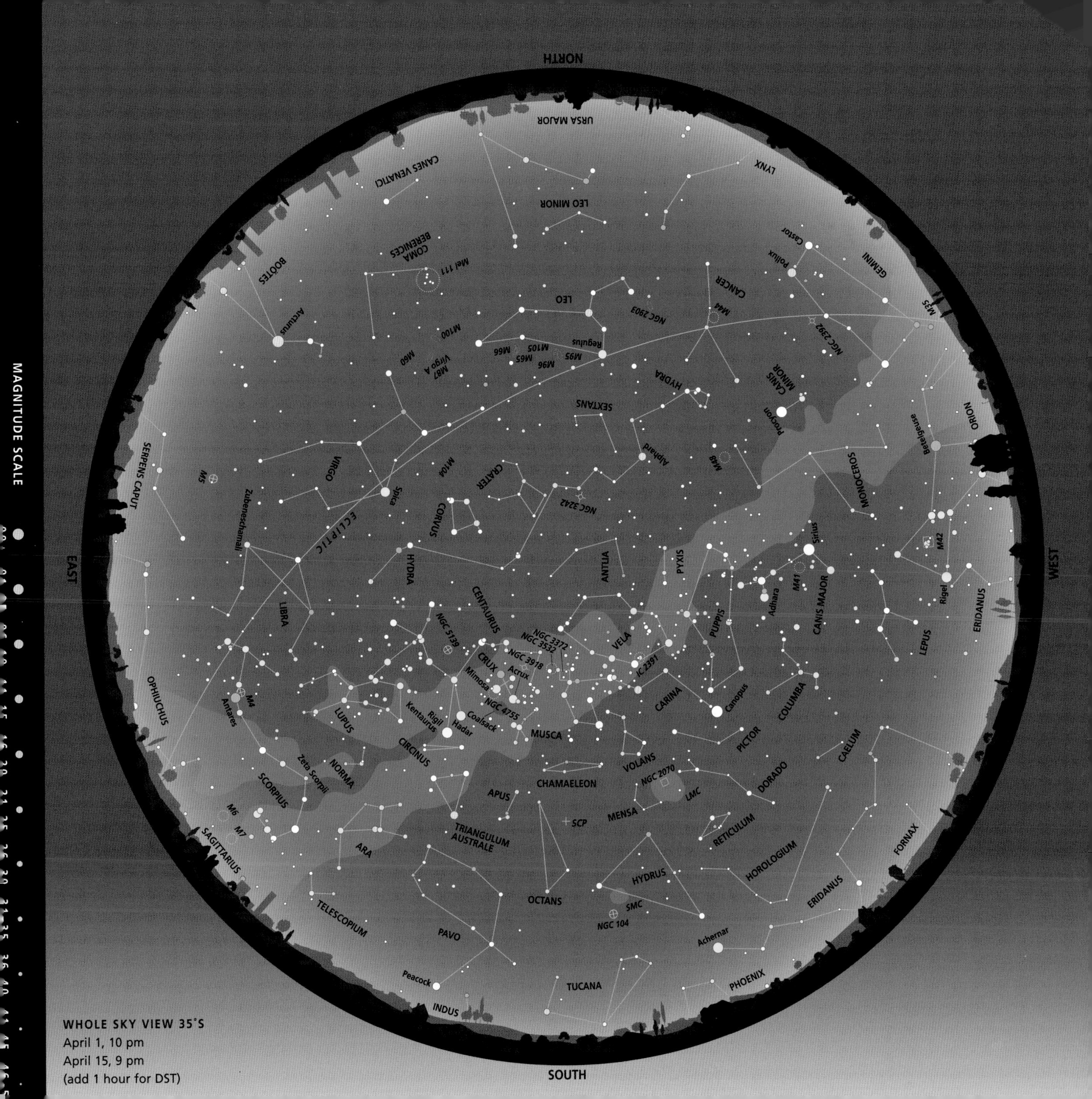
NORTH
SOUTH
EAST
WEST
MAGNITUDE SCALE
WHOLE SKY VIEW 35°S
April 1, 10 pm
April 15, 9 pm
(add 1 hour for DST)
URSA MAJOR
LYNX
CANES VENATICI
LEO MINOR
COMA BERENICES
BOÖTES
GEMINI
CANCER
LEO
ECLIPTIC
HYDRA
SEXTANS
CANIS MINOR
ORION
SERPENS CAPUT
VIRGO
CRATER
CORVUS
MONOCEROS
LIBRA
CENTAURUS
ANTLIA
PYXIS
CANIS MAJOR
ERIDANUS
LEPUS
OPHIUCHUS
LUPUS
CRUX
VELA
PUPPIS
CARINA
COLUMBA
MUSCA
CIRCINUS
NORMA
SCORPIUS
PICTOR
CAELUM
VOLANS
CHAMAELEON
DORADO
APUS
MENSA
RETICULUM
ARA
TRIANGULUM AUSTRALE
HYDRUS
HOROLOGIUM
FORNAX
SAGITTARIUS
TELESCOPIUM
OCTANS
PAVO
TUCANA
PHOENIX
INDUS
Arcturus
Castor
Pollux
Regulus
Procyon
Betelgeuse
Alphard
Spica
Zubeneschamali
Sirius
Rigel
Adhara
Canopus
Antares
Mimosa
Acrux
Rigil Kentaurus
Hadar
Coalsack
Zeta Scorpii
Achernar
Peacock
SCP
SMC
LMC
Mel 111
M100
M60
M87
Virgo A
M66
M65
M105
M96
M95
M44
NGC 2903
NGC 2392
M35
M48
M104
M5
NGC 3242
M42
M41
NGC 5139
NGC 3372
NGC 3532
NGC 3918
IC 2391
NGC 4755
M4
M6
M7
NGC 2070
NGC 104

May stars of the southern skies

Due north, the sky is a circus for the inverted northern figures of Virgo, Leo and Boötes, each marked by their brightest stars Spica, Regulus and Arcturus, respectively. Virgo and Leo are also great places to hunt galaxies. In the south, the brilliant stars Alpha and Beta Centauri are now the highest they will get. They will cross the sky in a small, slow arc from east to west as they rotate about the south pole. The Small Magellanic Cloud is rising again, while its larger counterpart sinks in sympathy toward the southern horizon.

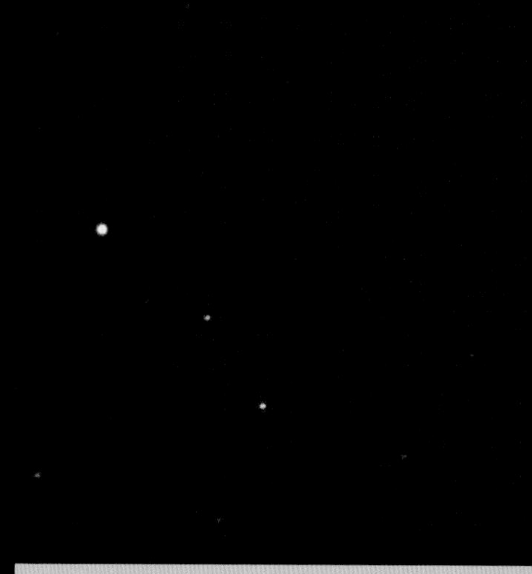

ZUBENESCHAMALI

Object type Colored star
Constellation Libra
Also known as Beta Librae
Description In most constellations it is the alpha star that is the most brilliant. But there are a few exceptions, and Libra the Scales (or the Balance) is one of them. Here, it is Zubeneschamali—Beta Librae—that is the brightest and northernmost star in the constellation. It is conspicuous because many observers claim it has a greenish color to it whether to the naked eye or with optical aid. This is interesting, because green stars cannot physically exist. Take a look and judge for yourself. Zubeneschamali is an Arabic term meaning "The Northern Claw," a name reflecting the possibility that this unusual star was part of a different constellation during Roman times.
Apparent magnitude 2.6
Spectral type B8 (blue)
Luminosity class V (dwarf)
Distance 160 light-years
Other interesting stars in this sky Alpha Centauri, Zeta Scorpii

JEWEL BOX CLUSTER (NGC 4755)

Object type Open cluster
Constellation Crux
Description The Jewel Box is a real beauty, despite its compactness—an aggregation of about 50 stars surrounding and including Kappa Crucis. At an age of just 7 million years, this is a very young cluster, containing many giant blue and white stars. You can see the Jewel Box unaided on a clear, dark night, but binoculars or a small telescope provide the best views. If you do use a telescope, you will need moderate magnification because the cluster is small. Don't magnify too much or your field of view will be too large to contain all the stars. In the center, a beautiful red supergiant contrasts strongly with the other blue and white cluster members, giving the cluster the appearance of a cornucopia of sparkling, colored gems—the brightest forming an "A" shape.
Apparent magnitude 4
Apparent size 10 arcminutes
Actual size 22 light-years
Distance 7600 light-years
Other clusters in this sky Omicrom Velorum Cluster, Omega Centauri (NGC 5139)

VIRGO A (M87)

Object type Elliptical galaxy
Constellation Virgo
Also known as NGC 4486
Description Virgo A is a giant elliptical galaxy with a faint outer edge that in deep photos extends for a tremendous half a billion light-years—a real monster. However, the brightest part of the galaxy spans about 120,000 light-years—still substantially larger than the Milky Way. Virgo A is the dominant cluster in the vast Coma–Virgo cluster of galaxies, and is famous, among other reasons, for a spectacular jet of charged particles that spurts from a supermassive black hole in its heart. In larger binoculars, Virgo A appears as a comet-like blur, and as a fuzzy ball in a telescope. Don't expect to see any details of the jet, though, which you can only glimpse in professional telescopes with objectives measuring several feet.
Apparent magnitude 8.6
Apparent size 7 arcminutes
Actual size 120,000 light-years
Distance 60,000,000 light-years
Other galaxies in this sky M100, M105

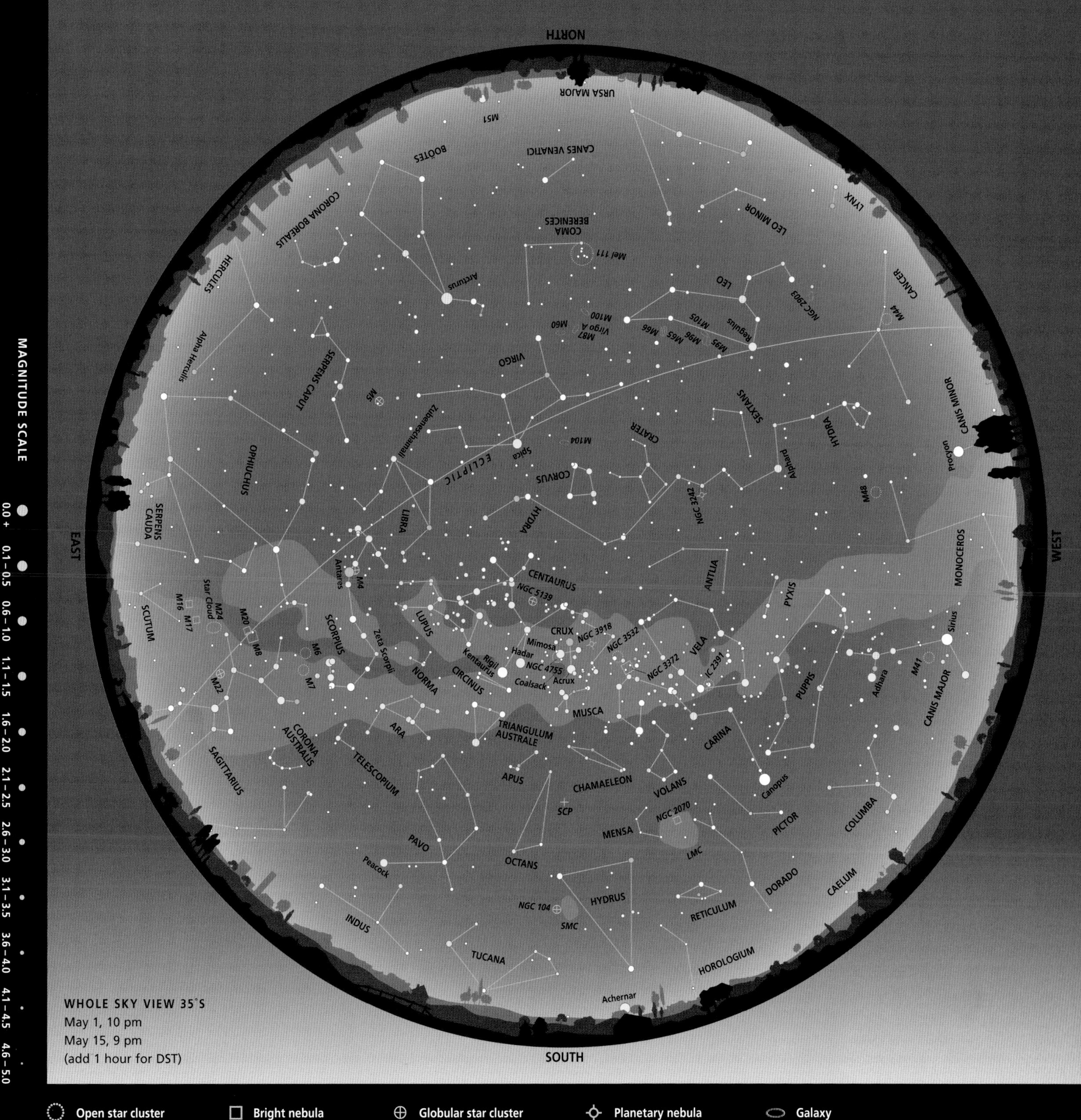

NORTH
EAST
WEST
SOUTH
MAGNITUDE SCALE
0.0 +
0.1–0.5
0.6–1.0
1.1–1.5
1.6–2.0
2.1–2.5
2.6–3.0
3.1–3.5
3.6–4.0
4.1–4.5
4.6–5.0
WHOLE SKY VIEW 35°S
May 1, 10 pm
May 15, 9 pm
(add 1 hour for DST)
Open star cluster
Bright nebula
Globular star cluster
Planetary nebula
Galaxy

June stars of the southern skies

As the southern winter begins to take hold, Orion and Sirius have at last departed the western skies, while Canopus and the Large Magellanic Cloud sink toward the southwest. Centaurus and Crux still shine down high in the south, while even higher overhead, around midnight, Scorpio crosses the zenith flagged by the brilliant red Antares. Later, as Scorpius moves aside, the dramatic star clouds of Sagittarius take its place. Now the northeastern skies see some new arrivals—the bright stars Vega in Lyra and Altair in Aquila.

LAGOON NEBULA (M8)

Object type Diffuse nebula
Constellation Sagittarius
Also known as NGC 6523
Description The Lagoon Nebula is a delightful object, easily ranking alongside the Orion Nebula and the Eta Carinae Nebula for its beauty. It is just visible to the keen unaided eye on a dark night, a bright patch superimposed on the rich Milky Way star clouds in Sagittarius. The Lagoon Nebula is a star factory. Through binoculars it becomes a bright smudge with a few faint stars within it—these are part of the open cluster NGC 6530, born from the nebula's gases and now residing in its eastern half. But with a telescope, the view is vastly improved. Many more stars dot the nebula, as do patches of darker dust including one long, curving filament that seems to split the nebula into two parts.
Apparent magnitude 5.8
Apparent size 90 x 40 arcminutes
Actual size 130 light-years
Distance 5200 light-years
Other nebulae in this sky Eagle Nebula (M16), Trifid Nebula (M20)

PTOLEMY'S CLUSTER (M7)

Object type Open cluster
Constellation Scorpius
Also known as The Scorpion's Tail, NGC 6475
Description This open cluster of about 80 stars was known by the astronomer Claudius Ptolemaeus (Ptolemy) as long ago as AD 130. At about 220 million years old, M7 is one of the most ancient star clusters. Compared to the Solar System and cosmic timescales in general, this is still extremely youthful, but older clusters are rare because they tend to disperse with age as they orbit the galaxy and experience the gravitational tugs of neighboring stars. You can easily see M7 with the naked eye as a misty blob about twice the size of the Moon, superimposed on a rich background of more distant stars. But as with many open clusters, it is best if you view it through binoculars or a low-power telescope.
Apparent magnitude 4
Apparent size 1.3 degrees
Actual size 21 light-years
Distance 900 light-years
Other clusters in this sky M22, Jewel Box Cluster (NGC 4755)

M60 GALAXY

Object type Elliptical galaxy
Constellation Virgo
Also known as NGC 4649
Description Like M87, also in Virgo, M60 is a giant elliptical galaxy belonging to the Coma–Virgo galaxy cluster. Through a low-power telescope, M60 is the easternmost galaxy in a chain 1.5 degrees long, which includes M58 (a spiral) and M59 (another elliptical). You can spot M60 with binoculars, but it will not impress. For a better view, use at least a 6-inch (150 mm) telescope. You should then be able to see that M60, a fuzzy blob, has a close companion galaxy—a face-on spiral with catalog number NGC 4647 (see image above). Astronomers suspect that M60 and NGC 4647 are at similar distances and so are physically interacting with each other, engaged in a gravitational tug-of-war.
Apparent magnitude 8.8
Apparent size 7 arcminutes
Actual size 120,000 light-years
Distance 60,000,000 light-years
Other galaxies in this sky M100, M105

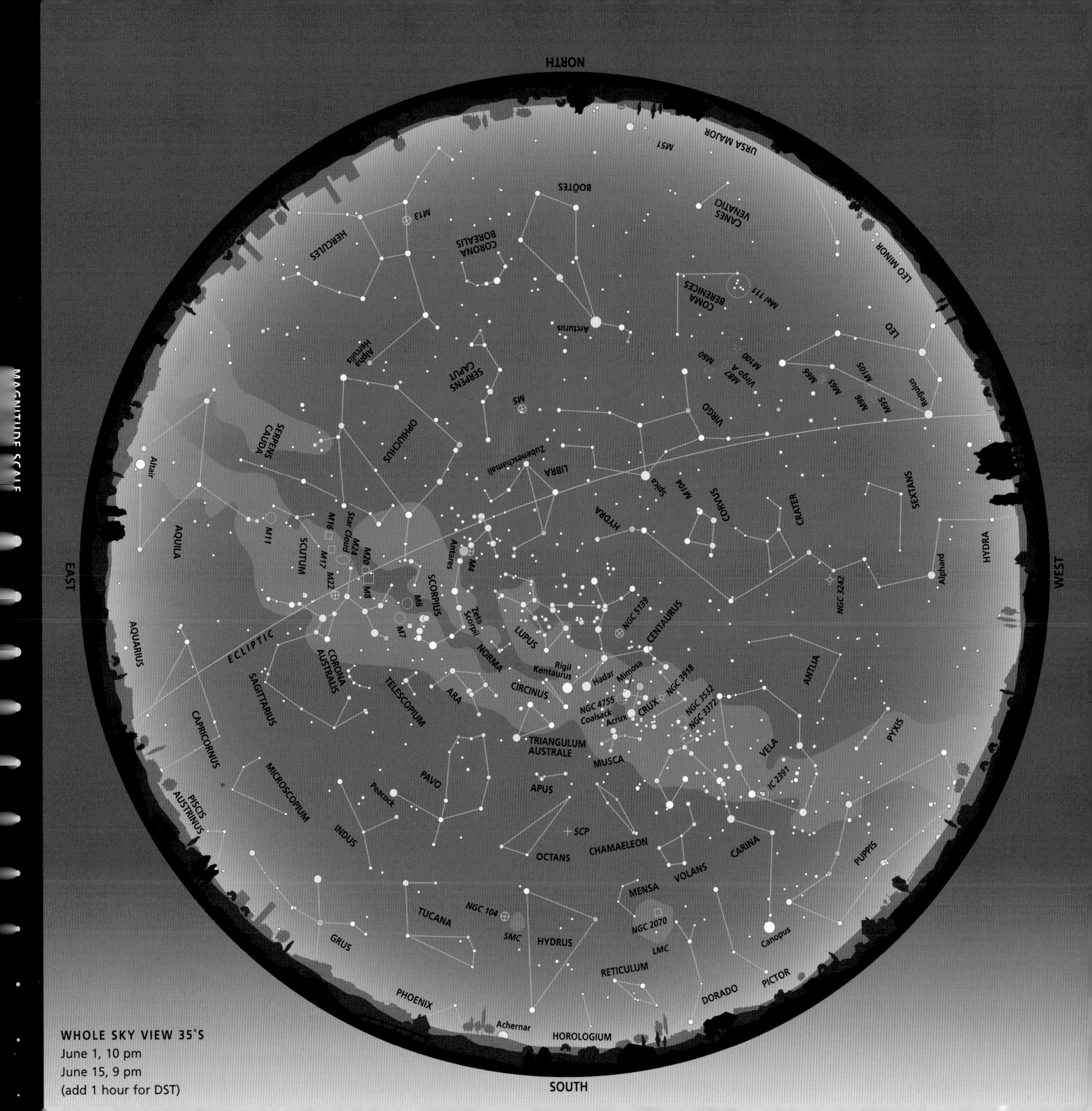

WHOLE SKY VIEW 35°S
June 1, 10 pm
June 15, 9 pm
(add 1 hour for DST)

July stars of the southern skies

The Summer Triangle—a symbol of the northern hemisphere's summer—is now becoming more prominent, heralding the contrasting southern winter. Altair in Aquila is the highest, shining down in the northern half of the sky from about 45 degrees. Vega in Lyra is lower down, and Deneb in Cygnus is around 10 degrees above the misty horizon. In the south, Crux is now pointing sideways, Carina kisses the horizon, and Alpha and Beta Centauri are fast descending. Leo hunts no more, and Boötes, too, has departed.

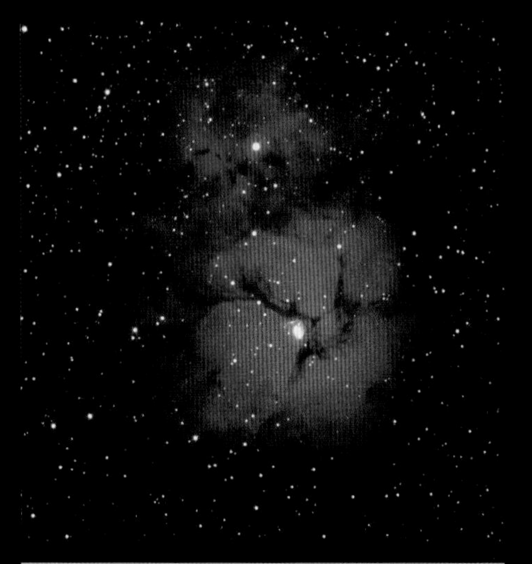

TRIFID NEBULA (M20)

Object type Diffuse nebula
Constellation Sagittarius
Also known as NGC 6514
Description Located about 1.5 Moon diameters north of the brighter and larger Lagoon Nebula, the Trifid is a beautiful but fairly faint nebula. Long-exposure photographs show a vibrant red cloud trisected by three lanes of dark gas that give the nebula the appearance of a flower. A blue component is also present, caused by starlight reflecting off dust, while at the heart of the Trifid lies a bright blob—a system of six stars, illuminating the surrounding nebulosity. In binoculars the Trifid shows up as little more than a blurry star. But with a 6-inch (150 mm) telescope you will be able to make out the famous lanes of dust. They look even better in a larger instrument.
Apparent magnitude 8
Apparent size 29 x 27 arcminutes
Actual size 45 light-years
Distance 5200 light-years
Other nebulae in this sky Lagoon Nebula (M8) Eagle Nebula (M16)

M22 CLUSTER

Object type Globular cluster
Constellation Sagittarius
Also known as NGC 6656
Description M22 was the first known globular cluster, discovered in 1665, possibly earlier, and is one of the sky's finest—and nearest. Its proximity makes it brighter than the famous M13 Hercules globular cluster—only Omega Centauri and 47 Tucanae are more brilliant. The cluster is also notable because it contains a planetary nebula. M22's brilliance makes it conspicuous to the naked eye and an easy target for binoculars. But to see this object at its best you need a telescope. Even a moderate 3-inch (75 mm) instrument will begin to resolve the cluster into a ball of stars, while a telescope with twice this aperture will take your breath away. You might also be able to discern the cluster's slightly oval shape.
Apparent magnitude 5.1
Apparent size 24 arcminutes
Actual size 73 light-years
Distance 10,400 light-years
Other clusters in this sky M4 Omega Centauri (NGC 5139)

RAS ALGETHI

Object type Variable star
Constellation Hercules
Also known as Rasalegti, Alpha Herculis
Description Ras Algethi is a cool red supergiant or possibly a bright giant, some 900 times more luminous than the Sun and several hundred times its diameter. The Arabic name means "The Kneeler's Head," describing the way that Hercules is often envisaged—as a kneeling man. As with most stars of this type, Ras Algethi is a variable. Its brightness changes by the best part of a magnitude over 50 to 150 days, the result of irregular stellar pulsations as the star "breathes" in and out. If you keep regular tabs on this star—you won't need any optical aid—you should be able to see how its brightness changes relative to the nearest bright star, Delta Herculis.
Apparent magnitude 3.1 to 3.9
Spectral type M5 (red)
Luminosity class Ib (underluminous supergiant) or II (bright giant)
Distance 250–500 light-years
Other interesting stars in this sky Beta Librae Alpha Centauri

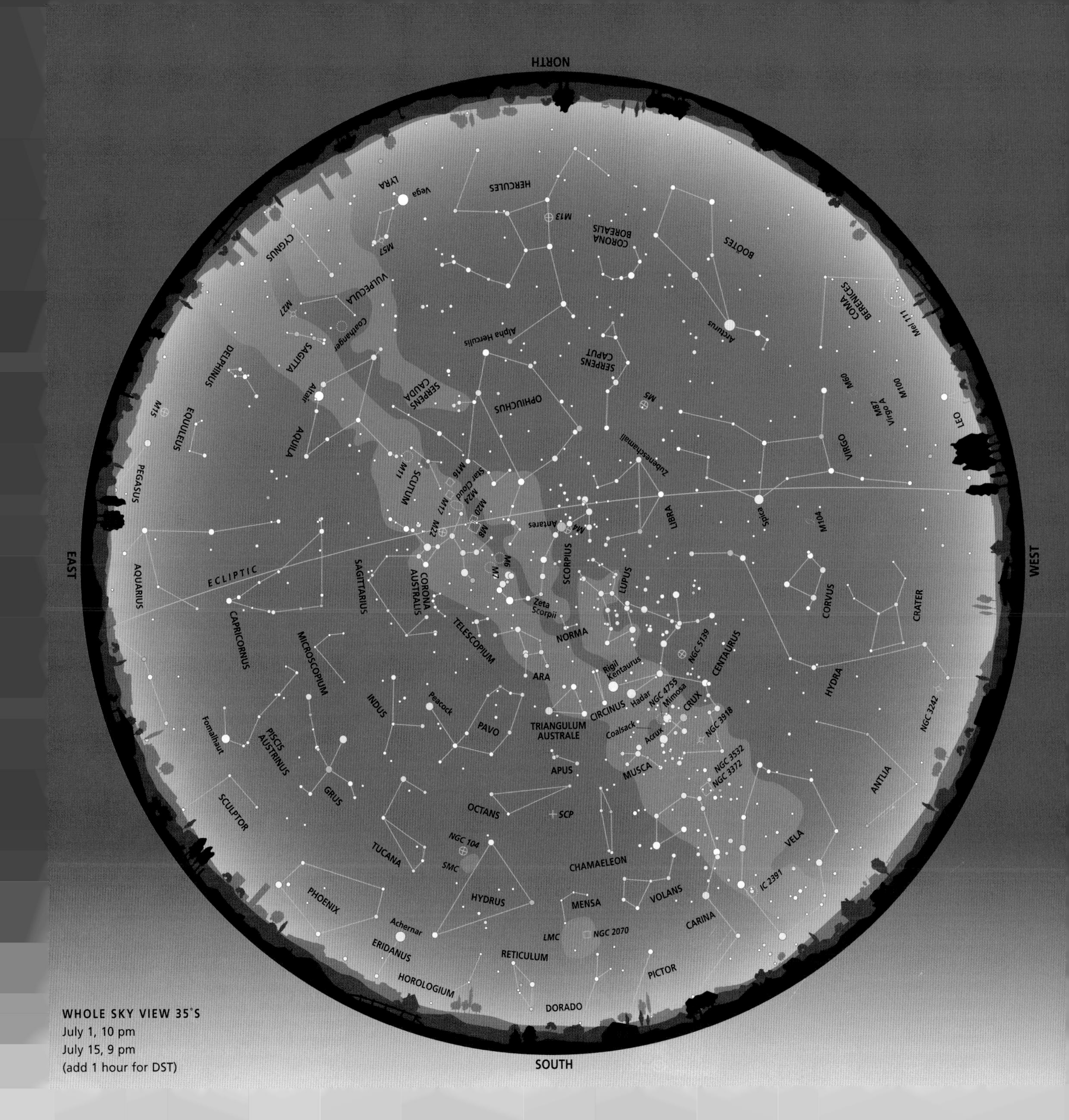
NORTH
SOUTH
EAST
WEST
LYRA
Vega
HERCULES
M13
CORONA BOREALIS
BOÖTES
CYGNUS
M57
VULPECULA
M27
Coathanger
SAGITTA
Alpha Herculis
Arcturus
COMA BERENICES
Mel 111
DELPHINUS
Altair
SERPENS CAUDA
SERPENS CAPUT
OPHIUCHUS
M5
M60
M100
M87
Virgo A
LEO
M15
EQUULEUS
AQUILA
Zubeneschamali
VIRGO
PEGASUS
M11
SCUTUM
M16
Star Cloud
M24
M17
M20
M22
M8
Antares
M4
LIBRA
Spica
M104
EAST
AQUARIUS
ECLIPTIC
SAGITTARIUS
CORONA AUSTRALIS
M6
M7
SCORPIUS
LUPUS
CORVUS
CRATER
CAPRICORNUS
Zeta Scorpii
NORMA
TELESCOPIUM
NGC 5139
CENTAURUS
MICROSCOPIUM
ARA
Rigil Kentaurus
HYDRA
INDUS
Peacock
PAVO
TRIANGULUM AUSTRALE
CIRCINUS
Hadar
NGC 4755
Mimosa
CRUX
NGC 3242
Fomalhaut
PISCIS AUSTRINUS
Coalsack
Acrux
NGC 3918
GRUS
APUS
MUSCA
NGC 3532
NGC 3372
ANTLIA
SCULPTOR
OCTANS
SCP
NGC 104
TUCANA
SMC
CHAMAELEON
VELA
PHOENIX
HYDRUS
MENSA
VOLANS
IC 2391
Achernar
ERIDANUS
LMC
NGC 2070
CARINA
RETICULUM
PICTOR
HOROLOGIUM
DORADO
WHOLE SKY VIEW 35˚S
July 1, 10 pm
July 15, 9 pm
(add 1 hour for DST)

August stars of the southern skies

In the west, Sagittarius and Scorpius are still high overhead. To their right, Altair shines down steadily from around 45 degrees and takes flight toward the west as the night progresses, with Deneb and Vega—the other two stars of the Summer Triangle—safely in tow.

Carina is below the southern horizon now. But Crux and Centaurus are still well placed in the southwest, at around the same height above the horizon as the now prominent Small Magellanic Cloud. Its larger cousin will make an appearance too, but not until after midnight.

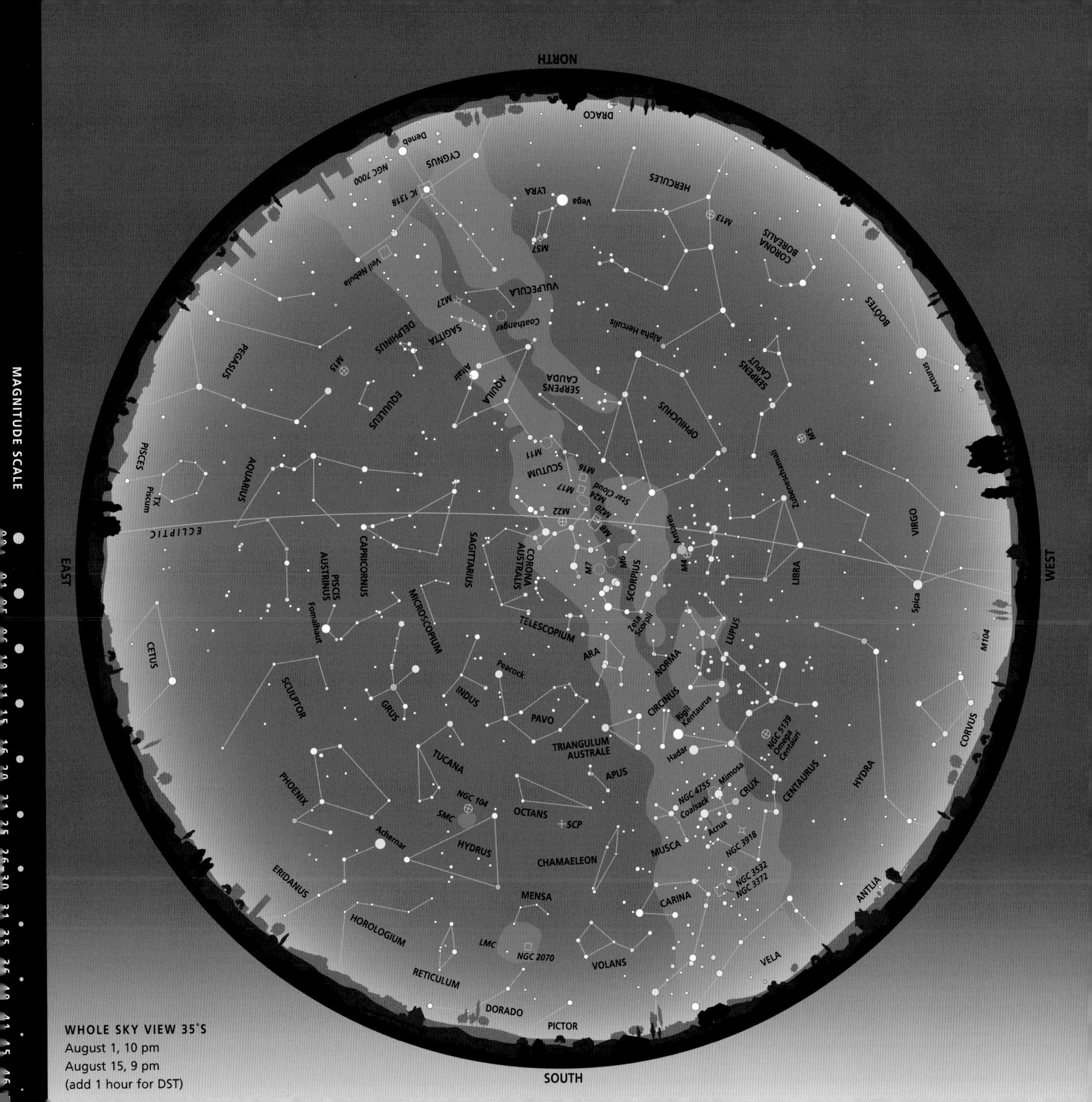

WHOLE SKY VIEW 35°S
August 1, 10 pm
August 15, 9 pm
(add 1 hour for DST)

September stars of the southern skies

Aquila still shines in the northwest, but Cygnus the Swan now skates by with a wingtip skimming the horizon, and Lyra is not much higher. The Milky Way is awesome this month. It emerges from the northeast below Cygnus and traverses the entire sky, slicing it in two. It flows past Aquila, then Scorpius, then into Sagittarius high overhead. It spills down toward Centaurus and Crux, before disappearing over the southern horizon. As the sky spins, the bright stars in the northwest make way for Pegasus, which heralds the arrival of spring.

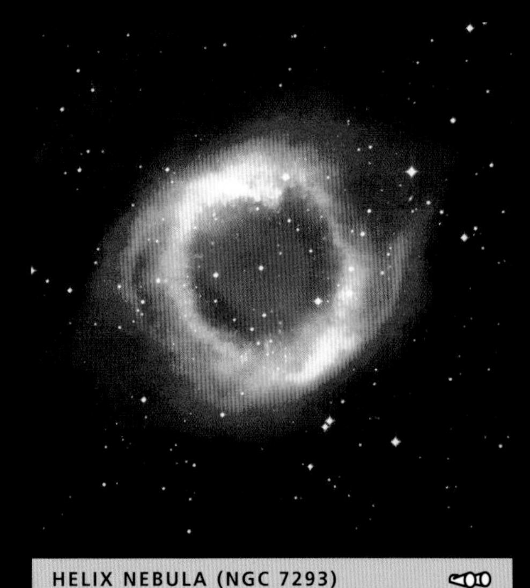

HELIX NEBULA (NGC 7293)

Object type Planetary nebula
Constellation Aquarius
Description This famous object is the closest known planetary nebula—the cast-off atmosphere of a star long dead—and subsequently has the largest apparent diameter as seen from Earth. Astronomers studied it in detail in 1996 using the Hubble Space Telescope, which uncovered thousands of comet-like knots of gas extending radially inward toward the central star. Scientists think these are the result of the interplay between two separate shells of stellar ejecta—a hot shell of fast gas overtaking a slower shell of cooler gas ejected earlier. You can see the Helix Nebula with binoculars as an oval spot about half the apparent size of the Moon. However, your best view will be gleaned with a small telescope using low magnification and a high-contrast filter.
Apparent magnitude 6.5
Apparent size 15 arcminutes, halo 28 arcminutes
Actual size 2 light-years, halo 3.7 light-years
Distance 450 light-years
Other nebulae in this sky Eagle Nebula (M16) Saturn Nebula (NGC 7009)

NGC 6752 CLUSTER

Object type Globular cluster
Constellation Pavo
Description Discovered in 1826 by Scottish astronomer James Dunlop, NGC 6752 is a vast globular cluster. It is comparable in size with the other southern hemisphere globulars 47 Tucanae and Omega Centauri. Despite its size, NGC 6752 is not as well known as these other two, and even the slightly fainter M13 in Hercules is a more familiar object to most. Still, NGC 6752 does not disappoint. You can just spot it with the naked eye, located 2 degrees above Lambda Pavonis, where it forms an elbow shape with Omega Pavonis off to the side. To obtain a decent view of this cluster, use binoculars or a low-power, small aperture telescope. As it's in a relatively dust-free patch of sky, the colors of its brighter red stars show up well in larger instruments.
Apparent magnitude 5.4
Apparent size 55 arcminutes
Actual size 210 light-years
Distance 13,000 light-years
Other clusters in this sky M4 M55

M74 GALAXY

Object type Spiral galaxy
Constellation Pisces
Also known as NGC 628
Description M74 is a large, beautiful open-armed spiral galaxy in Pisces, 2 degrees away from Eta Piscium. For size, it compares comfortably with the Milky Way. M74 is quite faint, and you will not see it without a moderate to large telescope—the smaller the scope, the darker the sky you will need. A 4-inch (100 mm) instrument will resolve the nucleus. Larger telescopes will reward you with a view of some of the spiral structure. But to see much more than this, you will need around 16 inches (400 mm) of aperture. With that kind of instrument, the galaxy takes on a knotty appearance, courtesy of numerous foreground stars and some nebulae within the galaxy itself.
Apparent magnitude 9.2
Apparent size 10 x 9 arcminutes
Actual size 105,000 light-years
Distance 35,000,000 light-years
Other galaxies in this sky M77 NGC 253

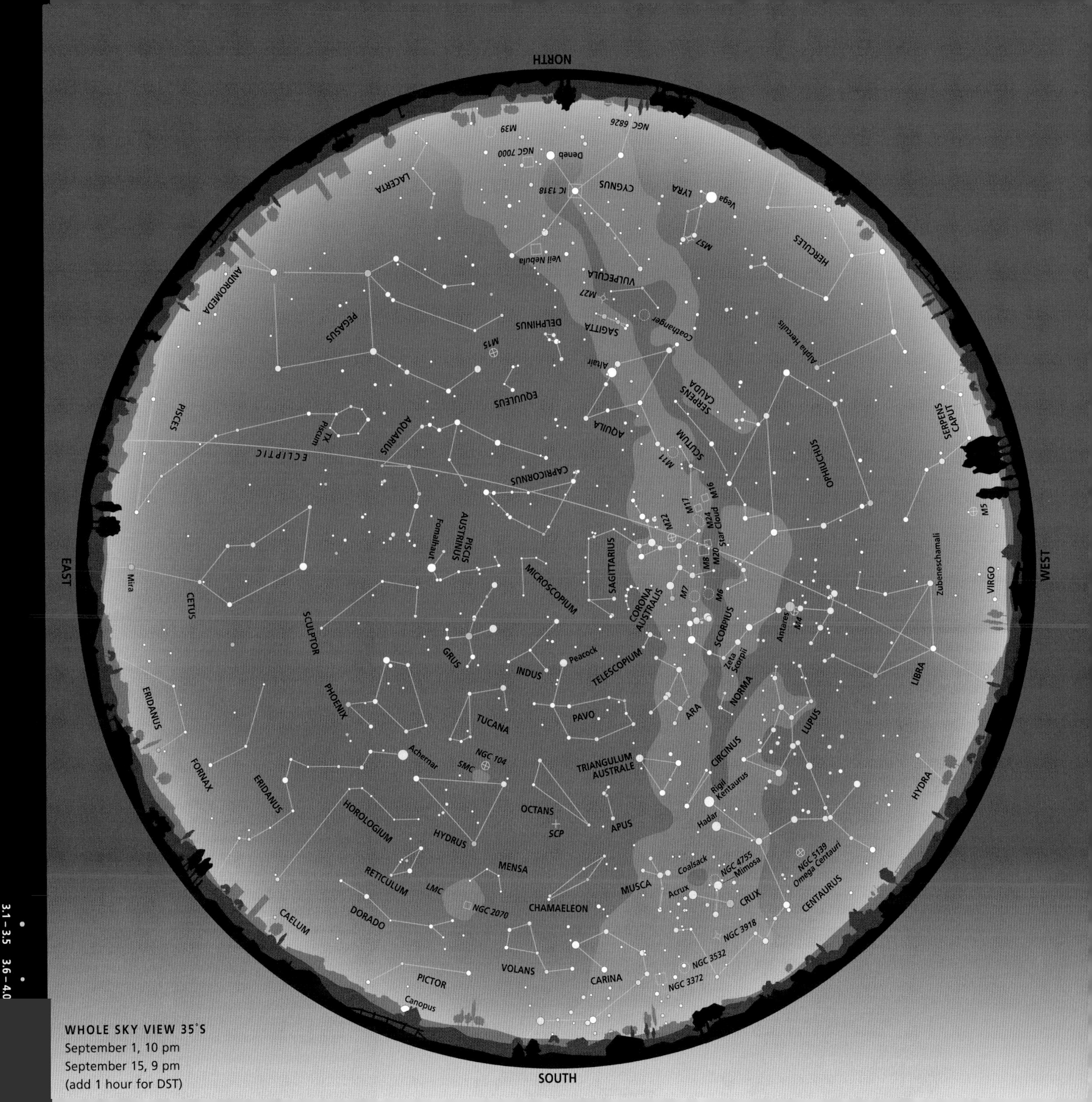

WHOLE SKY VIEW 35°S
September 1, 10 pm
September 15, 9 pm
(add 1 hour for DST)

October stars of the southern skies

Crux is now low down and sinks below the southern horizon before midnight. Alpha and Beta Centauri follow shortly afterward. The Milky Way remains prominent, while both Magellanic Clouds are now well placed for viewing just to the left of it in the south. Below them and farther left, Canopus is rising again and Sirius follows later. Lyra, Pegasus and Aquila still beg attention in the north and west. But otherwise the spring skies are a playground for a host of relatively dim constellations—the zodiacal Aquarius, Pisces and Capricornus.

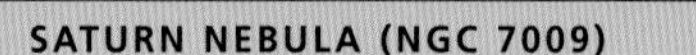

SATURN NEBULA (NGC 7009)

Object type Planetary nebula
Constellation Aquarius
Description This is a small planetary nebula that gets its name from the two spectacular jets that emerge symmetrically, one from either side. Lord Rosse, a famous nineteenth-century Irish astronomer, thought that the object bore a resemblance through his telescope to the ringed planet Saturn, seen edge-on. The Saturn Nebula lies at an uncertain distance of about 3000 light-years, and has a faint central star of magnitude 11.5. Through a small telescope of 3 inches (75 mm) with moderate magnification, the nebula appears as a greenish or bluish oval spot—similar in appearance to the above image. To see evidence of the jets that gave the object its name, though, you will need access to at least a 12-inch (300 mm) instrument.
Apparent magnitude 8.4
Apparent size 0.4 x 1.6 arcminutes
Actual size 1.4 light-years
Distance 3200 light-years
Other nebulae in this sky Trifid Nebula (M20) Helix Nebula (NGC 7293)

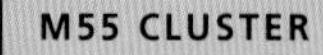

M55 CLUSTER

Object type Globular cluster
Constellation Sagittarius
Also known as NGC 6809
Description M55 is a globular cluster of moderate size and luminosity, and one of two clusters drifting beyond the rich star clouds of Sagittarius. It spans an area of sky roughly two-thirds the size of the Moon's disk but, unlike Sagittarius' other Messier globular cluster, M22, M55 is too faint to be seen without the aid of an optical instrument. To find it, point your binoculars about halfway between Tau Sagittarii and the double star Theta Sagittarii. Most globular clusters appear fairly nebulous with binoculars, but M55 has a very loose, grainy appearance, which is more akin to an open cluster than a globular. To get the best view, use a 4- to 6-inch (100–150 mm) telescope with low magnification.
Apparent magnitude 7
Apparent size 17 arcminutes
Actual size 97 light-years
Distance 18,000 light-years
Other clusters in this sky M30 NGC 6752

M77 GALAXY

Object type Spiral galaxy
Constellation Cetus
Also known as NGC 1068, Cetus A
Description M77 is a fairly large spiral galaxy, which was first thought to be a cluster, perhaps because of its foreground stars. It is the constellation's brightest, but like the others, M77 is still faint and has a small apparent size of just 6 arcminutes. M77 is a so-called Seyfert galaxy—a class of spiral that has a very active and bright nucleus, probably due to the presence of a supermassive black hole, sucking in stars and spewing out radiation. You will need a small telescope, 3 to 4 inches (75–100 mm) in aperture, to spot this object. You will then see the bright nucleus surrounded by a fuzzy ellipse of light. Don't expect much more with a larger telescope, as M77 has little of an impressive spiral structure.
Apparent magnitude 8.9
Apparent size 6 arcminutes
Actual size 105,000 light-years
Distance 60,000,000 light-years
Other galaxies in this sky M74 NGC 253

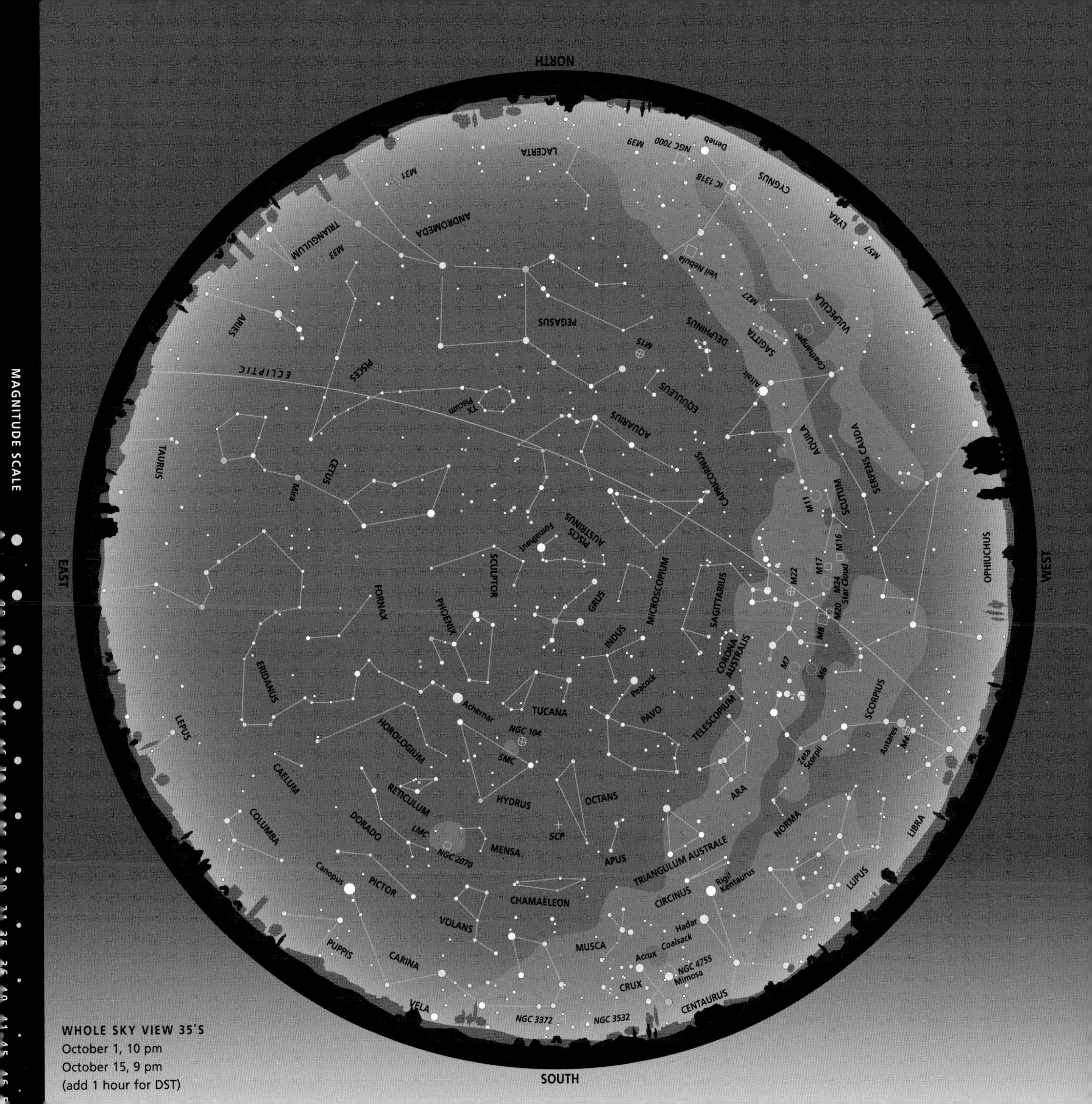
MAGNITUDE SCALE
NORTH
EAST
WEST
SOUTH
ECLIPTIC
LACERTA
M39
NGC 7000
Deneb
IC 1318
CYGNUS
LYRA
M57
M31
ANDROMEDA
TRIANGULUM
M33
Veil Nebula
M27
VULPECULA
PEGASUS
DELPHINUS
SAGITTA
Coathanger
ARIES
M15
PISCES
Altair
EQUULEUS
TX Piscum
AQUARIUS
AQUILA
SERPENS CAUDA
TAURUS
CETUS
Mira
CAPRICORNUS
M11
SCUTUM
PISCIS AUSTRINUS
Fomalhaut
M16
M17
OPHIUCHUS
SCULPTOR
MICROSCOPIUM
SAGITTARIUS
M22
M24
Star Cloud
M20
FORNAX
PHOENIX
GRUS
M8
INDUS
CORONA AUSTRALIS
M7
M6
ERIDANUS
Peacock
SCORPIUS
Achernar
TUCANA
PAVO
TELESCOPIUM
LEPUS
HOROLOGIUM
NGC 104
Antares
M4
Zeta Scorpii
SMC
CAELUM
RETICULUM
HYDRUS
OCTANS
ARA
COLUMBA
DORADO
LMC
SCP
NORMA
LIBRA
NGC 2070
MENSA
APUS
TRIANGULUM AUSTRALE
Canopus
PICTOR
LUPUS
CHAMAELEON
CIRCINUS
Rigil Kentaurus
VOLANS
Hadar
Coalsack
MUSCA
PUPPIS
CARINA
Acrux
NGC 4755
Mimosa
CRUX
CENTAURUS
VELA
NGC 3372
NGC 3532
WHOLE SKY VIEW 35°S
October 1, 10 pm
October 15, 9 pm
(add 1 hour for DST)

November stars of the southern skies

Summer is on its way. The Milky Way band of light encircles the sky low down, hugging the horizon all around and lost in its haze. The spectacular stars of Orion, Canis Major, Taurus and Carina shine brilliantly in the east, rising higher as the night deepens. But otherwise, this time of year is distinctive for its lack of celestial splendor. The sky is filled everywhere—except in the east and, later, in the north—with dim and unimportant constellations. Still, there is a consolation prize: this is a great season to spot striking galaxies.

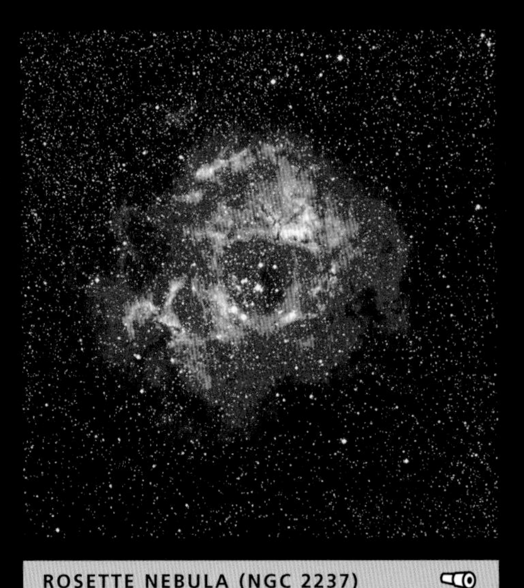

ROSETTE NEBULA (NGC 2237)

Object type Diffuse nebula
Constellation Monoceros
Description The Rosette Nebula is a great circular cloud of gas and dust that, at an estimated mass of about 10,000 Suns, is one of the most massive known. This object also appears large, with a diameter approaching three full Moons. In fact, it is so large that its brightest parts have actually been assigned four separate NGC numbers: 2237, 2238, 2239 and 2246. However, with its considerable angular size, it spreads its light thinly—you will spot it in small telescopes or binoculars but only if you have a dark sky. But more positively, what you will be able to see is the small cluster located at the center of the nebula, NGC 2244. A larger telescope of 8 to 10 inches (200–250 mm), combined with a deep sky filter, will reveal the striking nebulosity.
Apparent magnitude 5
Apparent size 80 x 60 arcminutes
Actual size 130 light-years
Distance 5500 light-years
Other nebulae in this sky Eight Burst Nebula (NGC 3132) Saturn Nebula (NGC 7009)

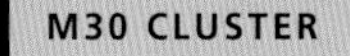

M30 CLUSTER

Object type Globular cluster
Constellation Capricornus
Also known as NGC 7099
Description M30 is another of Charles Messier's globular clusters, of average size and distance. It is located in Capricornus—an otherwise fairly insignificant constellation—about 2 degrees from Zeta Capricorni and outside Capricornus' distinctive triangle of stars. You can see M30 in binoculars or small telescopes, but as little more than a faint, "round nebula"—Messier's own description. A moderate instrument, about 3 to 4 inches (75–100 mm) of aperture, is needed to resolve the brightest stars and to demonstrate that M30 is an attractive object. Meanwhile, an even larger telescope reveals this globular cluster's dense and impressive central core.
Apparent magnitude 7.5
Apparent size 11 arcminutes
Actual size 83 light-years
Distance 26,000 light-years
Other clusters in this sky M41 NGC 3532

NGC 253 GALAXY

Object type Spiral galaxy
Constellation Sculptor
Also known as Silver Coin Galaxy, Sculptor Galaxy
Description NGC 253 is a bright spiral galaxy, one of the brightest beyond the Local Group of galaxies to which the Milky Way belongs. It is the largest of the Sculptor group—a galaxy cluster often referred to as the South Polar Group because of its location in that region. The Sculptor group is the closest galaxy cluster to the Local Group. Seen from our vantage point, NGC 253 is edge-on. Photographs easily show the extensive lanes of cosmic dust tracing the spiral arms. With binoculars, NGC 253 appears as an elongated band of light. It is best revealed in a small telescope, with which you will be able to discern some of the vast dust that gives the galaxy a patchy appearance.
Apparent magnitude 7.1
Apparent size 25 x 7 arcminutes
Actual size 73,000 light-years
Distance 10,000,000 light-years
Other galaxies in this sky M74, NGC 55

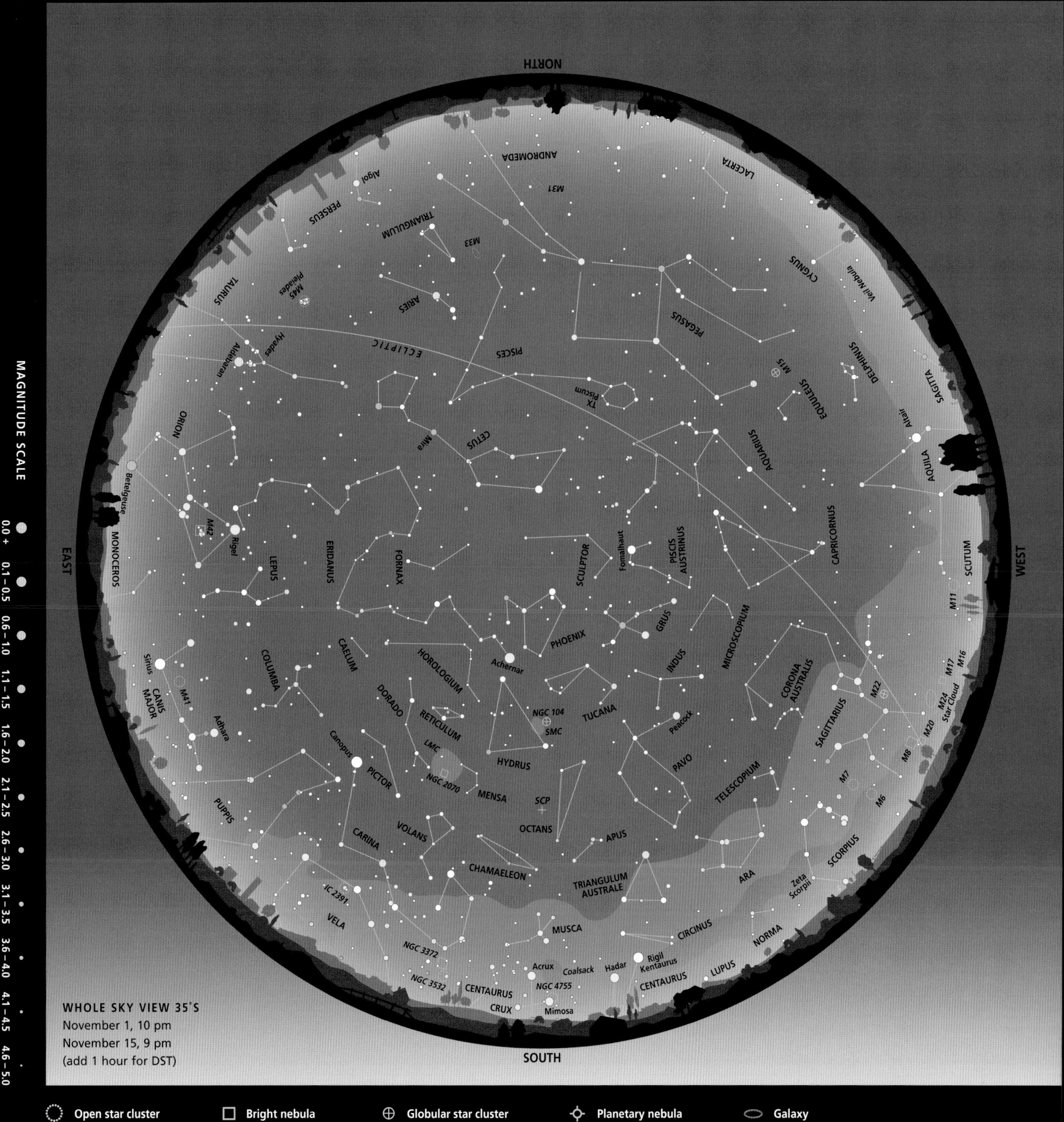

NORTH
SOUTH
EAST
WEST
MAGNITUDE SCALE
0.0 +
0.1 – 0.5
0.6 – 1.0
1.1 – 1.5
1.6 – 2.0
2.1 – 2.5
2.6 – 3.0
3.1 – 3.5
3.6 – 4.0
4.1 – 4.5
4.6 – 5.0
WHOLE SKY VIEW 35°S
November 1, 10 pm
November 15, 9 pm
(add 1 hour for DST)
Open star cluster
Bright nebula
Globular star cluster
Planetary nebula
Galaxy

December stars of the southern skies

Summer is now in full swing, and it is still the northeastern skies that hold the most interest. There Orion hunts in a handstand, high above the horizon, accompanied by Sirius and Canopus—the two brightest stars—Aldebaran in Taurus, the Gemini Twins Castor and Pollux and, later, bright Capella in Auriga. The western skies remain devoid of bright stars, but in the south, the Southern Cross returns, with the bright stars of Centaurus—Alpha and Beta Centauri—not far behind, rising out of the horizon and bringing the Milky Way with them.

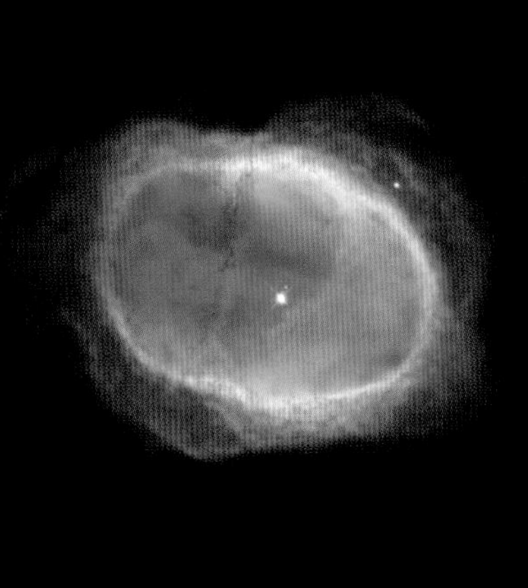

EIGHT BURST NEBULA (NGC 3132)

Object type Planetary nebula
Constellation Vela
Description The Eight Burst Nebula is a handsome planetary nebula on the Vela–Antila border, and it is considered by many to be the southern hemisphere's equivalent of the more famous Ring Nebula (M57). Indeed, amateurs often refer to NGC 3132 as the Southern Ring Nebula. The two objects do bear a significant resemblance to one another, except that NGC 3132 has a brighter central star. The Hubble Telescope photographed this nebula in 1998 and revealed odd filaments of dust bisecting it—which astronomers are trying to understand. You won't see the Eight Burst Nebula well with binoculars. With about 4 inches (100 mm) of aperture, you should be able to make out a fuzzy circular patch. With a larger instrument, the central hole will become apparent.
Apparent magnitude 8
Apparent size 48 arcseconds
Actual size 0.5 light-years
Distance 2000 light-years
Other nebulae in this sky Rosette Nebula (NGC 2237) Tarantula Nebula (NGC 2070)

M47 CLUSTER

Object type Open cluster
Constellation Puppis
Also known as NGC 2422
Description M47 is a fairly bright open cluster. It contains some 50 stars in a region up to 14 light-years across and, from Earth, spans the diameter of the full Moon. Overall, M47's stars are similar to those of the Pleiades. On a dark night, you will be able to spot M47 unaided as a nebulous patch, but binoculars or a small telescope are needed to see it at its best. With such an instrument, both M47 and M46—a neighboring cluster—fit nicely in the field of view and make a handsome pair. Meanwhile, if you are equipped with a larger telescope, increase the magnification and take a look at M46. You should be able to see a small circular patch. This is NGC 2438, a planetary nebula in the foreground.
Apparent magnitude 5
Apparent size 30 arcminutes
Actual size 14 light-years
Distance 1600 light-years
Other clusters in this sky M30 M41

NGC 55 GALAXY

Object type Irregular or spiral galaxy
Constellation Sculptor
Also known as The Cigar Galaxy
Description Located in Sculptor, NGC 55 is an irregular galaxy that may be an edge-on spiral, thought to be similar in structure to the Large Magellanic Cloud. Both NGC 55 and NGC 253, another edge-on galaxy in Sculptor, are members of the Sculptor Group, a galaxy cluster very close to our own Local Group. As such, NGC 55 is a galaxy essentially on our doorstep, at just 7 million light-years away. You can spot it in binoculars or a small telescope, but you need larger instruments to see it properly. It shows up well in an 8-inch (200 mm) telescope, where you will be able to discern that one end is brighter than the other. This is because the central bulge is not located centrally but is offset.
Apparent magnitude 8.8
Apparent size 30 x 6 arcminutes
Actual size 53,000 light-years
Distance 6,000,000
Other galaxies in this sky NGC 253 Small Magellanic Cloud

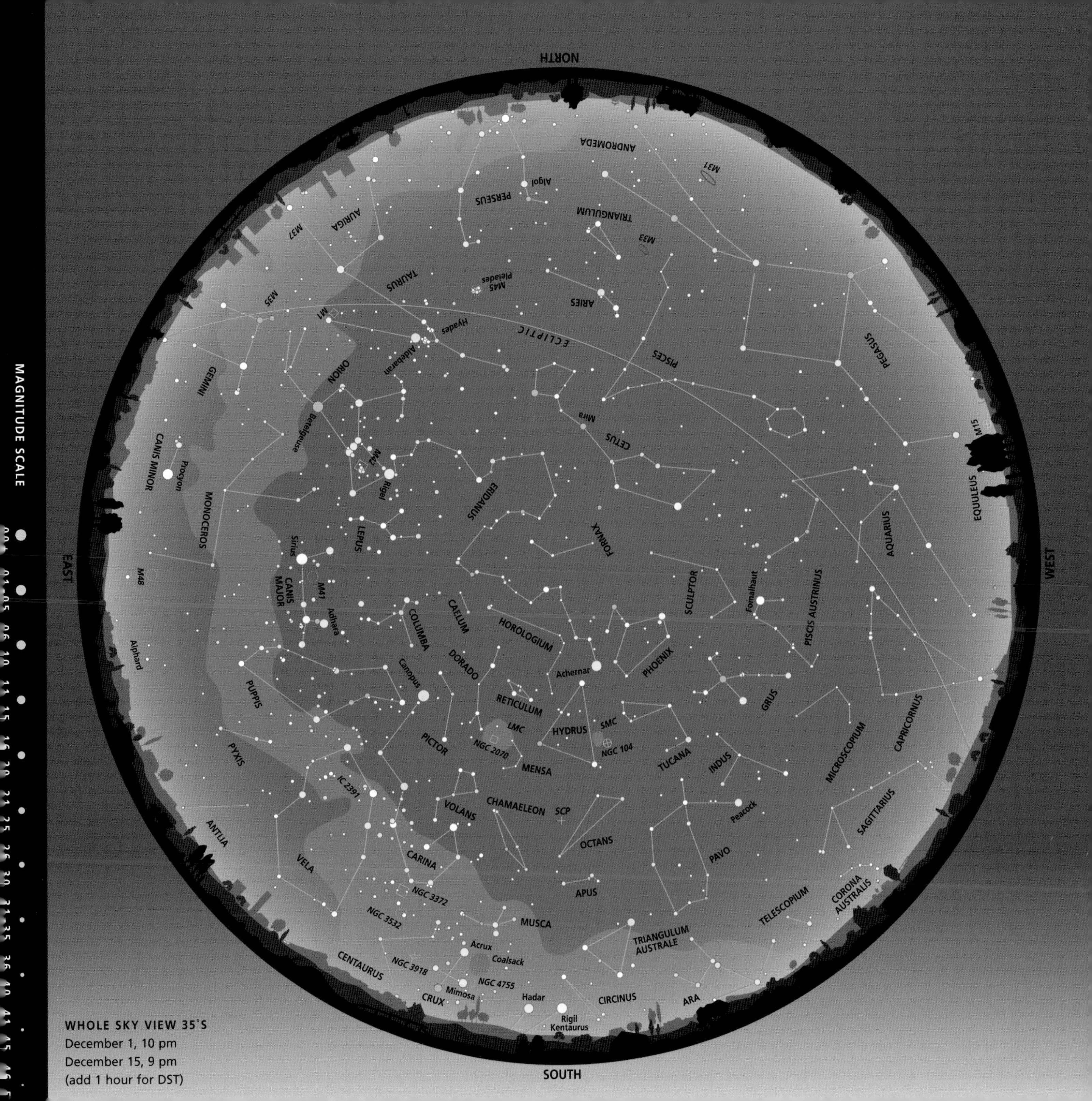

WHOLE SKY VIEW 35°S
December 1, 10 pm
December 15, 9 pm
(add 1 hour for DST)

The Universe

Previous page One small peek into the remote sky reveals the Universe at a distant time in the past.

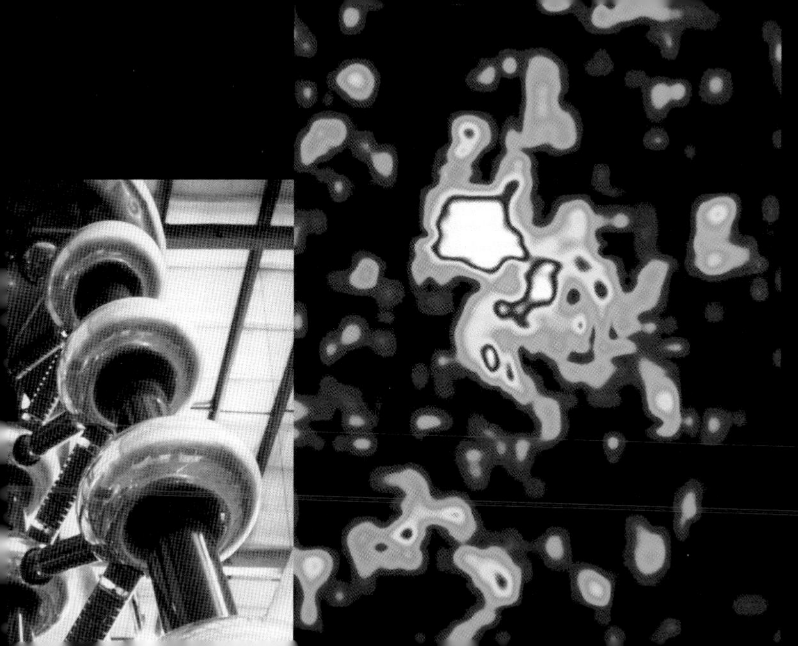

The Universe

Our understanding of the Universe today is very different from what it was just 10 years ago. The very fabric of space and time is now seen as inextricably interwoven with a mysterious "dark energy" about which we know next to nothing. And this is just one example of the many unknowns that astronomers struggle with daily.

Formation of the Universe

Hard as it may be to appreciate, there once existed a time, 13.9 billion years ago, when the Universe existed merely as a single point. Everything that it would later become—space, time, matter and energy—was squeezed into an infinitely hot and dense point. Suddenly, for reasons we may never know, this initial seed expanded in a fireball of creation—the "Big Bang." At an age of 10^{-35} seconds, the Universe increased its size from far smaller than a subatomic particle to about the diameter of a grapefruit, a period of exceedingly rapid expansion that cosmologists call inflation. The expansion continued after inflation, albeit more slowly. For three minutes the Universe cooled and cooked its broiling fog of subatomic particles. Then, at about T-plus three minutes, those particles bonded together to make the first nuclei of hydrogen and helium. It would be another 300,000 years, though, before electrons joined those nuclei to make stable atoms.

This young Universe consisted almost entirely of a hot bath of hydrogen and helium, its ambient temperature similar to the surface of a cool star. Light (in packets called photons) could now move freely, and the Universe became transparent. Gradually, matter coagulated to make vast gaseous structures that within a billion years would become the first stars and galaxies. As stars formed and exploded, the elements they had created in their cores were spread throughout space. The Universe became richer and its contents more diverse. Our Milky Way formed several billion years after the Big Bang, and the Solar System billions of years later still, about 4.6 billion years ago. Coincidentally, it was around this time that the Universe's expansion began to pick up again. And as far as we can tell, the expansion is still accelerating, showing no signs of stopping.

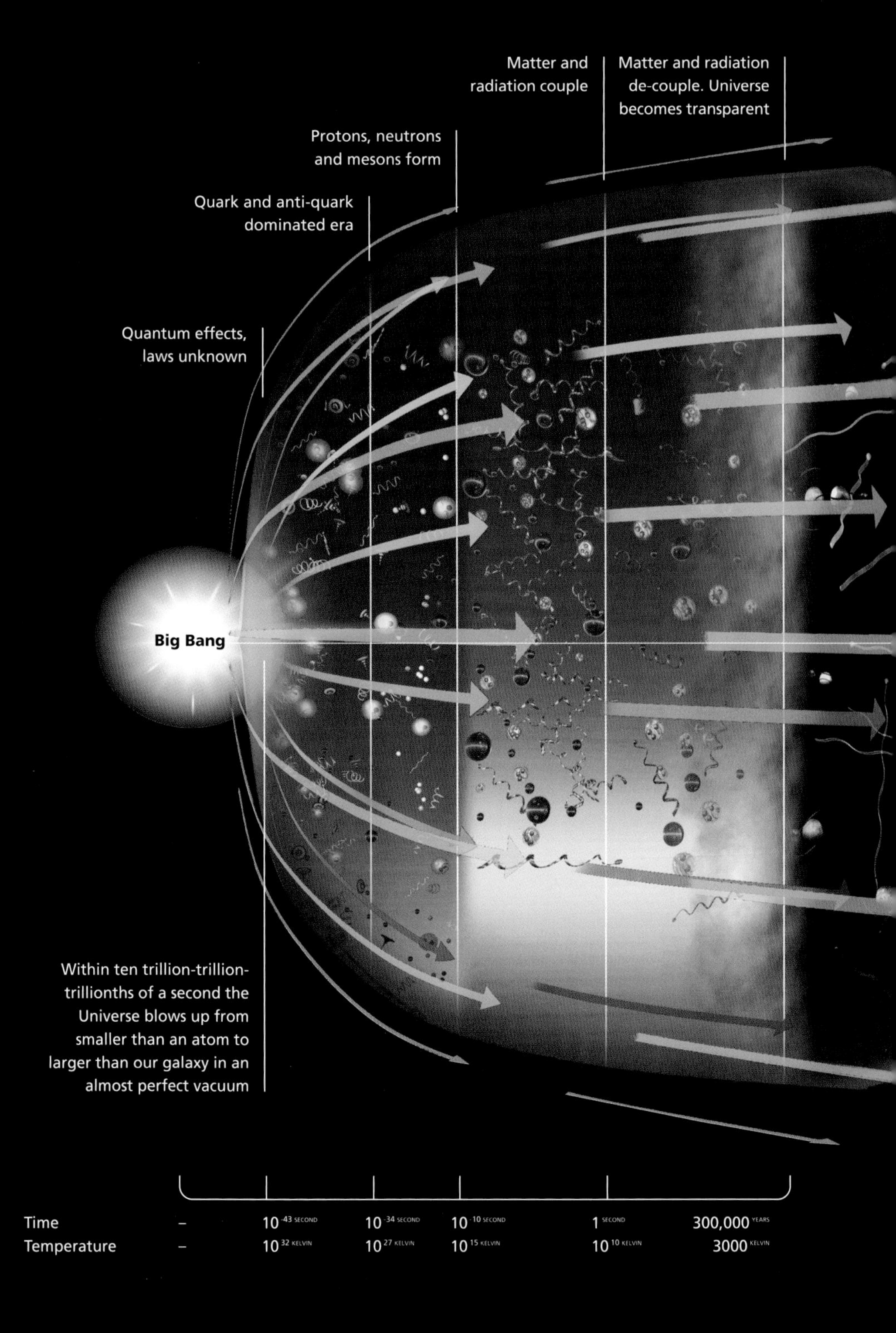

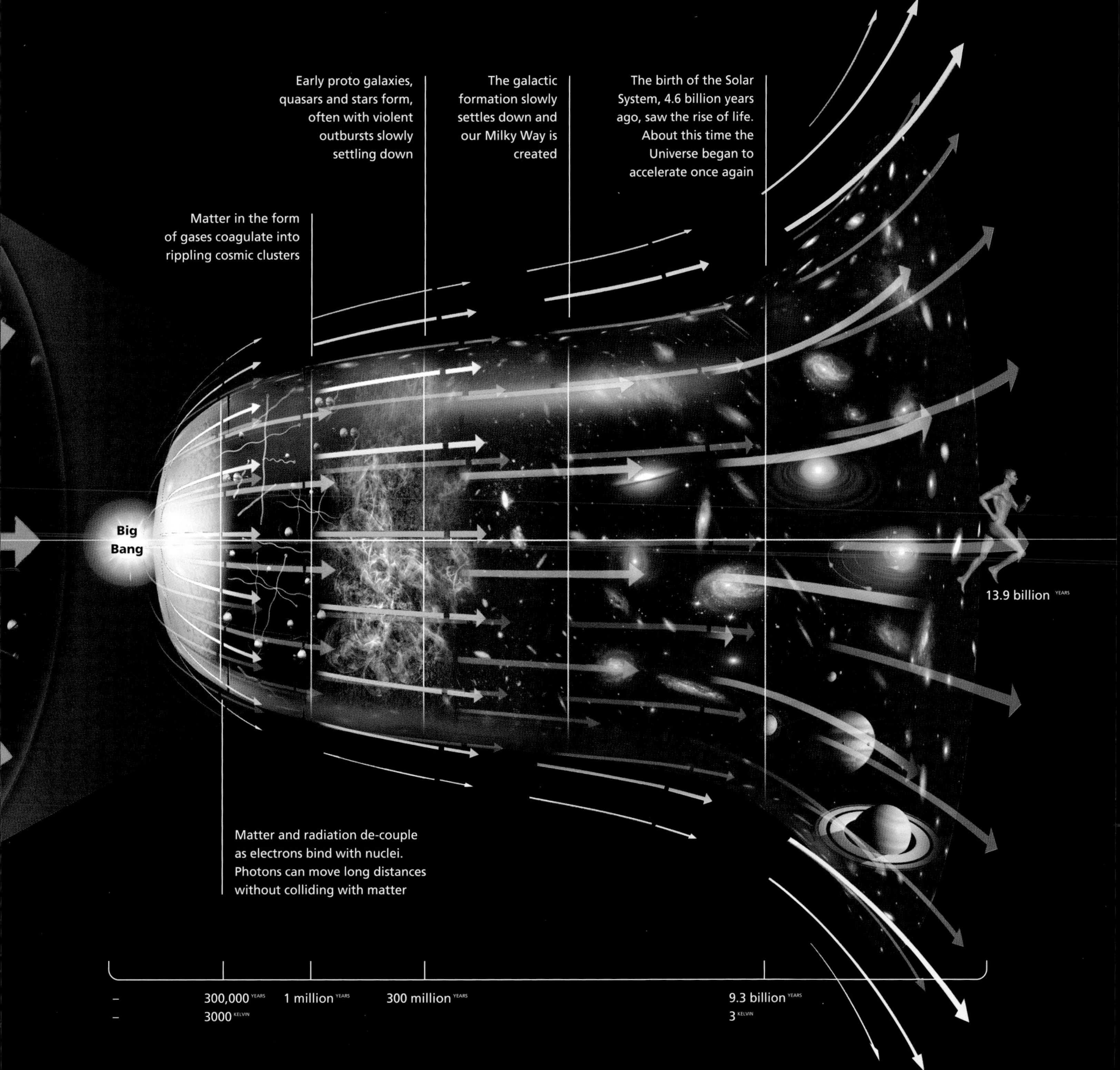

Early proto galaxies, quasars and stars form, often with violent outbursts slowly settling down
The galactic formation slowly settles down and our Milky Way is created
The birth of the Solar System, 4.6 billion years ago, saw the rise of life. About this time the Universe began to accelerate once again
Matter in the form of gases coagulate into rippling cosmic clusters
Big Bang
13.9 billion YEARS
Matter and radiation de-couple as electrons bind with nuclei. Photons can move long distances without colliding with matter
300,000 YEARS
3000 KELVIN
1 million YEARS
300 million YEARS
9.3 billion YEARS
3 KELVIN

Evidence for the Big Bang

The Big Bang is one of the most successful astronomical theories. Since its initial conception in the 1930s, it has survived every test it has been given. The first important piece of supporting evidence was the discovery of the so-called microwave background radiation. This had been predicted by the theory, but its eventual discovery was one of pure chance. In 1965, two communications scientists, Arno Penzias and Robert Wilson, locked onto a strange hiss in the microwave region of the radio spectrum that seemed to emanate from the entire sky. Despite their efforts to remove it, it persisted, and they eventually realized that they had discovered the background radiation. It is the remains of the energy left over from the bath of radiation that pervaded the Universe when it was just 300,000 years old. Most recently, the COBE satellite has further increased astronomers' faith in the theory.

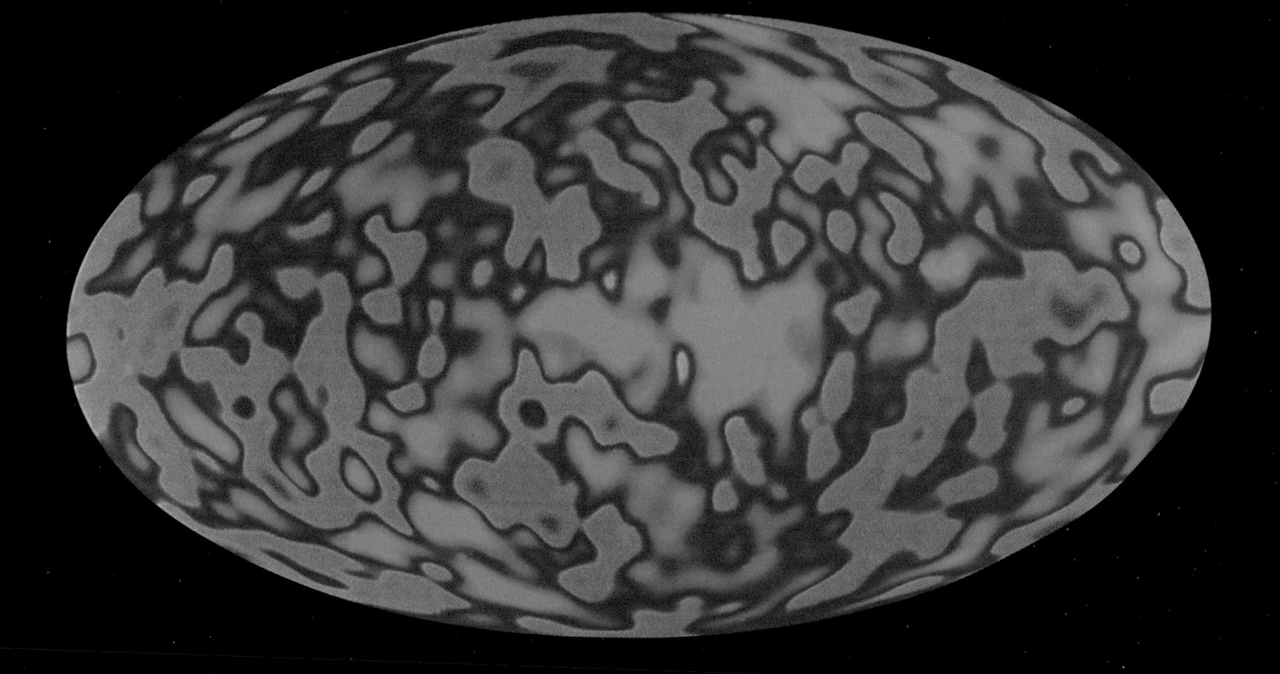

ECHOES OF THE EARLY UNIVERSE
In 1989, NASA launched the Cosmic Background Explorer (COBE). COBE was designed to observe, map and measure any inhomogeneities in the background radiation. As predicted by the Big Bang theory, the background radiation was shown to have a perfect blackbody spectrum, similar to that of a normal star. But COBE's biggest claim to fame was finding that the background radiation is very smooth, showing only tiny variations in intensity (blue and pink in the image above) across the Universe. This is expected according to the Big Bang model, since any non-uniformities in the young Universe ought to have been smoothed out by the rapid expansion.

← **Edwin Hubble's observations** were instrumental in showing that the Universe was expanding. Expansion, cosmologists came to reason, was caused by the Universe's birth in an escalating fireball of energy.

⇠ **Fred Hoyle disliked the Big Bang theory,** even though, ironically, he named it. He coined the term as a sarcastic rebuke at what he saw as an absurd theory.

→ **Arno Penzias and Robert Wilson** were working at Bell Laboratories in 1965 when the Horn Antenna (*pictured*) picked up the cosmic background radiation.

Our place in the Universe

For centuries, we humans have struggled to come to terms with our place in the Universe. For most of that time we knew only about the Moon, the planets and the stars beyond—unreachable and unknowable. Gradually, though, we have grasped the bigger picture of our position in the Universe—Earth is just one of nine planets that orbit the Sun. Other stars have planets, too, and together the vast ensemble of stars and planets forms the Milky Way Galaxy. Lastly, to remove Earth even farther from any point of significance, we have learned that the galaxies are as numerous as are the stars in the Milky Way. Today we have a much better understanding of our place in space, and there doesn't seem to be anything special about it.

↓ **Despite the overwhelming scale** of the Universe, you can still see some amazingly distant objects in the night sky, equipped only with the aid of a modest telescope—or even the naked eye.

THE SOLAR SYSTEM

THE LOCAL GROUP

THE MILKY WAY

THE SOLAR SYSTEM

Our most immediate place in space is the Solar System. The Solar System is just one of many planetary systems orbiting other stars. But whether or not any of these alien planets supports life is another matter. Earth is the largest of the terrestrial planets, but is dwarfed when compared to Jupiter—being over 11 times larger. Earth is the third of nine planets, about 93 million miles (150 million km) from the Sun. That may seen distant, but compared to the orbit of the outermost planet, Pluto, it's a walk to the local corner store. When farthest from the Sun, Pluto is fully 50 times more distant than Earth.

THE MILKY WAY

The Solar System, in turn, also has its own place in space. It can be found about two-thirds of the way out from the center of a spiral galaxy that we call the Milky Way. This gigantic, swirling city of stars and nebulae measures around 100,000 light-years from one end to the other—or alternatively, 600 million billion miles (950 million billion km). In that distance you could line up some 130 million entire Solar Systems, edge to edge. However, the Milky Way is an unremarkable galaxy, and like other spirals, is quite flat, with a pronounced central bulge where the density of stars is much higher.

THE LOCAL GROUP AND BEYOND

The Milky Way, not surprisingly, has its own backyard. It is just one of around 30 or so other galaxies that form the Local Group. The Milky Way and Andromeda are its two largest members, while the other galaxies are all small, mainly dwarf ellipticals or dwarf spheroidals. Beyond the Local Group can be found tens of millions of other clusters of galaxies, some thousands of times more massive than the Local Group. And on the grandest of scales, these clusters are in turn grouped into larger assemblages called superclusters. The supercluster to which the Local Group belongs is called the Local Supercluster.

Life in the Universe

There are few questions more fundamental: "Is there life elsewhere in the Universe?" The quick answer is that we just do not know. Our radio telescopes scan the heavens for signals from other stars, but the stars have so far stayed silent. And what little data we have obtained from the planets and moons on our doorstep, in the Solar System, says that they appear to be sterile. But consider this. The Sun is just one insignificant star in a swarming galaxy of 200 billion others. This galaxy, the Milky Way, is itself just one among at least 100 billion others. Lastly, we know now that planets around stars are common. At the time of writing, well over 100 nearby stars have been found to harbor planets. And so, given the vast number of potential planets in the Universe—billions of billions—it seems absurd to think that life only ever got going on one of them, a puny ball of rock and iron called Earth. Even if, as some biologists claim, the processes that led to the development of life were a mere fluke, there is a lot of scope out in the vastness of the Universe for other flukes!

THE SETI INSTITUTE

Founded in 1984 and situated in Silicon Valley, USA, the SETI Institute is the leading center for the Search for Extraterrestrial Intelligence (SETI). In 1995, the Institute began Project Phoenix, which continues to this day. Using the largest radio telescopes in the world, Project Phoenix scans a selected subset of nearby stars for potential signals from alien civilizations, scrutinizing 28 million radio bands simultaneously. So far the project has not yielded a positive detection, although a few interesting signals have arisen that required further work to eliminate them. Project Phoenix is entirely funded by private individuals and by grants from foundations.

← **Chairman Frank Drake and Jill Tarter** from the SETI Institute—stand underneath a radio telescope dish with which they scan the skies for signals from alien civilizations. In 1960, Frank Drake performed the first search for extraterrestrial intelligence.

↓ **Europa, one of Jupiter's four largest satellites,** is one possible abode of alien life in the Solar System. Its surface, seen here, is a cracked network of ice. But beneath, there may be an ocean of liquid water, laden with life. We will have to wait and see.

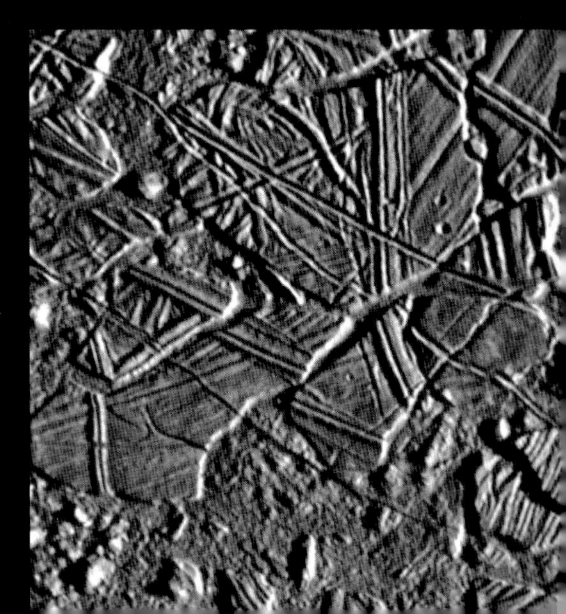

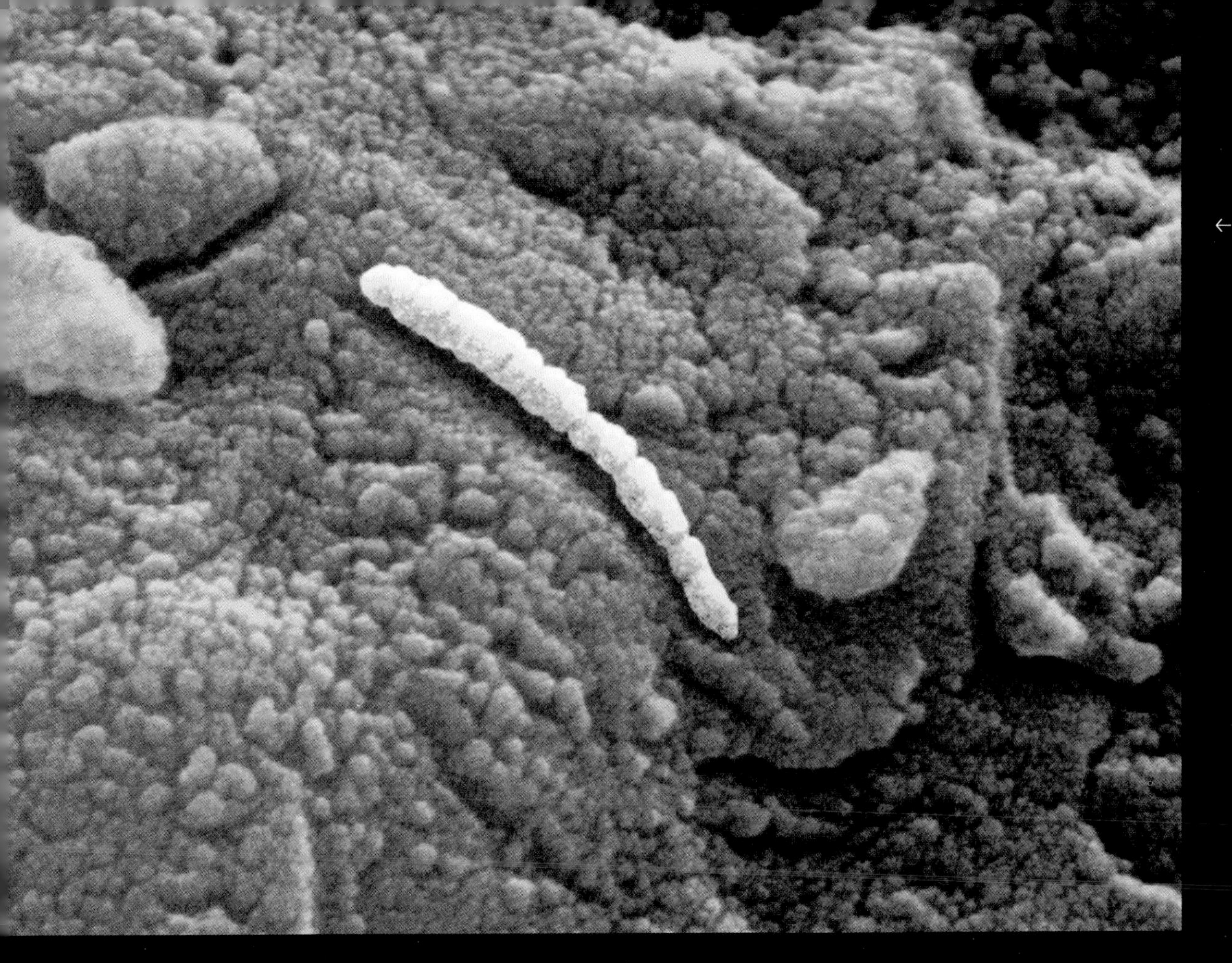

← **In 1996, scientists amazed the world** with a find that made front-page news: the first indications of life on another world. The world was Mars; the lifeforms, little more than primitive bacteria apparently thriving on that world in the distant past. The conclusion was based on the study of a meteorite that fell to Earth but which originated on Mars. Indeed, structures within the rock do resemble wormlike organisms. But since the find, researchers have become skeptical. Today, most of them believe that the formations are certainly natural but not a result of biology.

LIFE IN THE SOLAR SYSTEM

There is no doubt that life took hold on Earth as soon as it was physically able to. Within just a few hundred million years of its formation, the first self-replicating molecules gave way to the first bacterial spores. This rapidity is striking. Life is so complex that some scientists are amazed it developed at all in such a short time. They argue that life on Earth must have been given a head start, that the cloud of gas and dust from which the Solar System formed was already seeded with a rich cocktail of molecules that, given a planetary environment, would organize themselves into a form of life very quickly. If this is true, it seems odd that only one planet, Earth, ever developed life. So perhaps life did arise elsewhere in the Solar System, but has so far evaded detection. Today, the planet Mars and the Jovian satellite Europa are considered the most likely places to find life in the Solar System beyond the Earth.

→ **Stories about UFOs** and alien abductions abound. Many people genuinely believe they have experienced a close encounter with an alien intelligence. But the one vital thing lacking from countless testimonies is proof. There is not a single morsel to give a rational scientist pause to stop and consider.

Gamma ray bursters

Supernovae and quasars have had their share of the limelight: now, gamma ray bursters (GRBs) are thought to be the brightest objects in the Universe. Discovered in the 1960s, GRBs are intense releases of gamma rays lasting from less than a couple of seconds to more than several minutes. They occur all over the sky and at great distances, emanating from faraway galaxies. Even now, decades since their first sightings, astronomers are unsure how these mysterious cosmic blasts are created. But they have two good ideas. The current favorite is that GRBs are like supernovae but much more powerful—they have been called "hypernovae." A supermassive star, 20 to 30 times more massive than the Sun, collapsing at the end of its life, crumples in on itself to become a black hole. The other possibility is that GRBs are the result of a collision between two extremely dense, dead stellar cores called neutron stars, again leading to the formation of a black hole. Whichever is the correct theory, a black hole is thought to be the end product. The actual gamma rays are probably the results of shockwaves, successive waves of material ejected by the GRBs slamming into each other and producing a flood of high-intensity electromagnetic radiation.

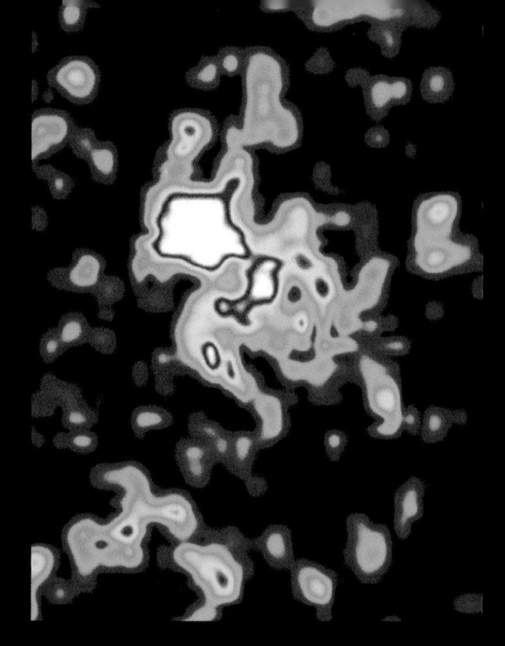

↑ **Hubble captured this supernova fireball** (white), which accompanied a gamma ray burst in 1997.

→ **A blast of intense radiation** signals a new black hole in this artist's impression. Gamma ray bursts provide more power than a million million million Suns, yet measure no larger than the Sun itself.

INSIDE A GAMMA RAY BURSTER

Gamma rays

Internal shockwaves

Internal shockwaves

Central black hole

Internal shockwaves

Optical, X-ray and radio afterglow

Material emitted by GRB collides with surrounding medium

GAMMA RAY BURSTER CROSS-SECTION

This is a cross-section through the funnel of material emitted by the black hole at the center of a GRB. Several waves of material are belched into space at great speed. When a later wave collides with an older one, emitted earlier, a shockwave forms. It is during these pile-ups that the GRB's distinctive gamma rays are produced. Later, as the ejected material collides with the interstellar medium—the thin gruel of gas between the stars—more shockwaves are produced, with lower energy, leading to the GRB's "afterglow" in the optical, X-ray and radio regions of the spectrum.

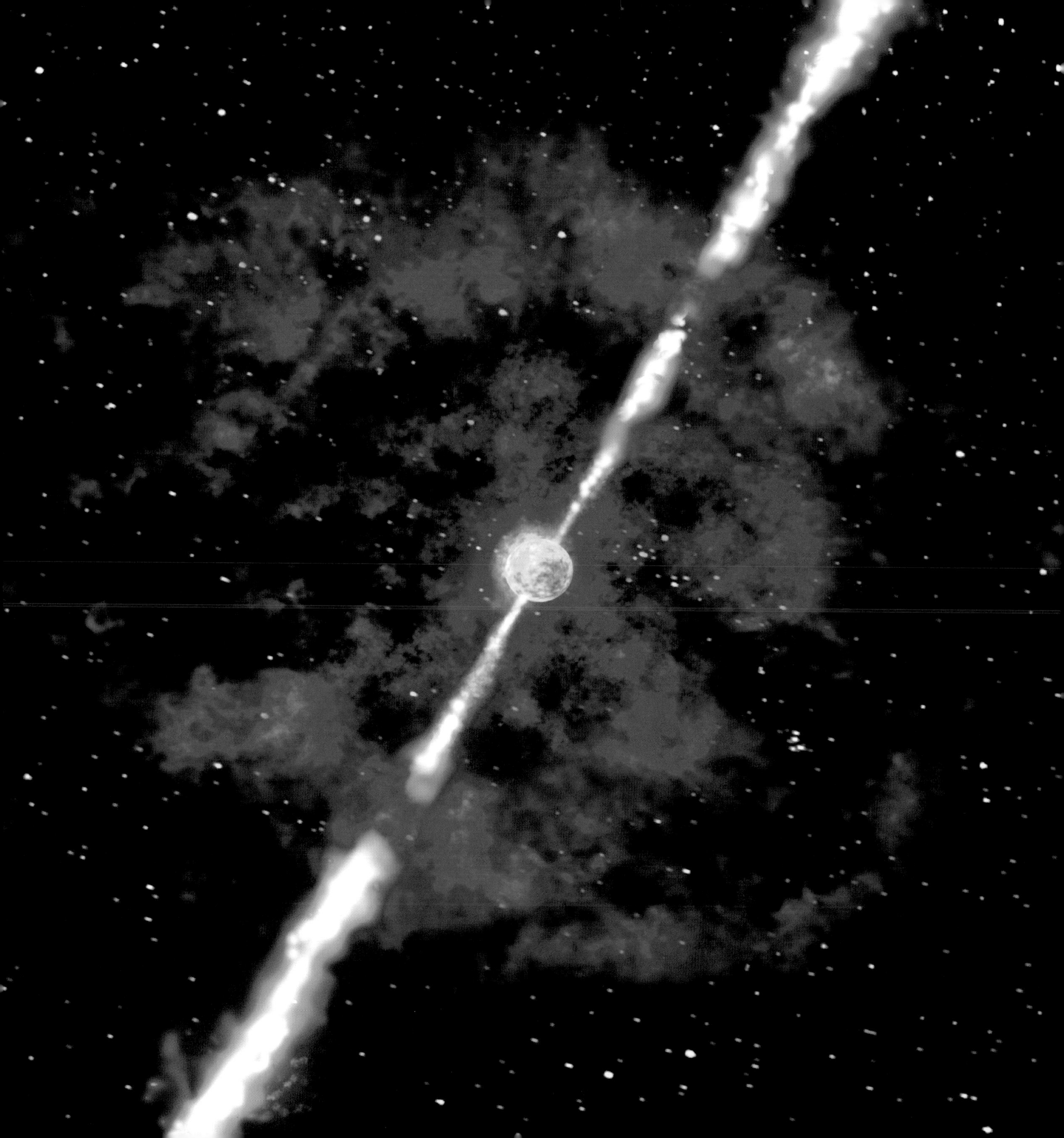

Wormholes

Think of an exotic astronomical object and a black hole will come to mind. But wormholes are even more fantastical. These are the super-highways through spacetime that, according to some theorists, might connect two different regions of the Universe. If the tunnel is wide enough, an object—even a person—could enter the wormhole at one end and emerge instantaneously at the other, at a different time, a different place, or even in another universe entirely.

It sounds like science fiction. But wormholes do exist in at least one sense—as mathematical solutions to a series of formulae known as the Einstein field equations. Indeed, many cosmologists take wormholes very seriously as an integral part of their research. Still, whether these objects can exist in physical reality is another matter entirely.

SPACE TRAVEL

If you were to enter one end of a wormhole, say in Paris, you could in theory emerge from the other side an arbitrary distance from your entry point, in Sydney perhaps. The effect is that you seem to travel faster than light—though you are not. In fact you are not traveling through space, but spacetime. As an analogy, think of space as a piece of paper. Fold it and stick a pencil through it. The pencil penetrates both halves of the folded paper and joins them. An ant could walk along the pencil to get to the other side of the paper much quicker than by traversing the paper itself. So the pencil is like a bridge connecting two distant parts of the paper, in the same way that a wormhole connects two otherwise remote regions of space.

← **Wormholes exist as a mathematical solution** to Einstein's famous field equations of relativity. In the 1930s, he (*left*) and another scientist, Nathan Rosen, found that a black hole might connect to another via a sort of tunnel through spacetime that has become known as an Einstein–Rosen bridge. Initially it was thought that wormholes could not exist in nature, but nobody has been able to prove this.

9/9
2004

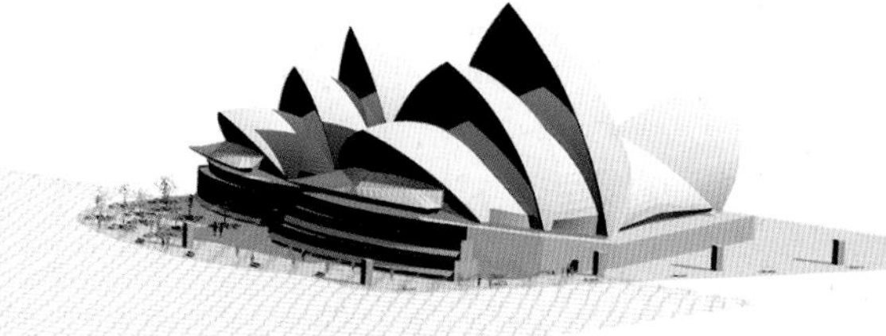

TIME TRAVEL

Just as wormholes enable instantaneous travel across vast distances, so they might also permit time travel. You could enter in 2004 and emerge in 2010. For you, no time will have passed, but the world will have aged. Imagine you take one mouth of a wormhole in a spaceship at great speed. The other mouth stays on Earth. As you are traveling quickly, time slows for you. This is a real phenomenon known as time dilation—accurate clocks carried in fast aeroplanes have been observed to slow. So when you return to Earth with your wormhole mouth, you and it will have aged very little, while decades may have passed on Earth. Thus the two wormhole mouths now connect two different times. If you entered the mouth you took with you, you could emerge from the other mouth, on Earth, when you left, traveling back in time!

Dark matter

In the 1970s, astronomers stumbled across an embarrassing fact: at least 90 percent of the Universe was hidden from them. They realized that stars and galaxies were not behaving as they would if they only contained as much material as we can see. Scratching their heads, astronomers were forced to conclude that there is a lot of material out there that, despite all efforts, simply cannot be seen directly. This became known as the "missing mass."

But how did they know it was there if they could not see it? It's a question of gravity. In the early 1600s, Johannes Kepler demonstrated, with simple mathematics, how to deduce the mass of an astronomical object by measuring the speed at which other bodies move around it in their orbits. That's how we know the mass of the Sun: from the orbits of the planets. When astronomers looked at galaxies and measured the speeds of the stars orbiting within them, they were puzzled. The stars were moving so quickly that the galaxies must have been much more massive than they appeared judging by the amount of light they emitted. Today, astronomers are more certain than ever that the Universe is replete with a mysterious "dark matter" that evades detection and yet which out-masses visible material at least ten to one.

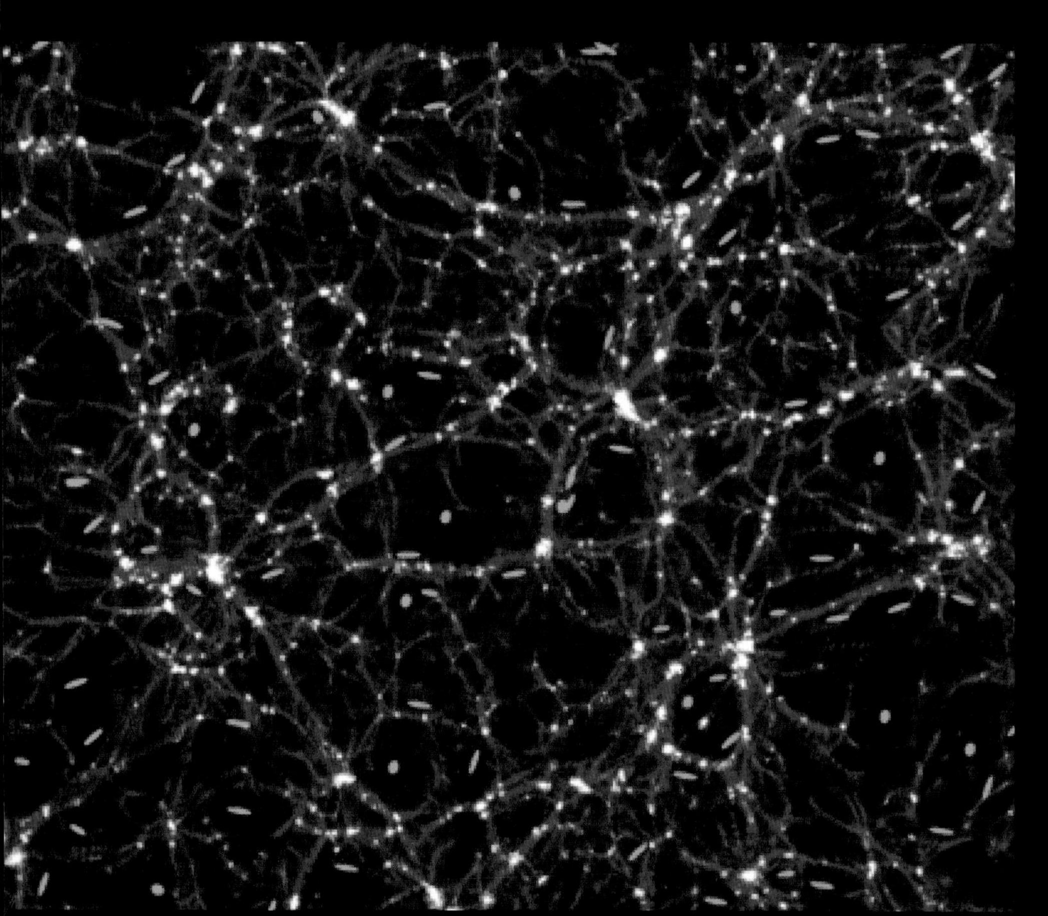

THE SHAPE OF DARK MATTER

Nobody is sure what form dark matter might take, but there are several ideas. Briefly, there are two classes of possible objects, whimsically named WIMPs and MACHOs. WIMPs, short for "weakly interacting massive particles," are hypothetical and exotic subatomic particles, having no electric charge, that we have not yet detected because they interact very weakly with the ordinary matter we can see. They simply pass right through normal matter. Indeed, if they exist, countless numbers of WIMPs could be moving through your body right now with virtually no interaction. If WIMPs do indeed contribute toward the missing mass, there must be an enormous number of them, considering that they are subatomic and yet outweigh normal matter by such large factors. MACHOs, on the other hand, are much bigger. They are "compact halo objects"—astronomical bodies that emit little or no light. MACHO candidates include supermassive black holes, free-floating planets (not bound to stars), and brown dwarfs—objects too puny to have become true stars.

↑ **Vera Rubin is the American astronomer** who first drew attention to the possible existence of dark matter—calling it, at the time, the "missing matter." Her conclusions, made in the 1970s, were based on research completed during the previous two decades.

← **For the first time,** scientists have been able to map the distribution of dark matter in the Universe (red in this image). The blue parts represent distant galaxies.

SEARCHING FOR DARK MATTER

Despite the fact that dark matter is, well, dark, that doesn't stop scientists from trying to see it—or at least detect it. Generally, astronomers conduct searches of the Universe looking for MACHOs, while particle physicists look for WIMPs in their particle experiments on Earth. On the astronomers' camp, the most promising detection method is to look for gravitational lensing events. Einstein's theory of general relativity shows how mass can bend rays of light, just like a lens, hence the name gravitational lensing. It occurs when an invisible, massive foreground object moves directly in front of another object and bends its light in a characteristic way. Many examples are known of distant galaxies being distorted by the gravity of foreground galaxies, so the effect has been confirmed. And many candidate MACHOs have been detected, with inferred masses between those of Jupiter and the Sun. Particle physicists have had less success, without a single WIMP detection under their belt. The search goes on . . .

← ↓ **To conduct their search for dark matter candidates** called WIMPs, particle physicists use equipment such as these particle accelerators at the Fermilab National Accelerator Laboratory, Illinois, USA. They attempt to create WIMPs inside these machines, and then try to find them as they interact with the atoms in a block of test material. It requires patience: only one WIMP in a million will typically hit an atom and emit a signal scientists can detect. So far, no WIMP particle has ever been confirmed.

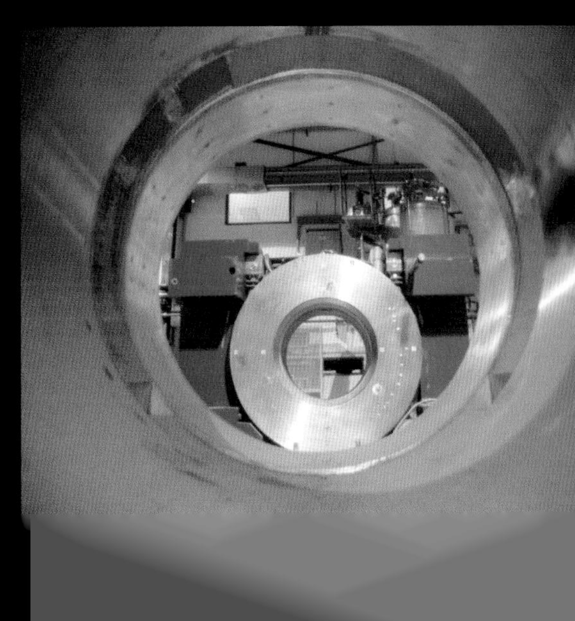

Dark energy

In the late 1990s, while observing very distant supernovae, astronomers stumbled across something that went against everything that they had learned about cosmology up until that point. We have known for 80 years that the Universe is expanding. However, what was recently discovered is that the rate at which it expands seems to be increasing. The Universe is accelerating. At first, it did not make sense. Since everything in space has gravity, every object pulls on its neighbor, and as a result the Universe's expansion ought to be slowing down or at least leveling off—certainly not speeding up. Cosmologists, struggling to explain this new discovery, came up with a possible solution. The Universe, they said, is pervaded by a mysterious "dark energy." Nobody is sure what its nature is, but its effect is to stretch the very fabric of the Universe apart. Dark energy has therefore been likened to an anti-gravity, although this is not strictly true. If the theory is correct, dark energy accounts for most of everything in the known Universe.

THE EXPANDING UNIVERSE

THE EXPANDING UNIVERSE
About 13.9 billion years ago, as near as we can tell, the Universe came into existence. The Big Bang created time, matter and energy where none had existed. For the first 300,000 years, the Universe existed as a hot fog of charged subatomic particles. But then, the background temperature dropped to the point where atoms could form, and the Universe finally become transparent. Stars formed quickly, and galaxies—all within the first billion years, and maybe much sooner. And as a direct result of its explosive birth, the Universe continued to expand, just like ripples on a pond. Astronomers think that, for most of its history, the Universe's expansion was actually slowing down, the mutual attraction of all the matter within it acting as a break that hindered the expansion. But in the fairly recent past, around two or three billion years ago, something happened that reversed this. Nowadays, all signs are that the Universe's expansion has picked up again and is actually accelerating, its galaxies growing not only farther apart, but taking comparatively less time to do so.

← **Supernovae, seen in an artist's impression,** are the explosions that result when a massive star runs out of fuel at the end of its life. It was by studying these events that astronomers realized that something was odd with the expansion of the Universe.

Big Bang

↓ **We have known for some decades** that there is more to the Universe than meets the eye—quite literally. Since the discovery of dark energy (if it exists), cosmologists have revamped the composition of the Universe. If we could convert the dark energy into pure matter, it would comprise about 65 percent of the known mass of the Universe. Dark matter makes up about 30 percent, leaving just a few percent for the bit of matter we can actually see.

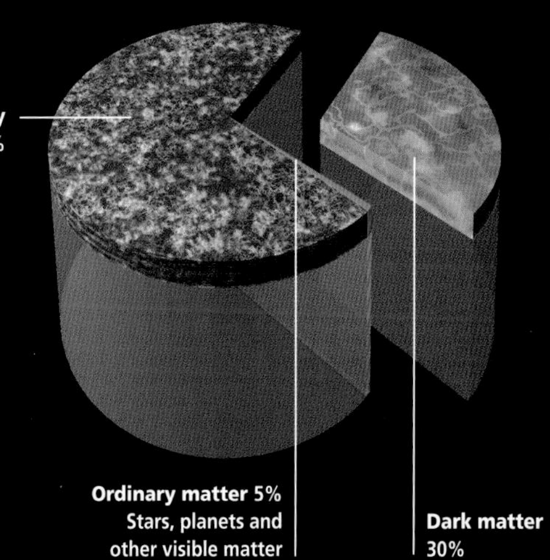

Future of the Universe

The fate of the Universe is perhaps equally as fundamental a question as its origins. Of course, we will never know how it all ends—or even if it does. But there are some clues. As far as we can see, the future depends on the density of space, on how closely packed galaxies are. Every object attracts everything else with the force of gravity. Space expands, but at the same time the gravity of its contents tries to pull everything back together again, just as a balloon resists when you inflate it. This tug-of-war between gravity and expansion can only have one outcome. If the cosmos is sparse, below a so-called critical density, the self-gravity of the Universe will never be able to overcome its expansion. It will expand forever. If the density is greater than critical, the expansion will slow and then, one day, it will reverse. The Universe will perhaps collapse to a point like the Big Bang in reverse—or rather, a "Big Crunch." The final alternative will transpire if the density of the Universe exactly matches the critical value. It will then expand forever, constantly slowing down but never quite stopping altogether.

Even before the discovery of dark energy—a mysterious sort of repulsive force that is causing the expansion of the Universe to accelerate—most astronomers were of the opinion that the Universe had a near-critical density and that it would expand forever. With the advent of dark energy, this seems even more likely. But while space and time might exist and expand forever, that is not to say that the contents of the Universe themselves are immortal.

THE DEAD UNIVERSE

Regardless of how densely packed the Universe is, the matter within it will one day cease to exist. As time goes on, more and more interstellar gas will be consumed to form new stars. Some of these will explode as supernovae or expand as planetary nebulae, thus spreading gas back into space. But some of that material will remain locked away in black holes, neutron stars or white dwarfs. Eventually, there must come a point where no more stars can form. The neutron stars and white dwarfs will persist, but eventually they will decay into radiation, since matter is, on an exceedingly long timescale, ultimately unstable. Even the black holes, as Stephen Hawking showed, must eventually evaporate. The future is a grim one: a Universe of time and space and energy, but no physical matter.

↓ **The density of the Universe** determines its geometry. A Universe with sub-critical density is positively curved, like the surface of a sphere, and is said to be "closed."

↓ **A Universe which has critical density** does not have net curvature and is said to be "flat." Its shape can be likened to the surface of a two-dimensional sheet.

↓ **Lastly, the Universe** could be negatively curved, like a saddle—or "open." In either of these last two possible scenarios, the volume of the Universe would be unbounded.

SPHERE UNIVERSE

FLAT UNIVERSE

SADDLE UNIVERSE

THE SHAPE OF THE FUTURE

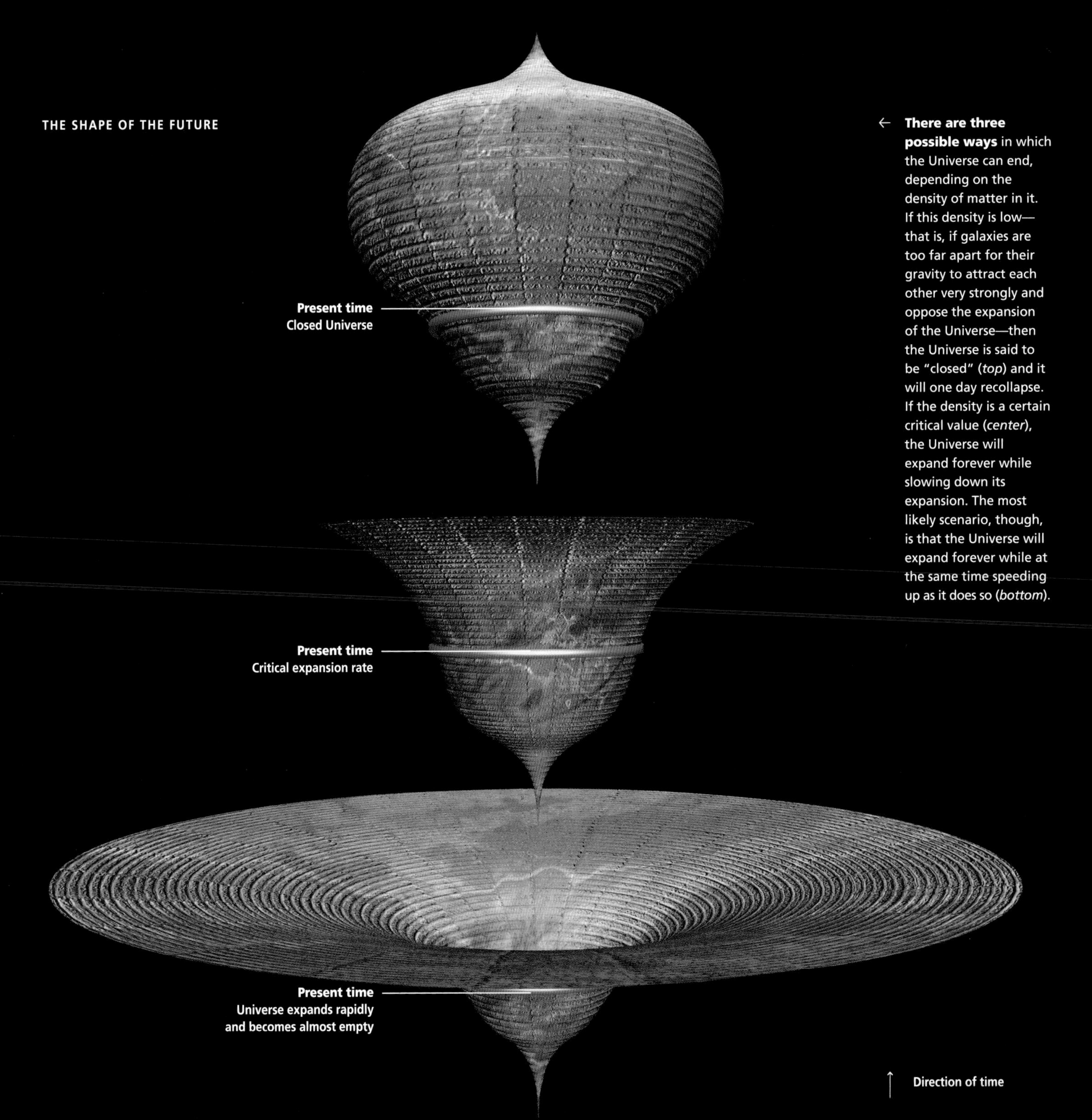

← **There are three possible ways** in which the Universe can end, depending on the density of matter in it. If this density is low—that is, if galaxies are too far apart for their gravity to attract each other very strongly and oppose the expansion of the Universe—then the Universe is said to be "closed" (*top*) and it will one day recollapse. If the density is a certain critical value (*center*), the Universe will expand forever while slowing down its expansion. The most likely scenario, though, is that the Universe will expand forever while at the same time speeding up as it does so (*bottom*).

Factfile

Mercury

Venus

Earth

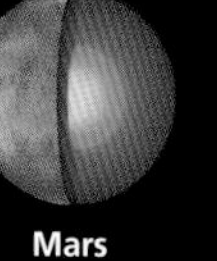
Mars

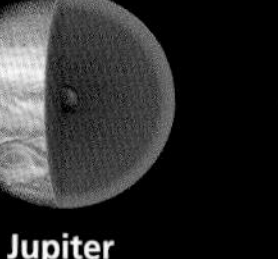
Jupiter

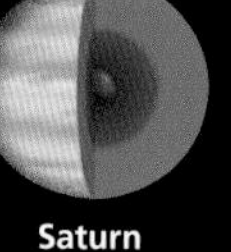
Saturn

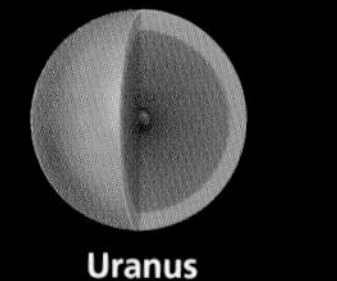
Uranus

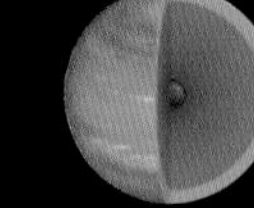
Neptune

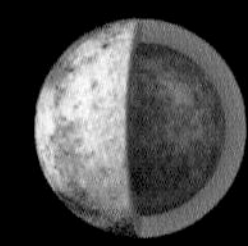
Pluto

MERCURY FACTS

Origin of name *Mercurius*, messenger of the Roman gods

Discovered Known since antiquity

Diameter 3031 miles (4878 km), 38.2% of Earth's

Mass 0.055 x Earth

Volume 0.06 x Earth

Farthest from Sun 43.4 million miles (69.8 million km)

Closest to Sun 28.6 million miles (46.0 million km)

Mean surface temperature 800°F (430°C)

Sunlight strength 450–1040% of Earth's

Apparent magnitude -2.0 to +3.0

How it can be seen Naked eye

Surface gravity 0.38 gee (38% of Earth's)

Magnetic field strength 0.003 gauss (1% of Earth's)

Number of satellites 0

VENUS FACTS

Origin of name *Venus*, Roman goddess of love and beauty

Discovered Known since antiquity

Diameter 7521 miles (12,104 km), 94.8% of Earth's

Mass 0.95 x Earth

Volume 0.86 x Earth

Farthest from Sun 67.7 million miles (108.9 million km)

Closest to Sun 66.8 million miles (107.5 million km)

Mean surface temperature 900°F (480°C)

Sunlight strength 190% of Earth's

Apparent magnitude -4.0 to -4.6

How it can be seen Naked eye

Surface gravity 0.9 gee (90% of Earth's)

Magnetic field strength <0.00002 gauss (<0.007% of Earth's)

Number of satellites 0

EARTH FACTS

Origin of name *Nerthus*, an ancient Germanic goddess

Diameter 7926 miles (12,756 km)

Mass 1.314×10^{25} pounds (5.973×10^{24} kg)

Volume 0.3 trillion cubic miles (1.1 trillion km^3)

Farthest from Sun 94.5 million miles (152.1 million km)

Closest to Sun 91.4 million miles (147.1 million km)

Length of year 365.22 days

Rotation time 23 hours 56 minutes

Mean surface temperature 72°F (22°C)

Highest peak 29,035 feet (8850 m), Mount Everest

Lowest trough -36,200 feet (-11,033 m), Marianus Trench

Surface gravity 1.0 gee

Magnetic field strength 0.31 gauss

Number of satellites 1

MARS FACTS

Origin of name *Mars Gradivus*, Roman god of war

Discovered Known since antiquity

Diameter 4221 miles (6794 km), 53.2% of Earth's

Mass 0.11 x Earth

Volume 0.15 x Earth

Farthest from Sun 154.8 million miles (249.1 million km)

Closest to Sun 128.4 million miles (206.7 million km)

Mean surface temperature -10°F (-23°C)

Sunlight strength 36–52% of Earth's

Apparent magnitude +1.8 to -2.6

How it can be seen Naked eye

Surface gravity 0.38 gee (38% of Earth's)

Magnetic field strength <0.0003 gauss (<0.1% of Earth's)

Number of satellites 2 (Phobos and Deimos)

JUPITER FACTS

Origin of name *Jove*, Roman god of heaven and Earth

Discovered Known since antiquity

Diameter 89,405 miles (143,884 km), 1127.9% of Earth's

Mass 317.7 x Earth

Volume 1323 x Earth

Farthest from Sun 506.9 million miles (815.7 million km)

Closest to Sun 460.4 million miles (740.9 million km)

Mean surface temperature -240°F (-150°C)

Sunlight strength 3–4% of Earth's

Apparent magnitude -1.2 to -2.5

How it can be seen Naked eye

Surface gravity 2.3 gee (230% of Earth's)

Magnetic field strength 4.28 gauss (1380% of Earth's)

Number of known satellites 61

SATURN FACTS

Origin of name *Saturnus*, Roman god of agriculture

Discovered Known since antiquity

Diameter 74,898 miles (120,536 km), 944.9% of Earth's

Mass 95.2 x Earth

Volume 752 x Earth

Farthest from Sun 933.9 million miles (1503 million km)

Closest to Sun 837.6 million miles (1348 million km)

Mean surface temperature -110°F (-80°C)

Sunlight strength 1% of Earth's

Apparent magnitude +0.6 to +1.5

How it can be seen Naked eye

Surface gravity 1.16 gee (116% of Earth's)

Magnetic field strength 0.22 gauss (71% of Earth's)

Number of known satellites 31

URANUS FACTS

Origin of name *Ouranos*, Greek primeval god of the sky

Discovered March 1781, by William Herschel

Diameter 31,763 miles (51,118 km), 400.7% of Earth's

Mass 14.5 x Earth

Volume 64 x Earth

Farthest from Sun 1866 million miles (3003 million km)

Closest to Sun 1702 million miles (2739 million km)

Mean surface temperature -355°F (-215°C)

Sunlight strength 0.2–0.3% of Earth's

Apparent magnitude +5.5 to +5.9

How it can be seen Telescope

Surface gravity 1.17 gee (117% of Earth's)

Magnetic field strength 0.23 gauss (75% of Earth's)

Number of known satellites 27

NEPTUNE FACTS

Origin of name *Neptunus*, Roman god of water

Discovered September 1846, by Johann Galle

Diameter 30,778 miles (49,532 km), 388.3% of Earth's

Mass 17.1 x Earth

Volume 54 x Earth

Farthest from Sun 2825 million miles (4546 million km)

Closest to Sun 2769 million miles (4456 million km)

Mean surface temperature -265°F (-220°C)

Sunlight strength 0.1% of Earth's

Apparent magnitude +7.9

How it can be seen Telescope

Surface gravity 1.77 gee (177% of Earth's)

Magnetic field strength 0.14 gauss (45% of Earth's)

Number of known satellites 13

PLUTO FACTS

Origin of name *Pluto*, Greek god of the underworld

Discovered February 1930, by Clyde Tombaugh

Diameter 1429 miles (2300 km), 18.0% of Earth's

Mass 0.002 x Earth

Volume 0.01 x Earth

Farthest from Sun 4586 million miles (7380 million km)

Closest to Sun 2763 million miles (4447 million km)

Mean surface temperature -380°F (-230°C)

Sunlight strength 0.04–0.1% of Earth's

Apparent magnitude +13.7

How it can be seen Telescope

Surface gravity 0.06 gee (6% of Earth's)

Magnetic field strength Unknown

Number of satellites 1 (Charon)

ASTEROID FACTS

Name	Discovery	Distance from the Sun	Orbit time	Rotation time	Diameter
1 Ceres	Piazzi, 1801	2.76 AU	4.6 yrs	9h 5m	567 miles (913 km)
2 Pallas	Olbers, 1802	2.77 AU	4.6 yrs	7h 49m	325 miles (523 km)
4 Vesta	Olbers, 1807	2.36 AU	3.6 yrs	5h 21m	323 miles (520 km)
10 Hygeia	De Gasparis, 1849	3.14 AU	5.6 yrs	27h 40m	267 miles (429 km)
511 Davida	Dugan, 1903	3.18 AU	5.7 yrs	5h 8m	209 miles (337 km)
704 Interamnia	Cerulli, 1910	3.06 AU	5.4 yrs	8h 44m	207 miles (333 km)
Juno	Harding, 1804	2.67 AU	4.4 yrs	7h 13m	166 miles (268 km)
Sylvia	Pogson, 1866	3.50 AU	6.6 yrs	5h 11m	162 miles (260 km)
Eunomia	De Gasparis, 1851	2.64 AU	4.3 yrs	5h 5m	159 miles (256 km)
Euphrosyne	Fergusson, 1854	3.14 AU	5.6 yrs	5h 32m	159 miles (256 km)
Europa	Goldschmidt, 1858	3.10 AU	5.5 yrs	5h 38m	157 miles (252 km)
253 Mathilde	Palisa, 1885	2.65 AU	4.3 yrs	17h 24m	41 miles (66 km)
243 Ida	Palisa, 1884	2.86 AU	4.9 yrs	4h 38m	37 miles (60 km)
433 Eros	Witt & Charlois, 1898	1.46 AU	1.8 yrs	5h 18m	21 miles (33 km)
951 Gaspra	Neujmin, 1916	2.21 AU	3.3 yrs	7h 3m	11 miles (18 km)

COMET FACTS

Name	First recorded	Farthest from the Sun	Closest to the Sun	Type	Orbit time	Orbit inclin.	Orbit eccent.
Encke	1786	4.1 AU	0.3 AU	short period	3.3 yrs	11.9°	0.85
Wirtanen	1954	5.1 AU	1.1 AU	short period	5.5 yrs	11.7°	0.65
Wild 2	1978	5.3 AU	1.6 AU	short period	6.4 yrs	3.2°	0.54
d'Arrest	1851	5.6 AU	1.4 AU	short period	6.5 yrs	19.5°	0.61
Biela	1772	0.86 AU	unknown	short period	6.62 yrs	12.55°	0.76
Crommelin	1818	0.75 AU	17.4 AU	short period	27.5 yrs	29.1°	0.92
Tempel-Tuttle	1866	19.5 AU	1.0 AU	short period	32.9 yrs	163°	0.90
Borsen-Metcalf	1847	0.48 AU	34.1 AU	short period	69.5 yrs	19.3°	0.97
Halley	239 BC	35.0 AU	0.6 AU	short period	76.0 yrs	162°	0.97
Swift-Tuttle	1862	52.3 AU	1.0 AU	short period	137.3 yrs	113°	0.96
Bennett	1970	281.8 AU	0.5 AU	long period	1678 yrs	90.0°	0.99
Donati	1858	311.6 AU	0.6 AU	long period	1950 yrs	117°	0.99
Hale-Bopp	1995	370.6 AU	0.9 AU	long period	2529 yrs	89.4°	0.99
Hyakutake	1996	2006 AU	0.2 AU	long period	31,781 yrs	125°	0.99
Ikeya-Seki	1965	4000 AU	0.01 AU	long period	89,443 yrs	129°	0.99

NOTES

PLANET FACTS

Apparent magnitude The visible brightness of a planet, star or other celestial object as seen from Earth. Brighter objects have smaller numbers.
Eccentricity of orbit How elliptical the planet's orbit is. The larger the number, the more elliptical the orbit.
Magnetic field strength A measurement of the magnetic force exerted by the planet. The unit is the gauss.
Surface gravity The strength of the planet's gravitational field at the surface compared to that of Earth.
Sunlight strength The brightness of the Sun at the planet in question, as compared to that from Earth.

ASTEROID FACTS

Distance from the Sun The asteroid's average distance from the Sun.
Orbit time All years listed are Earth years.

COMET FACTS

Name Comets are named for their discoverers.
Orbit time All years listed are Earth years.
Orbit inclin. The inclination of the comet's orbit relative to the plane of Earth's orbit.
Orbit eccent. The eccentricity of the comet's orbit. The larger the number, the more elliptical the orbit.

SATELLITE FACTS

Orbit time All days are Earth days.
Mass Due to large numbers, mass measurement can be given in scientific notation, such as 3.7×10^9. The superscript number after 10 denotes the number of zeros that follow the 1. Thus, 10^9 is 1,000,000,000, and 3.7×10^9 is 3,700,000,000, or 3700 million.
A tonne is a metric ton, equal to 1000 kg or about 2200 pounds.
A question mark indicates that the value is an unconfirmed estimate.
45 Eugenia and 243 Ida are asteroids with their own orbiting satellites.
The gas-giants of Jupiter, Saturn, Uranus and Neptune are like miniature Solar Systems in their own right, each controlling dozens of satellites, although most of them are puny. Jupiter alone has at least 61. As observational techniques improve, the number of known satellites grows steadily year by year.

GENERAL NOTES

1 AU (astronomical unit) is the average distance between Earth and the Sun, about 93 million miles (150 million km).
ly = light-year

SATELLITE FACTS

Planet	Satellite	Discovery	Distance from planet in miles (km)	Orbit time	Diameter in miles (km)	Mass in tons (tonnes)	Surface composition
Earth							
	The Moon	known since antiquity	238,856 (384,401)	27.3 days	2160 (3476)	8.0 x 10^{19} (7.3 x 10^{19})	anorthosite and basalt rock, dust
Mars							
	Phobos	Hall, 1877	5827 (9378)	0.3 day	17 (27)	1.1 x 10^{13} (9.6 x 10^{12})	carbon-rich rock, dust
	Deimos	Hall, 1877	14,577 (23,459)	1.3 days	9 (15)	2.0 x 10^{12} (1.8 x 10^{12})	carbon-rich rock, dust
45 Eugenia (asteroid)							
	S/1998(45)1	Merline, 1998	771 (1240)?	4.7 days	6 (10)?	unknown	rock?
243 Ida (asteroid)							
	Dactyl	Galileo spacecraft, 1993	56 (90)	unknown	1 (1.6)	4.1 x 10^{9} (3.7 x 10^{9})	carbon-rich rock, dust
Jupiter							
	Metis	Synnott, 1979	79,511 (127,960)	0.3 day	25 (40)	1.0 x 10^{14} (9.5 x 10^{13})	rock
	Adrastea	Jewitt & Danielson, 1979	80,145 (128,980)	0.3 day	16 (25)	2.1 x 10^{13} (1.9 x 10^{13})	rock
	Amalthea	Barnard, 1892	112,655 (181,300)	0.5 day	168 (270)	7.9 x 10^{15} (7.2 x 10^{15})	rock, sulfur coating?
	Thebe	Synnott, 1979	137,882 (221,900)	0.7 day	68 (110)	8.4 x 10^{14} (7.6 x 10^{14})	rock
	Io	Galileo, 1610	261,970 (421,600)	1.8 days	2264 (3643)	9.8 x 10^{19} (8.9 x 10^{19})	rock
	Europa	Galileo, 1610	416,880 (670,900)	3.5 days	1939 (3120)	5.3 x 10^{19} (4.8 x 10^{19})	rock
	Ganymede	Galileo, 1610	664,870 (1,070,000)	7.2 days	3273 (5268)	1.7 x 10^{20} (1.5 x 10^{20})	rock
	Callisto	Galileo, 1610	1,170,000 (1,883,000)	16.7 days	2983 (4800)	1.2 x 10^{20} (1.1 x 10^{20})	rock
	Leda	Kowal, 1974	6,893,500 (11,094,000)	238.7 days	10 (16)	6.3 x 10^{12} (5.7 x 10^{12})	carbon-rich dirt, ice
	Himalia	Perrine, 1904	7,133,300 (11,480,000)	250.6 days	116 (186)	1.0 x 10^{16} (9.5 x 10^{15})	carbon-rich dirt, ice
	Lysithea	Nicholson, 1938	7,282,500 (11,720,000)	259.2 days	22 (36)	8.4 x 10^{13} (7.6 x 10^{13})	carbon-rich dirt, ice
	Elara	Perrine, 1905	7,293,000 (11,737,000)	259.6 days	47 (76)	8.4 x 10^{14} (7.6 x 10^{14})	carbon-rich dirt, ice
	Euporie	Sheppard et al., 2001	11,994,000 (19,302,000)	550.7 days	1.2 (2)	unknown	unknown
	Thyone	Sheppard et al., 2001	13,012,000 (20,940,000)	627.3 days	2.4 (4)	unknown	unknown
	Ananke	Nicholson, 1951	13,173,100 (21,200,000)	631 days	19 (30)	4.2 x 10^{13} (3.8 x 10^{13})	carbon-rich dirt, ice
	Iocaste	Sheppard et al., 2000	13,216,000 (21,269,000)	631.5 days	3 (5)	unknown	unknown
	Carme	Nicholson, 1938	14,043,000 (22,600,000)	692 days	25 (40)	1.0 x 10^{14} (9.5 x 10^{13})	carbon-rich dirt, ice
	Eurydome	Sheppard et al., 2001	14,208,000 (22,865,000)	717.3 days	1.9 (3)	unknown	unknown
	Autonoe	Sheppard et al., 2001	14,316,000 (23,039,000)	762.7 days	2.4 (4)	unknown	unknown
	Taygete	Sheppard et al., 2000	14,515,000 (23,360,000)	732.2 days	3 (5)	unknown	unknown
	Pasiphae	Melotte, 1908	14,602,200 (23,500,000)	735 days	31 (50)	2.1 x 10^{14} (1.9 x 10^{14})	carbon-rich dirt, ice
	Kalyke	Sheppard et al., 2000	14,654,000 (23,583,000)	743.0 days	3 (5)	unknown	unknown
	Sinope	Nicholson, 1914	14,726,500 (23,700,000)	758 days	22 (36)	8.4 x 10^{13} (7.6 x 10^{13})	carbon-rich dirt, ice
	Megaclite	Sheppard et al., 2000	14,792,000 (23,806,000)	752.8 days	3.8 (6)	unknown	unknown
NB: Jupiter has at least 61 satellites—this list represents the planet's major moons.							
Saturn							
	Pan	Showalter, 1990	83,005 (133,583)	0.6 day	12 (20)	unknown	ice?
	Atlas	Terrile, 1980	85,544 (137,670)	0.6 day	24 (38)	unknown	dirty ice?
	Prometheus	Collins et al., 1980	86,590 (139,353)	0.6 day	87 (140)	1.5 x 10^{14} (1.4 x 10^{14})	ice?
	Pandora	Collins et al., 1980	88,048 (141,700)	0.6 day	68 (110)	1.4 x 10^{14} (1.3 x 10^{14})	ice?
	Epimetheus	Walker et al., 1966	94,089 (151,422)	0.7 day	87 (140)	6.0 x 10^{14} (5.5 x 10^{14})	dirty ice?
	Janus	Dollfus, 1966	94,120 (151,472)	0.7 day	137 (220)	2.2 x 10^{15} (2.0 x 10^{15})	dirty ice?
	Mimas	Herschel, 1789	115,280 (185,520)	0.9 day	244 (392)	4.2 x 10^{16} (3.8 x 10^{16})	ice
	Enceladus	Herschel, 1789	147,900 (238,020)	1.4 days	310 (500)	8.8 x 10^{16} (8.0 x 10^{16})	ice
	Tethys	Cassini, 1684	183,090 (294,660)	1.9 days	659 (1060)	8.4 x 10^{17} (7.6 x 10^{17})	ice
	Telesto	Smith et al., 1980	183,090 (294,660)	1.9 days	21 (34)	unknown	ice?
	Calypso	Pascu et al., 1980	183,090 (294,660)	1.9 days	21 (34)	unknown	ice?

Planet	Satellite	Discovery	Distance from planet in miles (km)	Orbit time	Diameter in miles (km)	Mass in tons (tonnes)	Surface composition
Saturn (continued)							
	Dione	Cassini, 1684	234,500 (377,400)	2.7 days	696 (1120)	1.2×10^{18} (1.1×10^{18})	dirty ice
	Helene	Laques & Lecacheux, 1980	234,500 (377,400)	2.7 days	22 (36)	unknown	dirty ice?
	Rhea	Cassini, 1672	327,490 (527,040)	4.5 days	950 (1528)	2.8×10^{18} (2.5×10^{18})	ice
	Titan	Huygens, 1655	759,210 (1,221,830)	16 days	3200 (5150)	1.5×10^{20} (1.4×10^{20})	liquid methane, ice
	Hyperion	Bond, 1848	920,310 (1,481,100)	21.3 days	218 (350)	1.9×10^{16} (1.7×10^{16})	dirty ice?
	Iapetus	Cassini, 1671	2,212,900 (3,561,300)	79.3 days	892 (1436)	2.1×10^{18} (1.9×10^{18})	ice, carbon-rich dirt
	Phoebe	Pickering, 1898	8,048,000 (12,952,000)	550.5 days	143 (230)	4.4×10^{14} (4.0×10^{14})	ice, carbon-rich dirt?
NB: Saturn has at least 31 satellites—this list represents the planet's major moons.							
Uranus							
	Cordelia	Voyager 2, 1986	30,926 (49,770)	0.3 day	16 (26)	unknown	carbon-rich dirt, ice?
	Ophelia	Voyager 2, 1986	33,424 (53,790)	0.4 day	19 (30)	unknown	carbon-rich dirt, ice?
	Bianca	Voyager 2, 1986	36,764 (59,166)	0.4 day	26 (42)	unknown	carbon-rich dirt, ice?
	Cressida	Voyager 2, 1986	38,388 (61,780)	0.5 day	39 (62)	unknown	carbon-rich dirt, ice?
	Desdemona	Voyager 2, 1986	38,948 (62,680)	0.5 day	34 (54)	unknown	carbon-rich dirt, ice?
	Juliet	Voyager 2, 1986	39,985 (64,350)	0.5 day	52 (84)	unknown	carbon-rich dirt, ice?
	Portia	Voyager 2, 1986	41,066 (66,090)	0.5 day	67 (108)	unknown	carbon-rich dirt, ice?
	Rosalind	Voyager 2, 1986	43,459 (69,940)	0.6 day	34 (54)	unknown	carbon-rich dirt, ice?
	Belinda	Voyager 2, 1986	46,762 (75,256)	0.6 day	41 (66)	unknown	carbon-rich dirt, ice?
	S/1986 U10	Karkoshka, 1999	47,483 (76,416)	0.6 day	25 (40)?	unknown	unknown
	Puck	Voyager 2, 1986	53,444 (86,010)	0.8 day	96 (154)	unknown	carbon-rich dirt, ice?
	S/2003 U1	Sheppard et al., 2003	60,729 (97,734)	unknown	6 (10)	unknown	unknown
	Miranda	Kuiper, 1948	80,399 (129,390)	1.4 days	301 (484)	7.6×10^{16} (6.9×10^{16})	ice
	Ariel	Lassell, 1851	118,694 (191,020)	2.5 days	720 (1158)	1.5×10^{18} (1.4×10^{18})	ice
	Umbriel	Lassell, 1851	165,471 (266,300)	4.1 days	728 (1172)	1.3×10^{18} (1.2×10^{18})	ice
	S/2001 U3	Showalter et al., 2001	2,660,000 (4,281,000)	266.6 days	8 (12)	unknown	unknown
	Titania	Herschel, 1787	270,862 (435,910)	8.7 days	982 (1580)	3.9×10^{18} (3.5×10^{18})	ice
	Oberon	Herschel, 1787	362,583 (583,520)	13.5 days	947 (1524)	3.3×10^{18} (3.0×10^{18})	ice
	Caliban	Gladman et al., 1997	4,455,000 (7,169,000)	579 days	37 (60)	unknown	unknown
	Stephano	Gladman et al., 1999	4,973,000 (8,004,000)	677.4 days	12 (20)	unknown	unknown
	Trinculo	Gladman et al., 2001	5,330,000 (8,578,000)	759.0 days	6 (10)	unknown	unknown
	Sycorax	Nicholson et al., 1997	7,589,000 (12,214,000)	1289 days	100 (160)	unknown	unknown
	Prospero	Holman et al., 1999	10,093,000 (16,243,000)	1977.3 days	19 (30)	unknown	unknown
	Setebos	Kavelaars et al., 1999	10,875,000 (17,501,000)	2234.8 days	19 (30)	unknown	unknown
	S/2001 U2	Sheppard et al., 2001	13,000,000 (21,000,000)	2823.4 days	12 (20)	unknown	unknown
NB: Uranus has at least 27 satellites—this list represents the planet's major moons.							
Neptune							
	Naiad	Voyager 2, 1989	29,967 (48,227)	0.3 day	36 (58)	unknown	carbon-rich dirt, ice?
	Thalassa	Voyager 2, 1989	31,112 (50,070)	0.3 day	50 (80)	unknown	carbon-rich dirt, ice?
	Despina	Voyager 2, 1989	32,638 (52,526)	0.3 day	92 (148)	unknown	carbon-rich dirt, ice?
	Galatea	Voyager 2, 1989	38,496 (61,953)	0.4 day	98 (158)	unknown	carbon-rich dirt, ice?
	Larissa	Reitsma & Voyager 2, 1989	45,701 (73,548)	0.6 day	129 (208)	unknown	carbon-rich dirt, ice?
	Proteus	Voyager 2, 1989	73,103 (117,647)	1.1 days	271 (436)	unknown	carbon-rich dirt, ice?
	Triton	Lassell, 1846	220,438 (354,760)	5.9 days	1681 (2706)	2.4×10^{19} (2.2×10^{19})	nitrogen and methane ice
	Nereid	Kuiper, 1949	3,425,900 (5,513,400)	360.1 days	211 (340)	unknown	carbon-rich dirt, ice?
NB: Neptune has at least 13 satellites—this list represents the planet's major moons.							
Pluto							
	Charon	Christy, 1978	12,201 (19,636)	6.4 days	737 (1186)	2.1×10^{18} (1.9×10^{18})	ice

Date	Type	Best seen from
Feb 5, 2000	partial	Antarctica
July 1, 2000	partial	southeast Pacific Ocean, Chile, Argentina
July 31, 2000	partial	Siberia, Arctic Ocean, northwest USA & Canada
Dec 25, 2000	partial	North and Central America
June 21, 2001	total	south Africa
Dec 14, 2001	partial	Pacific Ocean, Nicaragua
June 10, 2002	partial	Pacific Ocean
Dec 4, 2002	total	south Africa, Pacific Ocean & Australia
May 31, 2003	partial	central and east Europe, Asia
Nov 23, 2003	total	Antarctica
Apr 19, 2004	partial	south Africa
Oct 14, 2004	partial	northeast Asia, north Pacific Ocean
Apr 8, 2005	total	east Pacific Ocean, Colombia, Venezuela
Oct 3, 2005	partial	Spain, north and central Africa
Mar 29, 2006	total	west and north Africa, Turkey, south Russia
Sept 22, 2006	partial	Atlantic Ocean
Mar 19, 2007	partial	east Asia, Alaska
Sept 11, 2007	partial	South America, Antarctica
Feb 7, 2008	partial	Antarctica, east Australia, New Zealand
Aug 1, 2008	total	north Canada, Greenland, Siberia, Mongolia, China
Jan 26, 2009	partial	south Africa, Antarctica, southeast Asia, Australia
July 22, 2009	total	India, Nepal, China, central Pacific Ocean
Jan 15, 2010	partial	central Africa, India, Burma, China
July 11, 2010	total	south Pacific Ocean, Easter Island, Chile, Argentina
Jan 4, 2011	partial	Europe, north Africa, central Asia
June 1, 2011	partial	east Asia, Alaska, north Canada, Iceland
July 1, 2011	partial	south Indian Ocean
Nov 25, 2011	partial	south Africa, Antarctica, Tasmania, New Zealand
May 20, 2012	partial	China, Japan, Pacific Ocean, west USA
Nov 13, 2012	total	north Australia, south Pacific Ocean
May 10, 2013	partial	north Australia, Solomon Islands, Pacific Ocean
Nov 3, 2013	total	Atlantic Ocean, central Africa
Apr 29, 2014	partial	south Indian Ocean, Australia, Antarctica
Oct 23, 2014	partial	north Pacific Ocean, North America
Mar 20, 2015	total	northeast Atlantic Ocean
Sept 13, 2015	partial	south Africa, south Indian Ocean, Antarctica
Mar 9, 2016	total	south Pacific Ocean
Sept 1, 2016	annular	eastern Africa

Shower	Constellation	Date	Per hour	Parent object
Quadrantids	Boötes	Jan 3	40 meteors	unknown
Lyrids	Lyra	Apr 22	15 meteors	comet Thatcher
Eta Aquarids	Aquarius	May 5	20 meteors	comet Halley
Delta Aquarids	Aquarius	July 28	20 meteors	unknown
Perseids	Perseus	Aug 12	50 meteors	comet Swift-Tuttle
Orionids	Orion	Oct 22	25 meteors	comet Halley
Taurids	Taurus	Nov 3	15 meteors	comet Encke
Leonids	Leo	Nov 17	15 meteors	comet Tempel-Tuttle
Geminids	Gemini	Dec 14	50 meteors	asteroid 3200 Phaethon
Ursids	Ursa Minor	Dec 23	20 meteors	comet Tuttle

LUNAR ECLIPSES

Date	Type	Best seen from
Jan 21, 2000	total	North & South America
July 16, 2000	total	Australia, west Pacific Ocean
Jan 9, 2001	total	west Asia, Africa, Europe
July 5, 2001	partial	Australia, west Pacific Ocean
May 16, 2003	total	North & South America
Nov 9, 2003	total	Europe, west Africa, North & South America
May 4, 2004	total	west Asia, Africa, Europe
Oct 28, 2004	total	North & South America
Oct 17, 2005	partial	central Pacific Ocean
Sept 7, 2006	partial	Indian Ocean, Asia, east Africa
Mar 3, 2007	total	Africa, Europe
Aug 28, 2007	total	central Pacific, west North & South America
Feb 21, 2008	total	North & South America, west Europe
Aug 16, 2008	partial	west Asia, Europe, Africa
Dec 21, 2009	partial	Asia, Indian Ocean, Africa, Europe
June 26, 2010	partial	central Pacific Ocean, west North & South America
Dec 21, 2010	total	North America, west South America
June 15, 2011	total	southwest Asia, Africa, Indian Ocean
Dec 10, 2011	total	west Pacific Ocean, east Asia, Alaska, Yukon
June 4, 2012	partial	central Pacific Ocean, west North & South America
Apr 15, 2014	total	North America, west South America, Pacific Ocean
Oct 8, 2014	total	Pacific Ocean, west North & South America
Apr 4, 2015	total	Pacific Ocean, west North & South America
Sept 28, 2015	total	west Europe & Africa, North & South America

IMPORTANT UNMANNED MISSIONS

Planet	Mission	Country	Type	Launch date	Arrival date	Achievements
Moon						
	Luna 2	Soviet Union	impact	Sept 12, 1959	Sept 13, 1959	first impact
	Luna 3	Soviet Union	flyby	Oct 4, 1959	Oct 7, 1959	first farside images show terrain is mostly highlands
	Ranger 7	USA	impact	July 28, 1964	July 31, 1964	images surface until impact, finds many small craters
	Ranger 8	USA	impact	Feb 17, 1965	Feb 20, 1965	impacts in Mare Tranquillitatis, takes more than 7,000 photos
	Ranger 9	USA	impact	Mar 21, 1965	Mar 24, 1965	impacts in Alphonsus crater, finds volcanic vents
	Zond 3	Soviet Union	flyby	July 18, 1965	July 20, 1965	photographs lunar farside
	Luna 9	Soviet Union	lander	Jan 31, 1966	Feb 3, 1966	first soft landing, panoramic photos of surface
	Luna 10	Soviet Union	orbiter	Mar 31, 1966	Apr 3, 1966	first spacecraft to orbit the Moon
	Surveyor 1	USA	lander	May 30, 1966	June 2, 1966	first lander to make chemical measurements of surface
	Lunar Orbiter 1	USA	orbiter	Aug 10, 1966	Aug 14, 1966	photo-survey of potential Apollo landing sites
	Luna 11	Soviet Union	orbiter	Aug 24, 1966	Aug 28, 1966	photographs surface
	Luna 12	Soviet Union	orbiter	Oct 22, 1966	Oct 25, 1966	photographs surface
	Lunar Orbiter 2	USA	orbiter	Nov 6, 1966	Nov 10, 1966	photo-survey of potential Apollo landing sites
	Luna 13	Soviet Union	lander	Dec 21, 1966	Dec 24, 1966	panoramic photos, mechanical soil probe
	Lunar Orbiter 3	USA	orbiter	Feb 5, 1967	Feb 8, 1967	photo-survey of potential Apollo landing sites
	Surveyor 3	USA	lander	Apr 17, 1967	Apr 20, 1967	takes more than 6,000 photos, Apollo 12 later lands at site
	Lunar Orbiter 4	USA	orbiter	May 4, 1967	May 8, 1967	photo-survey of entire nearside hemisphere
	Lunar Orbiter 5	USA	orbiter	Aug 1, 1967	Aug 5, 1967	photo-survey of geologically interesting areas
	Surveyor 5	USA	lander	Sept 8, 1967	Sept 11, 1967	analyzes surface properties, takes more than 6,300 surface photos
	Surveyor 6	USA	lander	Nov 7, 1967	Nov 10, 1967	takes almost 30,000 surface photos
	Surveyor 7	USA	lander	Jan 7, 1968	Jan 10, 1968	lands near rim of Tycho crater, analyzes surface properties
	Zond 5	Soviet Union	flyby	Sept 15, 1968	Sept 18, 1968	flies around Moon, returns to Earth Sept 21
	Zond 6	Soviet Union	flyby	Nov 10, 1968	Nov 14, 1968	flies around Moon, returns to Earth Nov 17
	Zond 7	Soviet Union	flyby	Aug 7, 1969	Aug 11, 1969	flies around Moon, returns to Earth Aug 14
	Luna 16	Soviet Union	lander	Sept 12, 1970	Sept 20, 1970	collects rock sample and returns it to Earth
	Zond 8	Soviet Union	flyby	Oct 20, 1970	Oct 24, 1970	flies around Moon, returns to Earth Oct 27
	Luna 17	Soviet Union	rover	Nov 10, 1970	Nov 17, 1970	first robotic rover, drives 6 miles (10 km) on surface
	Luna 20	Soviet Union	lander	Feb 14, 1972	Feb 21, 1972	automatic sample return
	Luna 21	Soviet Union	rover	Jan 8, 1973	Jan 15, 1973	explores Posidonius crater, drives 23 miles (37 km)
	Luna 22	Soviet Union	orbiter	May 29, 1974	June 2, 1974	photo-survey from orbit
	Luna 24	Soviet Union	lander	Aug 9, 1976	Aug 14, 1976	lands in Mare Crisium, returns sample to Earth
	Hiten (Muses-A)	Japan	flyby & orbiter	Jan 24, 1990	Mar 19, 1990	flies past Moon and releases satellite
	Clementine	USA	orbiter	Jan 25, 1994	Feb 21, 1994	surveys surface mineralogy at high resolution
	Lunar Prospector	USA	orbiter	Jan 7, 1998	Jan 11, 1998	surveys composition, finds ice in polar craters
	Smart-1	European	orbiter	Sept 27, 2003	Sept 30, 2003	testing of an ion drive
	Lunar-A	Japan	orbiter	Aug, 2004	Aug, 2004	study interior and surface of Moon
	SELENE	Japan	orbiter	late 2005	late 2005	various goals
Mercury						
	Mariner 10	USA	flyby	Nov 3, 1973	Mar 29, 1974	first close-up images show cratered surface, detects large iron core
	Messenger	USA	orbiter and flyby	Mar 2004	Jul 2007	to study the environment
	BepiColonbo	European	orbiter	late 2011	mid 2014	mission still in planning
Venus						
	Mariner 2	USA	flyby	Aug 27, 1962	Dec 14, 1962	first flyby finds heavy atmosphere, hot surface
	Venera 4	Soviet Union	lander	June 12, 1967	Oct 18, 1967	measures atmosphere, fails on descent
	Mariner 5	USA	flyby	June 14, 1967	Oct 19, 1967	improves measurements of atmospheric pressure and temperature
	Venera 5	Soviet Union	lander	Jan 5, 1969	May 16, 1969	studies atmosphere
	Venus Venera 6	Soviet Union	lander	Jan 10, 1969	May 17, 1969	studies atmosphere
	Venera 8	Soviet Union	lander	Mar 27, 1972	July 22, 1972	sends back first data from surface
	Mariner 10	USA	flyby	Nov 4, 1973	Feb 5, 1974	flies past on way to Mercury, photographs swirling clouds

Planet	Mission	Country	Type	Launch date	Arrival date	Achievements
Venus (continued)						
	Venera 9	Soviet Union	orbiter & lander	June 8, 1975	Oct 22, 1975	first images of surface show volcanic rocks
	Venera 10	Soviet Union	orbiter & lander	June 14, 1975	Oct 25, 1975	photographs surface rocks and dirt
	Pioneer Venus Orbiter	USA	orbiter	May 20, 1978	Dec 4, 1978	first global radar map of landscape, studies clouds
	Pioneer Venus Probes	USA	entry probes	Aug 8, 1978	Dec 9, 1978	five probes sample atmosphere
	Venera 11	Soviet Union	orbiter & lander	Sept 9, 1978	Dec 25, 1978	photographs surface, analyzes atmosphere
	Venera 12	Soviet Union	orbiter & lander	Sept 14, 1978	Dec 21, 1978	photographs surface, analyzes atmosphere
	Venera 13	Soviet Union	orbiter & lander	Oct 30, 1981	Mar 1, 1982	photographs surface, analyzes atmosphere
	Venera 14	Soviet Union	orbiter & lander	Nov 4, 1981	Mar 5, 1982	photographs surface, analyzes atmosphere
	Venera 15	Soviet Union	orbiter	June 2, 1983	Oct 10, 1983	radar mapping of northern hemisphere
	Venera 16	Soviet Union	orbiter	June 7, 1983	Oct 14, 1983	radar mapping of northern hemisphere
	Vega 1	Soviet Union	lander & balloon	Dec 15, 1984	June 11, 1985	surveys atmosphere and winds with balloon
	Vega 2	Soviet Union	lander & balloon	Dec 21, 1984	June 16, 1985	surveys atmosphere and winds with balloon
	Magellan	USA	orbiter	May 4, 1989	Aug 10, 1990	surveys geology over most of Venus using radar
	Galileo	USA	flyby	Oct 18, 1989	Feb 10, 1990	flies past on way to Jupiter, studies clouds
	Cassini	USA	flyby	Oct 15, 1997	Apr 26, 1998	flies past on way to Saturn, studies clouds
	Cassini	USA	flyby	Oct 15, 1997	June 24, 1999	second Venus flyby, studies clouds
	Venus Express	European	orbiter	late 2005	early 2006	to study atmosphere and environment
	Planet-C	Japan	orbiter	Feb 2007	Sept 2009	to study atmosphere and environment
Mars						
	Mariner 4	USA	flyby	Nov 28, 1965	July 14, 1965	first close-up images show many craters, thin atmosphere
	Mariner 6	USA	flyby	Feb 24, 1969	July 31, 1969	returns 75 photos, increases geological knowledge
	Mariner 7	USA	flyby	Mar 27, 1969	Aug 5, 1969	returns 126 photos, increases geological knowledge
	Mars 3	Soviet Union	orbiter & lander	May 28, 1971	Dec 3, 1971	some data and few photos
	Mariner 9	USA	orbiter	May 30, 1971	Nov 12, 1971	first survey of entire surface, finds water channels, big volcanoes
	Mars 5	Soviet Union	orbiter	July 25, 1973	Feb 12, 1974	lasts a few days
	Mars 6	Soviet Union	orbiter & lander	Aug 5, 1973	Mar 12, 1974	little data return
	Mars 7	Soviet Union	orbiter & lander	Aug 9, 1973	Mar 9, 1974	little data return
	Viking 1	USA	orbiter & lander	Aug 20, 1975	June 19, 1976	geological survey from orbit, unsuccessful search for life on surface
	Viking 2	USA	orbiter & lander	Sept 9, 1975	Aug 7, 1976	geological survey from orbit, unsuccessful search for life on surface
	Mars Global Surveyor	USA	orbiter	Nov 7, 1996	Sept 12, 1997	maps entire planet at high resolution
	Mars Pathfinder	USA	lander & rover	Dec 2, 1996	July 4, 1997	explores geology of a once-flooded landscape
	Nozomi (Planet-B)	Japan	orbiter	July 4, 1998	Dec 2003	studies upper atmosphere, magnetosphere, solar wind
	Mars Climate Orbiter	USA	orbiter	Dec 11, 1998	Sept 23, 1999	crashes on arrival, no data returned
	Mars Polar Lander	USA	lander	Jan 3, 1999	Dec 3, 1999	fails upon arrival, no data returned
	Deep Space 2	USA	impactor	Jan 3, 1999	Dec 3, 1999	travels with Polar Lander, searches for subsurface ice
	Mars Odyssey	USA	orbiter and lander	Apr 7, 2001	Oct 24, 2001	studies atmosphere and environment
	Beagle 2	European	orbiter and lander	Jun 2, 2003	Dec 26, 2003	various tasks
	Spirit	USA	orbiter and lander	Jun 10, 2003	Jan 4, 2004	studies climate, atmosphere, and searches for life
	Opportunity	USA	orbiter and lander	Jul 8, 2003	Jan 25, 2004	studies climate, atmosphere, and searches for life
	Mars Surveyor	USA	orbiter	Aug 2005	Mar 2006	to study surface from orbit
Asteroids						
	Galileo	USA	flyby	Oct 18, 1989	Oct 29, 1991	first close-up images of an asteroid, 951 Gaspra, show craters
	Galileo	USA	flyby	Oct 18, 1989	Aug 28, 1993	asteroid 243 Ida: survey of cratered surface, finds moon, Dactyl
	NEAR	USA	flyby	Feb 17, 1996	June 27, 1997	asteroid 253 Mathilde: reveals big craters, low density
	NEAR	USA	flyby	Feb 17, 1996	Dec 23, 1998	asteroid 433 Eros: survey of craters and features
	Deep Space 1	USA	flyby	Oct 24, 1998	July 29, 1999	asteroid 9969 Braille: study of magnetic field, surface composition
	NEAR	USA	orbiter	Feb 17, 1996	Feb 14, 2000	asteroid 433 Eros: first spacecraft to orbit an asteroid
	Dawn	USA	orbiter	May 2006	July 2010	rendezvous with Vesta
	Dawn	USA	orbiter	May 2006	Aug 2014	rendezvous with Ceres

Planet	Mission	Country	Type	Launch date	Arrival date	Achievements
Jupiter						
	Pioneer 10	USA	flyby	Mar 3, 1972	Dec 3, 1973	first detailed study of a gas-giant planet
	Pioneer 11	USA	flyby	Apr 6, 1973	Dec 3, 1974	studies polar regions of Jupiter, magnetic environment
	Voyager 1	USA	flyby	Sept 5, 1977	Mar 5, 1979	first detailed images of planet and moons, discovers ring system
	Voyager 2	USA	flyby	Aug 20, 1977	July 9, 1979	follow-up on Voyager 1's discoveries
	Galileo	USA	orbiter & entry	Oct 18, 1989	Dec 7, 1995	first atmosphere probe, orbital tour of planet and moons
Saturn						
	Pioneer 11	USA	flyby	Apr 6, 1973	Sept 1, 1979	first spacecraft visit, discovers new rings
	Voyager 1	USA	flyby	Sept 5, 1977	Nov 12, 1980	detailed portraits of clouds, rings, and moons
	Voyager 2	USA	flyby	Aug 20, 1977	Aug 25, 1981	follow-up on Voyager 1's discoveries
	Cassini	USA	flyby	Oct 15, 1997	July 2004	orbital tour of planet and moons, lander for Titan
Uranus						
	Voyager 2	USA	flyby	Aug 20, 1977	Jan 24, 1986	first spacecraft visit, studies rings and moons, finds new moons
Neptune						
	Voyager 2	USA	flyby	Aug 20, 1977	Aug 25, 1989	first spacecraft visit, studies storms, finds geysers on Triton
Pluto						
	New Horizons	USA	flyby	Jan 2006	July 2015	to image and study Pluto and Charon
Comets						
	Vega 1	Soviet Union	flyby	Dec 15, 1984	Mar 6, 1986	study of Venus and Comet Halley
	Vega 2	Soviet Union	flyby	Dec 21, 1984	Mar 9, 1986	study of Venus and Comet Halley
	Giotto	European	flyby	July 2, 1985	Mar 14, 1986	photographs nucleus of comet Halley
	Suisei	Japan	flyby	Aug 18, 1985	Mar 1986	UV photos of comet Halley
	Sakigake	Japan	flyby	Jan 1986	Mar 8, 1986	various, Comet Halley
	Stardust	USA	flyby	Feb 7, 1999	Jan 2, 2004	sample collection & return from comet Wild 2
	Deep Space 1	USA	flyby	Oct 24, 1998	Sept 22, 2001	flyby of comets Wilson-Harrington and Borrelly
	Rosetta	European	flyby and lander	Feb 2004	Nov 2014	study of Comet Churyumov-Gerasimenko
	Deep Impact	USA	flyby and lander	Dec 2004	Jul 2005	study of Comet Tempel 1

CONSTELLATION FACTS

Constellation	Meaning	Hemisphere	Highlights
Andromeda	the Princess	Northern	spiral galaxy M31 (Andromeda), double star Gamma Andromedae, open cluster NGC 752
Antlia	the Air Pump	Southern	planetary nebula NGC 3132, spiral galaxy NGC 2997
Apus	the Bird of Paradise	Southern	double star Delta Apodis, variable star S Apodis, variable star Theta Apodis
Aquarius	the Water Carrier	equatorial	globular cluster M2, planetary nebulas NGC 7009 (Saturn) and NGC 7293 (Helix)
Aquila	the Eagle	equatorial	bright star Altair, variable stars Eta Aquilae and R Aquilae, open cluster NGC 6709
Ara	the Altar	Southern	open cluster NGC 6193, globular cluster NGC 6397
Aries	the Ram	Northern	double stars Gamma Arietis, Lambda Arietis, and Pi Arietis
Auriga	the Charioteer	Northern	bright star Capella, open clusters M36, M37, and M38, double star Omega Aurigae
Boötes	the Herdsman	Northern	bright star Arcturus, double star Mu Boötis
Caelum	the Chisel	Southern	double star Gamma Caeli, variable star R Caeli
Camelopardalis	the Giraffe	Southern	double star Beta Camelopardalis, variable star VZ Camelopardalis, star cluster NGC 1502
Cancer	the Crab	Northern	open clusters M44 (Beehive) and M67, double stars Zeta Cancri and Iota Cancri
Canes Venatici	the Hunting Dogs	Northern	spiral galaxies M51 (Whirlpool) and M94, globular cluster M3

CONSTELLATION FACTS

Constellation	Meaning	Hemisphere	Highlights
Capricornus	the Sea Goat	Southern	double stars Alpha Capricorni and Beta Capricorni, globular cluster M30
Carina	the Keel	Southern	bright star Canopus, emission nebula Eta Carinae, open clusters NGC 3532 and IC 2602
Cassiopeia	the Queen	Northern	variable star Gamma Cassiopeiae, open cluster M52
Centaurus	the Centaur	Southern	bright star Alpha Centauri, globular cluster Omega Centauri, elliptical galaxy NGC 5128
Cepheus	the King	Northern	variable stars Delta Cephei and Mu Cephei
Cetus	the Sea Monster	equatorial	variable star Mira Ceti, Seyfert galaxy M77
Chamaeleon	the Chameleon	Southern	double star Delta Chamaeleontis, planetary nebula NGC 3195
Circinus	the Drawing Compass	Southern	double star Alpha Circini
Columba	the Dove	Southern	variable star T Columbae, globular cluster NGC 1851
Coma Berenices	Berenice's Hair	Northern	globular cluster M53, spiral galaxies M64 (Black-eye) and NGC 4565
Corona Australis	the Southern Crown	Southern	double star Kappa Coronae Australis, globular cluster NGC 6541
Corona Borealis	the Northern Crown	Northern	double star Nu Coronae Borealis, variable star R Coronae Borealis
Corvus	the Crow	Southern	variable star R Corvi, galaxies NGC 4038 and NGC 4039 (Antennae galaxies)
Crater	the Cup	Southern	double stars Gamma and Iota Crateris
Crux	the Southern Cross	Southern	bright star Acrux, open cluster NGC 4755 (Jewel Box), dark nebula the Coalsack
Cygnus	the Swan	Northern	bright star Deneb, double star Beta Cygni, emission nebula NGC 7000 (North America)
Delphinus	the Dolphin	Northern	double star Gamma Delphini
Dorado	the Goldfish	Southern	irregular galaxy the Large Magellanic Cloud, emission nebula NGC 2070 (Tarantula)
Draco	the Dragon	Northern	double stars Psi Draconis and 39 Draconis, planetary nebula NGC 6543
Equuleus	the Little Horse	Northern	double star Gamma Equulei
Eridanus	the River	Southern	bright star Achernar, triple star Omicron 2 Eridani
Fornax	the Furnace	Southern	barred spiral galaxy NGC 1365 (Great Barred Spiral)
Gemini	the Twins	Northern	bright stars Castor and Pollux, open cluster M35, planetary nebula NGC 2392 (Clownface)
Grus	the Crane	Southern	double stars Delta Gruis and Mu Gruis
Hercules	the Strongman	Northern	variable star Alpha Herculis, globular clusters M13 (Hercules) and M92
Horologium	the Clock	Southern	variable stars R Horologii and TW Horologii
Hydra	the Sea Serpent	equatorial	double star 27 Hydrae, open cluster M48, planetary nebula NGC 3242, spiral galaxy M83
Hydrus	the Water Snake	Southern	double star Pi Hydri
Indus	the Indian	Southern	double star Theta Indi
Lacerta	the Lizard	Northern	BL Lacertae (the prototype active galaxy), open clusters NGC 7243 and NGC 7209
Leo	the Lion	equatorial	bright star Regulus, double star Gamma Leonis, spiral galaxies M65 and M66
Leo Minor	the Little Lion	Northern	red supergiant variable Mira
Lepus	the Hare	Southern	double star Gamma Leporis, globular cluster M79
Libra	the Scales	Southern	double star Alpha Librae, variable star Delta Librae
Lupus	the Wolf	Southern	open cluster NGC 5822, globular cluster NGC 5986
Lynx	the Lynx	Northern	a faint and distant globular cluster, known as the Intergalactic Tramp (NGC 2419)
Lyra	the Lyre	Northern	bright star Vega, variable star Beta Lyrae, planetary nebula M57 (Ring)
Mensa	the Table	Southern	dwarf star Alpha Mensae
Microscopium	the Microscope	Southern	double star Alpha Microscopii
Monoceros	the Unicorn	equatorial	open cluster M50, nebulas NGC 2237 (Rosette) and NGC 2264 (Cone)
Musca	the Fly	Southern	double star Theta Muscae, globular clusters NGC 4372 and NGC 4833
Norma	the Level	Southern	open clusters NGC 6067 and NGC 6087
Octans	the Octant	Southern	pole star Sigma Octantis, double star Lambda Octantis
Ophiuchus	the Serpent Carrier	equatorial	open clusters IC 4665 and NGC 6633, globular clusters M10 and M12
Orion	the Hunter	equatorial	bright stars Betelgeuse and Rigel, emission nebula M42 (Orion), dark nebula IC 434 (Horsehead)
Pavo	the Peacock	Southern	variable star Kappa Pavonis, globular cluster NGC 6752
Pegasus	the Flying Horse	Northern	double star Epsilon Pegasi, globular cluster M15
Perseus	the Hero	Northern	variable star Algol (Beta Persei), open clusters NGC 869 and NGC 884 (Double Cluster)
Phoenix	the Firebird	Southern	double and variable star Zeta Phoenicis
Pictor	the Painter's Easel	Southern	Beta Pictoris (which hosts a protoplanetary disk), Kapteyn's Star
Pisces	the Fishes	equatorial	double star Rho and 94 Piscium, spiral galaxy M74

Contellation	Meaning	Hemisphere	Highlights
Piscis Austrinus	the Southern Fish	Southern	bright star Fomalhaut
Puppis	the Stern	Southern	variable star L2 Puppis, open clusters M46 and M47
Pyxis	the Compass	Southern	a recurrent nova, T Pyxidis, which reaches a peak magnitude of +7 every 12 to 25 years
Reticulum	the Reticle	Southern	double star Zeta Reticuli
Sagitta	the Arrow	Northern	globular cluster M71
Sagittarius	the Archer	Southern	globular clusters M22 and M23, emission nebulas M8 (Lagoon), M17 (Omega), M20 (Trifid)
Scorpius	the Scorpion	Southern	bright star Antares, double star Beta Scorpii, open cluster M7, globular clusters M4 and M80
Sculptor	the Sculptor	Southern	spiral galaxies NGC 253 and NGC 55
Scutum	the Shield	Northern	open cluster M11 (Wild Duck)
Serpens	the Serpent	equatorial	double star Nu Serpentis, emission nebula M16 (Eagle), globular cluster M5
Sextans	the Sextant	equatorial	double star 17 and 18 Sextantis, elliptical galaxy NGC 3115 (Spindle)
Taurus	the Bull	equatorial	bright star Aldebaran, open clusters M45 (Pleiades) and Hyades, supernova remnant M1 (Crab)
Telescopium	the Telescope	Southern	double star Delta Telescopii
Triangulum	the Triangle	Northern	spiral galaxy M33 (Pinwheel)
Triangulum Australe	the Southern Triangle	Southern	variable star R Trianguli Australe, open cluster NGC 6025
Tucana	the Toucan	Southern	double star Beta Tucanae, globular cluster 47 Tucanae, irregular galaxy the Small Magellanic Cloud
Ursa Major	the Big Bear	Northern	double star Mizar and Alcor, spiral galaxies M81 and M101, peculiar galaxy M82
Ursa Minor	the Little Bear	Northern	Polaris the Pole Star, double star Gamma and 11 Ursae Minoris
Vela	the Sails	Southern	double star Gamma Velorum, open clusters IC 2391 and NGC 2547
Virgo	the Maiden	equatorial	bright star Spica, elliptical galaxies M47, M87, and M104 (Sombrero), quasar 3C 273
Volans	the Flying Fish	Southern	double star Gamma Volantis, barred spiral galaxy NGC 2442
Vulpecula	the Little Fox	Northern	planetary nebula M27 (Dumbbell)

BRIGHTEST STAR FACTS

Star	Constellation	Color	Type	Companion stars	Apparent magnitude	Absolute magnitude	Diameter (Sun = 1)	Distance from Earth
Sirius	Canis Major	white	main sequence	1 companion	–1.46	+1.4	1.7	8.6 ly
Canopus	Carina	white	bright giant	0 companions	–0.72	–2.5	unknown	74 ly
Rigil Kentaurus	Centaurus	yellow	main sequence	2 companions	–0.27	+4.4	1.2	4.3 ly
Arcturus	Boötes	orange	giant	0 companions	–0.04	–0.2	25	34 ly
Vega	Lyra	white	main sequence	0 companions	+0.03	+0.6	2.0	25 ly
Capella	Auriga	yellow	giant	1 companion	+0.08	–0.4	13	41 ly
Rigel	Orion	blue-white	supergiant	2 companions	+0.12	–8.1	63	1400 ly
Procyon	Canis Minor	yellow-white	subgiant	1 companion	+0.38	+2.6	2.0	11.4 ly
Achernar	Eridanus	blue-white	main sequence	0 companions	+0.46	–1.3	5.0	69 ly
Betelgeuse	Orion	red	supergiant	0 companions	+0.50 variable	–7.2	226	1400 ly
Hadar	Centaurus	blue-white	giant	1 companion	+0.61 variable	–4.4	unknown	320 ly
Acrux	Crux	blue-white	main sequence	1 companion	+0.76	–4.6	2.2	510 ly
Altair	Aquila	white	main sequence	0 companions	+0.77	+2.3	1.6	16 ly
Aldebaran	Taurus	orange	giant	0 companions	+0.85 variable	–0.3	46	60 ly
Antares	Scorpius	red	supergiant	1 companion	+0.96 variable	–5.2	510	520 ly
Spica	Virgo	blue-white	main sequence	1 companion	+0.98 variable	–3.2	6.6	220 ly
Pollux	Gemini	orange	giant	0 companions	+1.14	+0.7	10	40 ly
Fomalhaut	Pisces Austrinus	white	main sequence	0 companions	+1.16	+2.0	1.5	22 ly
Becrux / Mimosa	Crux	blue-white	giant	0 companions	+1.25	–4.7	unknown	460 ly
Deneb	Cygnus	white	supergiant	0 companions	+1.25 variable	–7.2	unknown	1500 ly

UNIVERSAL RECORDS

HOTTEST PLANET SURFACE IN THE SOLAR SYSTEM

The surface of Venus, 880°F (470°C). At its hottest, Mercury comes close: 800°F (427°C). Venus's thick atmosphere traps the Sun's heat, so midnight temperatures are as hot as those at noontime. (And the rocks are hot enough to glow dull red!)

COLDEST RECORDED SURFACE IN THE SOLAR SYSTEM

Triton, the largest satellite of Neptune. When the Voyager 2 probe passed this world in 1989, it found a frigid surface with a temperature of -391°F (-235°C).

BIGGEST CRATER IN OUR SOLAR SYSTEM

The Aitken Basin on the Moon's south pole, 1600 miles (2500 km) in diameter. This ancient impact scar is so heavily marked with smaller craters that it was not discovered until the Clementine probe visited the Moon in 1994. Scientists used data from Clementine to carefully map the Moon's surface. This mapping revealed the basin, a broad depression in the lunar farside that is more than 7 miles (12 km) deep.

TALLEST MOUNTAIN IN THE SOLAR SYSTEM

Olympus Mons on Mars, rising 15 miles (24 km) above its base. The second tallest is Maxwell Montes on Venus, which rises 7 miles (11 km) above the planet's average surface. Earth's officially tallest peak is Mount Everest, 5.5 miles (8.8 km) above sea level. However, Hawaii's Mauna Kea can also claim to be the tallest, since it rises about 5.6 miles (9 km) above the ocean floor it stands on.

BIGGEST CANYON IN THE SOLAR SYSTEM

Valles Marineris on Mars, roughly 2500 miles (4000 km) long, with a maximum width of about 370 miles (600 km) and a maximum depth of 5 miles (8 km). If it were in the United States, this canyon could extend from San Francisco on the west coast to the Appalachian Mountains in Virginia near the east coast. In Europe, it would stretch from Paris to Russia's Ural Mountains.

LARGEST PLANET IN THE SOLAR SYSTEM

Jupiter, with 317.8 times the mass of Earth, and about 11 times its diameter. Jupiter contains more mass than all the rest of the planets, satellites, comets, and asteroids.

LARGEST KNOWN PLANET

An unnamed planet orbiting the star HD 114762. This planet appears to have 11 times the mass of Jupiter, but some astronomers think it may actually be a brown dwarf, an object that is like a small, dim, cool star. If it is a brown dwarf, then the most massive planet would be one with 6.6 times Jupiter's mass that orbits the star 70 Virginis.

LARGEST SATELLITE IN THE SOLAR SYSTEM

Jupiter's Ganymede, 3273 miles (5268 km) in diameter. If Ganymede orbited the Sun instead of Jupiter, it would easily qualify as a planet. It is larger than either Mercury or Pluto.

GREATEST METEOR SHOWER

The Leonids on November 13, 1833, with up to 200,000 meteors per hour. Onlookers said that the meteors "fell like snowflakes," while many thought the world was about to come to an end. The remarkable display helped astronomers realize that meteors were entering Earth's atmosphere from outer space, and were not just an Earth-based event like rain.

LARGEST METEORITE

Hoba meteorite in Namibia, weighing 65 tons (60 tonnes)—about as heavy as nine elephants! Discovered in 1920, this iron meteorite almost 10 feet (3 m) long still lies in the ground where it landed. It was originally even larger—part of the meteorite has weathered away.

LARGEST ASTEROID

1 Ceres, 567 miles (913 km) in diameter. This largest of all asteroids was also the first to be found—and its discovery came on the first day of the 19th century: January 1, 1801. It was discovered by Giuseppe Piazzi at the Palermo Observatory.

LARGEST KUIPER-BELT OBJECT (KBO)

Quaoar (pronounced KWAH-o-WAH), 800 miles (1300 km) across, or half the size of Pluto. This world orbiting beyond Neptune in the Kuiper Belt is the largest known minor planet. It is bigger than all the asteroids in the Asteroid Belt combined. (Some astronomers consider Pluto itself a KBO rather than a planet, in which case it is the largest known.)

CLOSEST COMET TO EARTH

Comet Lexell in 1770, at a distance of 1.4 million miles (2.2 million km) from Earth—less than six times the distance to the Moon. Despite coming so close, this comet never developed much of a tail and its head looked no bigger than five times the size of the Moon in our night sky.

LONGEST COMET TAIL

Great Comet of March 1843, 190 million miles (300 million km) long. This tail was long enough to reach from the Sun to well past the orbit of Mars.

BROADEST STAR IN OUR NIGHT SKY

Betelgeuse in Orion, about 800 times the Sun's diameter. If it replaced the Sun in the Solar System, this bloated red supergiant star would reach past the orbit of Jupiter.

MOST MASSIVE STAR

Eta Carinae, approximately 150 times as massive as the Sun. Astronomers are not certain if Eta Carinae is one star or two.

LEAST MASSIVE STAR

Gliese 105C, about 10 percent as massive as the Sun. This is about as small as a star can be and still be a true star (an object that fuses hydrogen into helium).

NEAREST STAR

Proxima Centauri, third member of the Alpha Centauri system. This cool red dwarf star lies about 4.2 light-years away, about 0.1 light-year closer to us than the other two stars in the system.

GLOBULAR STAR CLUSTER WITH THE MOST STARS

Omega Centauri, with 1.1 million stars. This globular cluster measures about 180 light-years in diameter.

MOST MASSIVE GALAXY

Giant elliptical M87 in the constellation of Virgo, with at least 800 billion Suns' worth of mass. M87 is a member of the Virgo cluster of galaxies.

LEAST MASSIVE GALAXY

The Pegasus II dwarf elliptical, about 10 million solar masses. Smaller galaxies may exist, but as they are not very luminous, astronomers cannot detect them unless they lie close to us.

NEAREST GALAXY

The Canis Major dwarf galaxy. This galaxy in Canis Major is 25,000 light-years away from the Solar System and 42,000 light-years from the center of the Milky Way. It is the current record holder, but surveys find new dwarf elliptical galaxies every year or so, and an even closer galaxy may yet be found.

MOST DISTANT OBJECT VISIBLE TO THE NAKED EYE

Andromeda galaxy (M31), 2.9 million light-years away. When you look at this galaxy, you are seeing light that left the galaxy when the most recent great Ice Ages were beginning on Earth. The spiral galaxy M33 in Triangulum is farther and fainter, and may be visible to the very keenest naked eye.

MOST DISTANT OBJECT DETECTED

An unnamed galaxy in Ursa Major, 12.6 billion light-years away. This galaxy may not hold the record for very long. Astronomers working with giant telescopes on Earth and the Hubble Space Telescope find a new and more distant record-holder once or twice a year.

STAR CLUSTER FACTS

Cluster	Constellation	Type	Number of stars	Apparent magnitude	Diameter	Distance from Earth	What you need to see it
Hyades	Taurus	open	100	+0.8	17 ly	150 ly	naked eye
Pleiades	Taurus	open	several hundred	+1.6	13 ly	375 ly	naked eye
Beehive	Cancer	open	50	+3.9	15 ly	590 ly	binoculars
Double Cluster	Perseus	open	350	+4.3	61 ly	7000 ly	binoculars
47 Tucanae	Tucana	globular	460,000	+4.4	125 ly	16,000 ly	binoculars
Omega Centauri	Centaurus	globular	1,100,000	+4.5	180 ly	17,000 ly	binoculars
Jewel Box	Crux	open	50	+5.2	24 ly	6800 ly	binoculars
Hercules cluster (M13)	Hercules	globular	220,000	+6.4	110 ly	21,000 ly	binoculars
M15	Pegasus	globular	410,000	+7.0	120 ly	34,000 ly	binoculars
M4	Scorpius	globular	44,000	+7.1	50 ly	14,000 ly	binoculars

GALAXY FACTS

Cluster	Constellation	Type	Galaxy cluster	Apparent magnitude	Diameter	Distance from Earth	What you need to see it
M81	Ursa Major	spiral	Coma-Sculptor cloud	+7.9	30,000 ly	4,500,000 ly	telescope
M83	Hydra	spiral	Coma-Sculptor cloud	+8.2	52,000 ly	15,000,000 ly	telescope
M51 (Whirlpool)	Canes Venatici	spiral	Coma-Sculptor cloud	+9.0	50,000 ly	15,000,000 ly	telescope
M101	Ursa Major	spiral	Coma-Sculptor cloud	+7.9	120,000 ly	17,500,000 ly	telescope
NGC 6946	Cepheus	spiral	Coma-Sculptor cloud	+8.9	78,000 ly	18,000,000 ly	telescope
NGC 5128 (Centaurus A)	Centaurus	giant elliptical	Coma-Sculptor cloud	+7.0	138,000 ly	26,000,000 ly	telescope
M87	Virgo	giant elliptical	Virgo cluster	+8.6	147,000 ly	55,000,000 ly	telescope
NGC 1365 (Great Barred Spiral)	Fornax	barred spiral	Fornax cluster	+9.5	157,800 ly	55,000,000 ly	telescope
M104 (Sombrero)	Virgo	spiral	Virgo cluster	+8.3	160,000 ly	65,000,000 ly	telescope
NGC 4038/4039 (Antennae)	Corvus	spirals	Crater cloud	+10.7	220,000 ly	82,800,000 ly	telescope
NGC 1275	Perseus	Seyfert	Perseus cluster	+11.6	175,000 ly	230,000,000 ly	telescope
3C 273	Virgo	quasar	unknown	+12.0	unknown	1,900,000,000 ly	telescope

NEBULA FACTS

Cluster	Constellation	Type	Apparent magnitude	Diameter	Distance from Earth	What you need to see it
NGC 7293 (Helix nebula)	Aquarius	planetary	+6.5	2 ly	600 ly	telescope
M27 (Dumbbell nebula)	Vulpecula	planetary	+8.1	2 ly	815 ly	telescope
M57 (Ring nebula)	Lyra	planetary	+9.0	0.4 ly	1140 ly	telescope
IC 434 (Horsehead nebula)	Orion	dark	—	20 ly?	1200 ly?	telescope
M42 (Orion nebula)	Orion	emission	+4.0	40 ly	1500 ly	naked eye
NGC 2264 (Cone nebula)	Monoceros	dark	—	50 ly	3000 ly	telescope
M20 (Trifid nebula)	Sagittarius	emission/reflection	+8.5	40 ly	5000 ly	telescope
M8 (Lagoon nebula)	Sagittarius	emission	+5.8	130 ly	5200 ly	telescope
M1 (Crab nebula)	Taurus	supernova remnant	+8.4	13.7 ly	6520 ly	telescope
M16 (Eagle nebula)	Serpens	emission	+6.0	315 ly	7000 ly	telescope

Glossary

Absolute magnitude How bright a star would appear, in magnitudes, at a standard distance of 10 parsecs (32.6 light-years) from Earth.

Accretion disk A swirling disk of matter that often surrounds a black hole, either a small one in a binary star system or a supergiant at the center of an active galaxy.

Active galaxy A galaxy with a central black hole that is emitting a large amount of radiation which is non-stellar in origin.

Altazimuth mounting A simple telescope mounting that swings from side to side and up and down.

Aperture The diameter of a telescope's main light-collecting optics. Also, the diameter of a binocular lens.

Aphelion The farthest distance from the Sun in the elliptical orbit of a comet, asteroid or planet.

Apparent magnitude The visible brightness of a star or other celestial object as seen from Earth.

Arcminute A unit of angular measure equal to $\frac{1}{60}$ of a degree; the Moon and Sun are about 30 arcminutes across.

Arcsecond A unit of angular measure equal to $\frac{1}{60}$ of an arcminute; Jupiter averages some 44 arcseconds across.

Asteroid Also called a minor planet, a small stony and/or metallic object with a diameter of less than 600 miles (1000 km) orbiting the Sun, usually in the asteroid belt.

Asteroid Belt A reservoir of asteroids orbiting the Sun in the ecliptic between the orbits of Mars and Jupiter.

Astronomical unit (AU) The average distance between Earth and the Sun, about 93 million miles (150 million km).

Astrophysics The study of the dynamics, chemical properties and evolution of celestial bodies.

Atmosphere A layer of gases attached to a planet or moon by the body's gravity.

Aurora Curtains and arcs of light in the sky over middle and high latitudes. They are caused by particles from the Sun hitting Earth's atmosphere and causing some of its gases to glow.

Axis The imaginary line through the center of a planet, star or galaxy around which it rotates; also, a shaft around which a telescope mounting pivots.

Big Bang The eruption of a small, very hot lump of matter about 14 billion years ago that marked the birth of the Universe, according to our best cosmological theory.

Binary star Two stars linked by mutual gravity and revolving around a common center of mass (see *double star*).

Black hole An object so dense that no light or other radiation can escape it.

Blazar A type of quasar (quasi-stellar radio source) that has distinct characteristics.

Brown dwarf A starlike object not large enough to start hydrogen fusion reactions. They are one possible candidate for a form of dark matter.

Catadioptric telescope A telescope, such as a Schmidt-Cassegrain, that uses both mirrors and lenses to form an image.

Celestial equator The imaginary line encircling the sky midway between the celestial poles.

Celestial poles The imaginary points on the sky where Earth's rotation axis, extended infinitely, would touch the imaginary celestial sphere.

Celestial sphere The imaginary sphere enveloping Earth upon which the stars, galaxies and other objects all appear to lie.

Cepheid variable A variable star whose brightness varies with a period of a few days. The brightness is directly related to the period of pulsation, making Cepheids indicators of distance in astronomy.

Chromosphere In the Sun, the thin layer of atmosphere lying just above the photosphere (visible surface) and below the corona.

Cluster A group of stars or galaxies held together by their gravity.

Comet A small body composed of ice and dust that orbits the Sun on an elongated path.

Conjunction The moment when a given two celestial objects lie closest together in the sky.

Constellation One of the 88 official patterns of stars that divide the sky into sections.

Convection A heat-driven process that causes hotter material to move upward while lighter material sinks.

Corona The high-temperature, outermost atmosphere of the Sun, visible from Earth only during a total solar eclipse.

Dark energy A mysterious type of anti-gravity that theorists now think pervades the Universe and is what is making its expansion accelerate.

Dark matter A form of matter which cannot be seen via the usual methods. Dark matter far outweighs normal, visible matter.

Dark nebula A cloud of interstellar dust that blocks the light of stars and nebulae behind it, appearing in silhouette.

Declination The angular distance of a celestial object north or south of the celestial equator.

Deepsky object An object located beyond the Solar System.

Degree A unit of angular measure equal to $\frac{1}{360}$ of a circle.

Dobsonian mount A simple type of altazimuth mount.

Double star Two stars that appear close together in the sky. Optical doubles are chance alignments of the stars; binary or multiple systems are linked by gravity.

Dust lane The thin disk of a spiral galaxy has a layer of gas and dust, that, when seen edge-on, looks like a thin lane among the stars of the galaxy.

Eclipse When one celestial body passes in front of another, dimming or obscuring its light.

Ecliptic The apparent path of the Sun around the celestial sphere; marks the plane of Earth's orbit.

Electromagnetic spectrum The name given to the entire range of radiation that includes radio waves, infrared, optical light, ultraviolet light, X-rays and gamma rays.

Ellipse The oval, closed path followed by a celestial object moving under the influence of gravity.

Emission nebula A cloud of gas glowing as the gas re-emits energy absorbed from a nearby hot star.

Equator The imaginary line on a celestial body that lies halfway between its two poles.

Equatorial mount A telescope mount with one axis parallel to Earth's rotational axis, so the motion of the heavens can be followed with a single movement.

Equinox The moment when the Sun appears to stand directly above a planet's equator.

Escape velocity The minimum speed an object (such as a rocket) must attain in order to travel from the surface of a planet, moon or other body, and into space. If the speed is too low, gravity will pull the object back down.

Event horizon The region surrounding a black hole, beyond which no signal emitted either by the hole or an object falling into it can ever reach an outside observer. Beyond the event horizon, the escape velocity exceeds the speed of light.

Finderscope Also called a finder, a small, low-power telescope attached to and aligned with a larger one that allows you to locate the general area of sky.

Galaxy A huge gathering of stars, gas and dust, bound by gravity and having a mass ranging from 100,000 to 10 trillion times that of the Sun.

Gamma rays Radiation with a wavelength shorter than X-rays.

Gas-giant planet A planet whose composition is dominated by hydrogen (Jupiter, Saturn, Uranus and Neptune).

Gibbous The phase of a moon or a planet when it appears greater than a half disk, but less than a full disk.

Globular star cluster A spherical cluster that may contain over a million stars, most of them old and red.

Infrared (IR) Radiation with wavelengths just longer than those of visible light.

Interferometry A technique for linking two or more telescopes together.

Kuiper Belt The region of the Solar System, outside the orbits of the planets, that contains icy planetesimals.

Light-year The distance that light travels in one year, about 6 trillion miles (9.5 trillion km).

Local Group A gathering of about 30 nearby galaxies, including the Milky Way.

M objects Star clusters, nebulas and galaxies in the Messier catalog.

Magnetic field A region surrounding a magnetic object, within which an iron-rich body will experience a magnetic force.

Magnitude A logarithmic unit used to measure the brightness of celestial objects. Apparent magnitude describes how bright a star looks from Earth, while absolute magnitude is its brightness if placed at a distance of 32.6 light-years. The lower the magnitude, the brighter the star.

Main sequence The longest-lived period in the lifecycle of a star, during which it converts hydrogen nuclei into helium nuclei. This generates the outward pressure that counterbalances the inward pull of the star's self-gravity.

Mare (plural Maria) A plain of congealed lava on the surface of the Moon, darker than the surrounding areas.

Meridian An imaginary line on the sky that runs due north and south, passing through the zenith.

Meteor The bright, transient streak of light produced by a meteoroid, a piece of space debris, burning up as it enters Earth's atmosphere at high speed.

Meteorite The name given to any piece of interplanetary debris that reaches Earth's surface intact.

Meteoroid Any small debris traveling through space, usually from a comet or asteroid.

Multiple star Three or more stars linked by gravity.

Nebula A cloud of gas or dust in space; may be dark or luminous.

Nebulosity The presence of faint gas.

Neutron star A massive star's collapsed remnant, consisting almost wholly of very densely packed neutrons. May be visible as a pulsar.

NGC objects Galaxies, star clusters and nebulae listed in the New General Catalog.

Nova A white dwarf star in a binary system that brightens suddenly by several magnitudes as gas pulled away from its companion star explodes in a thermonuclear reaction.

Nucleus The central core of a galaxy or comet.

Oort Cloud A swarm of billions of comets extending out about 2 light-years from the Kuiper Belt.

Open star cluster A group of a few hundred relatively young stars bound together by gravity.

Opposition The point in a planet's orbit when it appears opposite the Sun in the sky.

Optical telescope Any telescope that collects visible light

Orbit The path of an object as it moves through space under the control of another's gravity.

Parallax The apparent change in position of a nearby object relative to a more distant background when viewed from different points; the term is used to determine distances to nearby stars.

Parsec A unit of distance related to the radius of Earth's orbit. It is equivalent to 3.26 light-years.

Penumbra The outer part of an eclipse shadow; and the lighter area surrounding a sunspot.

Perihelion The closest distance to the Sun in the elliptical orbit of a comet, asteroid or planet.

Photosphere The visible surface of the Sun or any other star.

Planetary nebula A shell of gas blown off by a low-mass star when it runs out of fuel in its core.

Planetesimal A small, rocky body; one of the small bodies that coalesced to form the planets.

Precession A slow, periodic wobble in Earth's axis caused by the pull of the Sun and Moon.

Prominence A cloud of cooler gas lying above the Sun's surface.

Pulsar An old, rapidly spinning star that flashes bursts of radio (and occasionally optical) energy.

Pulsating variable A star that changes its brightness as it expands and shrinks regularly.

Quasar Short for quasi-stellar radio source, quasars are thought to be the active nuclei of very distant galaxies.

Radio galaxy A type of active galaxy that is a strong source of radio energy

Red giant A large, cool, red star in a late stage of its life.

Reflection nebula A cloud of dust or gas visible because it reflects light from nearby stars.

Reflector A telescope that forms an image using mirrors.

Refractor A telescope that forms an image using lenses.

Resolving power The ability of a telescope to show two closely spaced objects as separate.

Retrograde motion The apparent backward (westward) motion of a celestial body relative to the stars as Earth overtakes it because of its greater orbital speed.

Right ascension (RA) The celestial coordinate analogous to longitude on Earth.

Rotation The spin of a planet, satellite or star on its axis.

Satellite Any small object orbiting a larger one, although the term is most often used for rocky or artificial objects orbiting a planet.

Seyfert galaxy A type of active galaxy with unusual, often violent core activity.

Singularity A point in space-time where our laws of mathematics and physics cannot operate. Singularities exist at the centers of black holes.

Solar flare A sudden release of magnetic energy in or near the Sun's corona emitting radiation into space.

Solar wind A ceaseless, but variable, high-speed stream of extreme charged particles flowing out into space from the Sun.

Solstice The moment when a planet's pole tilts most directly toward (or away from) the Sun, marking the beginning of summer and winter.

Spectroscopy The analysis of light to determine, by studying the spectral lines, the chemical composition and conditions of the object producing it, as well as that object's motion and velocity toward or away from Earth.

Sunspot A dark, highly magnetic region on the Sun's surface that is cooler than the surrounding area.

Supercluster A cluster of clusters: a vast assemblage of entire clusters of galaxies.

Supernova The explosion of a massive star, briefly outshining a galaxy, that occurs when the star reaches the end of its fuel supply.

Supernova remnant An expanding cloud of gas thrown into space by a supernova explosion.

Terrestrial planet A planet with a mainly rocky composition (Mercury, Venus, Earth and Mars).

Transit The passage of an astronomical body in front of another. Often, Venus and Mercury can be seen silhouetted against the Sun during transits.

Umbra The dark, inner part of an eclipse shadow. Also, the dark central part of a sunspot.

Variable star Any star whose brightness appears to change.

Wavelength The distance between two successive crests or troughs in a wave.

White dwarf The small, very hot but faint remnant of a star that remains after the red giant stage.

Wormholes Theoretical tunnels through hyperspace. They could provide a short cut to a distant place in our Universe or to another universe entirely. They might also permit time travel.

X-rays Radiation with wavelengths between ultraviolet and gamma rays.

Zenith The point directly over an observer's head, 90 degrees perpendicularly from the horizon

Zodiac Name given to the 12 constellations that lie along the path of the Sun on the sky. .

Index

A

B

C

H

I

J

K

L

M

N

T

U

V

W

X

Y

Z

Credits

PHOTOGRAPHS

t=top; b=bottom; c=center; l=left; r=right
AAO = Anglo–Australian Observatory; AAP = Australian Associated Press; APL = Australian Picture Library; AURA = Association of Universities for Research in Astronomy; CIT = California Institute of Technology; DS = Digital Stock; ESA = European Space Agency; ESO = European Southern Observatory; GPL = Galaxy Picture Library; GRIN = Great Images in NASA; HST = Hubble Space Telescope; ING = Isaac Newton Group of Telescopes; IPAC = Infrared Processing and Analysis Center (via NASA); ISP = Institute for Solar Physics; JPL = Jet Propulsion Laboratory; MIC = Meade Instruments Corporation; NASA = National Aeronautics and Space Administration; NOAO = National Optical Astronomy Observatory; NRAO = National Radio Astronomy Observatory; PD = Photodisc; SOHO = Solar and Heliospheric Observatory; SPL = Science Photo Library; TPL = photolibrary.com

Front cover Courtesy Space Telescope Science Institute
Front flap Human Space Flight/NASA
1 NASA/ESA & J. Hester (Arizona State University) **2** APL/Corbis **4–5** APL/Corbis **6–7** APL/Corbis **8–9** ACS Science & Engineering Team/NASA **10** l PD; c R. Albrecht (ESA/ESO), NASA; r T.A. Rector (NRAO/AUI/NSF & NOAO/AURA/NSF) & B.A. Wolpa (NOAO/AURA/NSF) **11** l, r APL/Corbis; c Robert Williams/Hubble Deep Field Team (STScI)/NASA **12** APL/Corbis **14–15** PD **16** l, c APL/Corbis **17** l NASA/GRIN; c APL/Corbis **18** l Fergus O'Brien/Getty Images; c Harvard College Observatory/TPL/SPL; r NASA/JPL **19** l NASA/TPL/SPL; c APL/Corbis; r AAO/David Malin **20** l, r APL/Corbis **21** t APL/Corbis; b Northwind Picture Archives; **22** APL/Corbis **23** t, b APL/Corbis **24** tl APL/Corbis; bl TPL/SPL **25** br APL/Corbis; c APL/British Museum, London/The Bridgeman Art Library **26** l TPL/SPL; c, r APL/Corbis **27** l, r APL/Corbis; c NASA/HST **28** t APL/Corbis; b Dr Fred Espenak/TPL/SPL **29** l, r APL/Corbis **30** l APL/Corbis; c David Parker/TPL/SPL; r David Nunuk/TPL/SPL **31** l Dr Ian Robson/TPL/SPL; c, r APL/Corbis **32** APL/Corbis **33** tl Ben Johnson/Ferranti Astron/TPL/SPL; bl, r APL/Corbis; **34** NASA/ESA & J. Hester (ASU) **35** t, b NASA/HST **36** bl, cl NRAO/AUI/NSF/TPL/SPL; t APL/Corbis **38** bl Dr Fred Espenak/TPL/SPL; tr APL/Corbis **39** cr Dr Kurt Weiler/TPL/SPL **40** tl, bl NASA/TPL/SPL; br APL/Corbis **42** cl Daniel Wang, North Western University/TPL/SPL; bl, tr APL/Corbis **43** r NASA/TPL/SPL **44** l, r, c APL/Corbis **45** l NASA/GRIN; c APL/Corbis; r NASA/TPL/SPL **46** Novosti/TPL/SPL **47** tl NASA/GRIN; b, tr NASA/TPL/SPL **48** tl, bl, br APL/Corbis; tr AAP/Tyler Morning Telegraph **49** br APL/Corbis **50** APL/Corbis **51** tl, c, bl APL/Corbis **52** t NASA; b NASA/TPL/SPL; **53** tl, bl APL/Corbis; tr NASA/TPL/SPL **54** l APL/Corbis; c NASA/JPL; r NASA/TPL/SPL **55** l, r APL/Corbis; c NASA **56** bl APL/Corbis **57** t, b NASA/TPL/SPL **58** APL/Corbis **59** t APL/Corbis; b Novosti/TPL/SPL **60** l APL/Corbis; tr NASA **61** tl Stockbyte; tr NASA/TPL/SPL **64–65** R. Albrecht (ESA/ESO), NASA **66** l NASA/TPL/SPL; c DS **67** l Royal Swedish Academy of Sciences; c NASA/GRIN **76** bl NASA/JPL/North Western University; r U.S. Geological Survey/TPL/SPL **78** tc NASA/JPL/TPL/SPL **79** t Royal Swedish Academy of Sciences **80** bl APL/Corbis; cr NASA/JPL **82** tr NASA/JPL **83** t NASA/TPL/SPL; b NASA/JPL **84** bl Jacques Descloitres, MODIS Land Rapid Response Team, NASA/GSFC; cr NSSDC/NASA **86** t Pekka Parviainen/TPL/SPL; b APL/Corbis **87** b NASA/GSFC/NOAA; cr APL/Corbis **88** NASA **89** t DS; b NASA **90** bl APL/Corbis/Sygma; cr NASA/U.S. Geological Survey **92** t APL/Corbis **93** t APL/Corbis; b NASA/JPL/MSSS **94** APL/Corbis **95** NASA **96** bl NASA; cr NASA/TPL/SPL **98** bl APL/Corbis Images–Bettmann **99** t APL/Corbis; r NASA/TPL/SPL **100** NASA/TPL/SPL **101** t APL/Corbis ; bl–br: NASA/TPL/SPL; NASA/DLR; APL/Corbis; NASA/DLR **102** NASA/TPL/SPL **104** NASA/JPL **105** DS **106** NASA/TPL/SPL **107** tl, tr, bl–br APL/Corbis **108** bl Space Telescope Science Institute/NASA/TPL/SPL; cr NASA/U.S. Geological Survey **111** t NASA/TPL/SPL; b APL/Corbis **112** APL/Corbis **113** t, bl–br APL/Corbis **114** bl NASA/TPL/SPL; cr NASA/JPL **116** APL/Corbis **117** b APL/Corbis **118** APL/Corbis **119** t, b NASA/TPL/SPL **120** bl APL/Corbis **122** bl, cr NASA/SOHO **123** cl NASA/SOHO **124** bl NASA/SOHO; tr NASA/SOHO **125** bl PD; r Royal Swedish Academy of Sciences **128** bl PD; cr APL/Corbis **130** t NASA/TPL/SPL **131** t NASA/TPL/SPL; b NASA **132** l, r Yann Arthus-Bertrand/APL/Corbis **134** bl, br APL/Corbis; tr NASA **135** tl, tr APL/Corbis **136** t Dr Fred Espenak/TPL/SPL; b Pekka Parviainen/TPL/SPL **137** tr, cr APL/Corbis **138** t Pekka Parviainen/TPL/SPL; b APL/Corbis **139** George East/TPL/SPL **140** bl APL/Corbis **141** t Pekka Parviainen/TPL/SPL; bl Dan Schechter/TPL/SPL **142** bl, br APL/Corbis; tr NASA/TPL/SPL **144** t Lowell Observatory/NOAO/AURA/NSF **145** APL/Corbis **146** t TPL/SPL; b APL/Corbis **147** NASA/HST **148–149** NOAO/AURA/NSF **150** l, c Celestial Image Co./TPL/SPL; r Max-Planck-Institut Fur Extraterrestrische Phusick/TPL/SPL **151** l Nuffield Radio Astronomy Laboratories, Jodrell Bank/TPL/SPL; c Celestial Image Co./TPL/SPL **152** NASA/HST **153** l George Greaney; cr Celestial Image Co./TPL/SPL; br Dr Rudolph Schild/TPL/SPL; **154** bl Extreme Ultraviolet Imaging Telescope Consortium; r Space Telescope Science Institute/NASA/TPL/SPL **156** Celestial Image Co./TPL/SPL **158** tr Space Telescope Science Institute/NASA/TPL/SPL **160** bl AAO/ David Malin; br NOAO/ AURA/NSF; tr Celestial Image Co./TPL/SPL **161** J-C Cuillandre/Canada-France-Hawaii Telescope/TPL/SPL **162** bl AAP/NASA; tr Celestial Image Co./TPL/SPL **163** NASA & The Hubble Heritage Team (STScI/AURA)/C.R. O'Dell (Vanderbilt University) **164** NASA, J.J. Hester, Arizona State University **165** tl NASA/The Hubble Heritage Team/STScI/AURA; r ESO/TPL/SPL **166** t Space Telescope Science Institute/NASA/TPL/SPL **167** bl NASA/TPL/SPL **169** tl Chris Butler/TPL/SPL; bl Marcella Carollo (ETHZ)/The Hubble Heritage Team (STScI/AURA)/NASA; tr, br NOAO/AURA/NSF **170** l NASA, Goddard Space Flight Center/TPL/SPL; r NASA/The Hubble Heritage Team (STScI/AURA)/N. Scoville (CIT) & T. Rector (NOAO) **171** NOAO/AURA/NSF **172** l Max-Planck-Institut Fur Extraterrestrische Physik/Rosat X-ray Telescope/TPL/SPL; r Tod R. Lauer/NASA/HST **173** AAO/David Malin **174** l, r NOAO/AURA/NSF **175** Space Telescope Science Institute/NASA/TPL/SPL**176** l Dr Jean Lorre/TPL/SPL; r NASA/CXC/SAO/PSU/CMU **177** Celestial Image Co./TPL/SPL **179** tl TPL/SPL; bl NRAO/AUI/NSF/SPL; br Fred Espenak/TPL/SPL **180** bl Nuffield Radio Astronomy Laboratories, Jodrell Bank/TPL/SPL **181** l NOAO/AUI/NSF/TPL/SPL; cr, br Dr David Roberts/TPL/SPL **183** t NASA, A. Fruchter & the ERO Team, STScI, ST-ECF; b Dr Rudolph Schild/TPL/SPL **184** t R. Giacconi et al./JHU/AUI/NASA; b J. Baum & N. Henbest/TPL/SPL **186** tr Space Telescope Science Institute/NASA/TPL/SPL; b John Dubinski, University of Toronto & San Diego Supercomputing Center/GPL **187** t Space Telescope Science Institute/NASA/TPL/SPL; b NASA/The Hubble Heritage Team/STScI/AURA **188–189** APL/Corbis **190** l Greg Pease/Getty Images; c Marcelo Bass, CTIO/NOAO/AURA/NSF **191** c Pekka Parviainen/TPL/SPL **192** tl APL/Corbis–Bettmann; bl R. Williams (STScI)/HDF-S Team/NASA; tr Robin Scagell/TPL/SPL **193** br AAO/David Malin **194** tl Mary Evans Picture Library; tr Observatorie de Paris/Bulloz **196** bl–br APL/Corbis **197** Pekka Parviainen/TPL/SPL **198** APL/Corbis **199** APL/Corbis **200** cr Derke/O'Hara/Getty Images; b Oliver Strewe **201** t Craig Mayhew & Robert Simmon, NASA/GSFC; b Greg Pease/Getty Images **202** tl, tr MIC **203** bl, bc, br NOAO/AURA/NSF; c MIC; tr Tenmon Guide, Japan **204** MIC **205** l, r MIC **210** l NOAO/AURA/NSF; c Martin Altman/Observatorium Hoher List of the University Bonn, Daun, Germany; r MPIA-HD, Birkle, Slawik/TPL/SPL **212** l ESO/TPL/SPL; c Till Checher; r AAO/David Malin **214** l NASA; c NOAO/AURA/NSF; r Martin Altman/Observatorium Hoher List of the University Bonn, Daun, Germany **216** l John Sanford/TPL/SPL; c Infrared Processing & Analysis Centre & University of Massachusetts; r NOAO/TPL/SPL **218** l Richard Crisp; c Naoyuki Kurita; r NOAO/TPL/SPL **220** l Sean Walker/John Bodreau; c Chris Cook; r Jean-Charles Cuillandre/Canada-France-Hawaii Telescope/TPL/SPL **222** l MPIA-HD, Birkle, Slawik/TPL/SPL; c J-C Cuillandre/Canada-France-Hawaii Telescope/TPL/SPL; r John Sanford/TPL/SPL **224** l NOAO/TPL/SPL; c J-C Cuillandre/Canada-France-Hawaii Telescope/TPL/SPL; r George Jacoby, Bruce Bohannan, Mark Hannal/NOAO/AURA/NSF **226** l Tony Hallas/TPL/SPL; c IPAC/NASA; r Robin Scagell/GPL **228** l Celestial Image Co./TPL/SPL; c Andy Steele; r Simon Driver (St. Andrews)/ING **230** l NASA/HST; c Space Telescope Science Institute/NASA/TPL/SPL; r Tony Hallas/TPL/SPL **232** l John Bodreau; c MPIA-HD, Birkle, Slawik/TPL/SPL; r Naoyuki Kurita **234** l Celestial Image Co./TPL/SPL; c NOAO/TPL/SPL; r Royal Observatory, Edinburgh/AATB/TPL/SPL **236** l Space Telescope Science Institute/NASA/TPL/SPL; c Akira Fuji; r Mount Stromlo and Siding Spring Observatories/TPL/SPL **238** l Dr Fred Espenak/TPL/SPL; c Celestial Image Co./TPL/SPL; r NOAO/AURA/NSF **240** l Dr. Luke Dodd/TPL/SPL; c Royal Observatory, Edinburgh/TPL/SPL; r AAO **242** l Robin Scagell/GPL; c Mount Stromlo and Siding Spring Observatories/TPL/SPL; r AAO **244** l MPIA-HD, Birkle, Slawik/TPL/SPL; c, r NOAO/AURA/NSF **246** l NOAO/TPL/SPL; c Kim Gordon/TPL/SPL; r Arpad Kovacsy **248** l, r Celestial Image Co./TPL/SPL; c Royal Observatory, Edinburgh/TPL/SPL **250** l Celestial Image Co./TPL/SPL; c David Malin/AAO; r Jean-Charles Cuillandre/Canada-France-Hawaii Telescope **252** l Jan Wisniewski; c, r NOAO/AURA/NSF **254** l Celestial Image Co./TPL/SPL; c NOAO/AURA/NSF; r J-C Cuillandre/Canada-France-Hawaii Telescope/TPL/SPL **256** l NASA/HST; c NOAO/AURA/NSF; r T.A.Rector/NOAO/AURA/NSF **258–259** Robert Williams/Hubble Deep Field Team (STScI)/NASA **260** l NASA; c The Archives, CIT **261** l APL/Corbis; c Space Telescope Science Institute/NASA/TPL/SPL **264** l, r APL/Corbis **265** t NASA/TPL/SPL; b APL/Corbis **266** TPL/SPL/Larry Landolfi **268** l APL/Corbis; r Arizona State University/NASA **269** t NASA/TPL/SPL; b APL/Corbis **270** tr Space Telescope Science Institute/NASA/TPL/SPL **271** NASA **272** The Archives, CIT **274** bl Yannick Mellier/IAP/TPL/SPL; tr John Irwin Collection/American Institute of Physics/SPL **275** l, r APL/Corbis **280–281** APL/Corbis
Back cover tl SOHO/EIT 28/ESA/NASA; tc SOHO/NASA; tr SOHO/LASCO C2/ESA/NASA

ILLUSTRATIONS

Peter Bull Art Studio 144
David Carroll 202b, 204l
Andrew Davies/Creative Communication 30, 68bl, 70bl, 72bl, 76tl, 80tl, 84tl, 90tl, 96tl, 102bl, 108tl, 114tl, 120tl, 126bl, 128tl, 140tl
Désirée DeKlerk Endpapers
Luigi Gallante 24tr, 24br
Mark A. Garlick/space-art.co.uk 74tr, 74cr, 74br, 75, 77, 81, 85, 86, 88, 91, 97, 99, 103, 104, 108cr, 109, 110b, 115, 116, 120c, 120br, 121, 123, 127, 129, 136, 139, 143, 145, 147, 155, 157, 158, 164, 166, 178, 182, 185, 266, 282
David Hardy/Wildlife Art Ltd 74bc, 78, 82, 92, 130, 132, 133
Moonrunner Design back cover, cover spine, 16, 22, 32, 34, 36, 37, 39, 40, 41, 43, 46, 50, 56, 58, 61, 62, 63, 66, 68c, 70c, 72c, 79, 94, 100, 106, 110c, 112, 118, 156, 159, 167, 168, 180, 190, 194, 195, 196, 197, 199, 260, 262, 270, 272, 276, 277, 278, 279
Wil Tirion (star maps) 191, 206, 207, 208 209, 211, 213, 215, 217, 219, 221, 223, 225, 227, 229, 231, 233, 235, 237, 239, 241, 243, 245, 247, 249, 251, 253, 255, 257
David Wood 204r

Published by Firefly Books Ltd. 2004

Conceived and produced by Weldon Owen Pty. Ltd.
59 Victoria Street, McMahons Point, Sydney, NSW 2060, Australia

First printing

Publisher Cataloging-in-Publication Data (U.S.)
Garlick, Mark A.
Astronomy : a visual guide / Mark A. Garlick. —1st ed.
[304] p. : col. ill., photos. ; cm. (Visual Guides)
Includes index.
Summary: A comprehensive guide to understanding and observing the night sky, including monthly star maps.
ISBN 1-55297-958-X.
1. Astronomy. I. Title. II. Series
522 22 QB44.3.G37 2004

National Library of Canada Cataloguing in Publication
Garlick, Mark A. (Mark Antony), 1968-
Astronomy : a visual guide / Mark A Garlick.
Includes index.
ISBN 1-55297-958-X
1. Astronomy--Popular works. 2. Astronomy--Observers' manuals. I. Title.
QB44.3.G37 2004 520 C2004-900138-8

Published in the United States in 2004 by
Firefly Books (U.S.) Inc.
P.O. Box 1338, Ellicott Station
Buffalo, New York 14205

Published in Canada in 2004 by
Firefly Books Ltd.
66 Leek Crescent
Richmond Hill, Ontario L4B 1H1

Illustrators: Mark A. Garlick and Malcolm Godwin/Moonrunner Design
Designers: Jo Raynsford and Steven Smedley
Cover Designer: Jacqueline Hope Raynor

Color reproduction by Chroma Graphics (Overseas) Pte. Ltd.
Printed by SNP Leefung Printers Ltd
Printed in China

Captions

page 1 Some 5500 light-years from the Solar System, the Swan Nebula (M17) is a glowing hotbed of star formation.

page 2–3 Bruce McCandless, shuttle astronaut, hovers above the air in his Manned Maneuvering Unit (MMU).

page 4–5 Edward White was the first American to "walk" in space, tethered to his Gemini 4 spacecraft in June 1965.

page 6–7 Craters and jagged peaks stretch across the lunar surface in this photo taken in the late twentieth century.

page 8–9 Hubble took this image of the Omega Nebula, also known as the Swan Nebula or the Horseshoe Nebula, thousands of light-years away in Sagittarius.

page 12–13 Space shuttle Discovery, the first shuttle to fly after the Challenger disaster of 1986, blasts off from Cape Kennedy in 1993.